Mastering

Advanced Pure Mathematics

Macmillan Master Series

Accounting
Advanced English Language
Advanced Pure Mathematics
Arabic
Banking
Basic Management
Biology
British Politics
Business Administration
Business Communication
Business Law
C Programming
Chemistry
COBOL Programming
Commerce
Communication
Computers
Databases
Economic and Social History
Economics
Electrical Engineering
Electronic and Electrical Calculations
Electronics
English as a Foreign Language
English Grammar
English Language
English Literature
English Spelling
French
French 2
German
German 2
Human Biology
Italian
Italian 2
Japanese
Manufacturing
Marketing
Mathematics
Mathematics for Electrical and Electronic Engineering
Modern British History
Modern European History
Modern World History
Pascal Programming
Philosophy
Photography
Physics
Psychology
Science
Social Welfare
Sociology
Spanish
Spanish 2
Spreadsheets
Statistics
Study Skills
Word Processing

Mastering

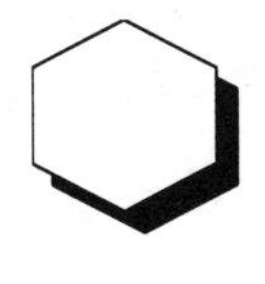

Advanced Pure Mathematics

Geoff Buckwell

First published 1996 by
MACMILLAN PRESS LTD
Houndmills, Basingstoke, Hampshire RG21 6XS
and London
Companies and representatives
throughout the world

ISBN 0–333–62049–6

A catalogue record for this book is available from the British Library.

10 9 8 7 6 5 4 3 2 1
05 04 03 02 01 00 99 98 97 96

Copy-edited and typeset by Povey–Edmondson
Okehampton and Rochdale, England

Printed in Malaysia

Contents

Introduction

Mastering Advanced Pure Mathematics is not just another A-level text book. It contains full explanations so that the book can be used at home as a support in the study of a wide range of syllabuses based on the common core A-level topic list. Throughout the book, there are a number of boxes headed DO YOU KNOW? These contain brief reminders of ideas. If you are happy with the ideas covered in DO YOU KNOW?, then you should be able to continue without any problems. Any ideas in the DO YOU KNOW sections you are not sure about can be studied in *Mastering Mathematics* (also in this series). So often, people try to study topics without the necessary background knowledge. *Mastering Advanced Pure Mathematics* helps you to avoid making that sort of mistake.

You will also notice boxes throughout the book headed MEMORY JOGGER. These points are extremely important. Try to remember them, as they will increase your level of understanding. In order to check that you have understood the work, you can try the exercises: these have been arranged so that you can check your progress as you go along. There are also miscellaneous examples at the end of each chapter that you should be able to complete on first reading. The revision problems at the ends of the chapters are more difficult and should not be attempted until you are familiar with most of the book, because often, questions in a chapter will contain topics covered in other chapters.

Scattered throughout the book are activities designed to broaden your knowledge. It is hoped that these will help your enjoyment of mathematics.

Acknowledgements

The authors and publishers are grateful to the following for permission to reproduce copyright material:

The Associated Examining Board, Northern Examinations and Assessment Board (incorporating Northern Examining Association and the Joint Matriculation Board), Oxford and Cambridge Schools Examination Board, University of Cambridge Local Examinations Syndicate, and University of London Examinations and Assessment Council, for A-Level questions from past examination papers. Every effort has been made to trace all the copyright-holders, but if any have been inadvertently overlooked the publishers will be pleased to make the necessary arrangement at the first opportunity.

The author thanks Josie Buckwell for her help in preparing the book, Matthew Buckwell for the computer-generated diagrams, and Colin Prior for providing the sketches.

1 Number

In order to study advanced mathematics, it is important that you have good skills in algebraic manipulation. The purpose of this chapter is to deal with those aspects of algebra concerned with powers. You should work through this chapter carefully, and any points listed in the DO YOU KNOW? section that you have not already covered can be found in *Mastering Mathematics*. Remember, people have been studying algebra for hundreds of years. Don't expect to master it overnight.

DO YOU KNOW?

(i) The meaning of an index number:

$x^4 = x \times x \times x \times x$ (x to the power 4)

$x^4 \times x^3 = x^{4+3} = x^7$ (add the powers when multiplying)

$t^6 \div t^4 = t^{6-4} = t^2$ (subtract the powers when dividing).

(ii) The general rules for indices that are **positive** integers:

$$x^m \times x^n = x^{m+n} \quad \text{(P1)}$$

$$x^m \div x^n = x^{m-n} \quad \text{(P2)}$$

$$(x^m)^n = x^{mn} \quad \text{(P3)}$$

$$x^0 = 1 \text{ whatever the value of } x \quad \text{(P4)}$$

$$(xy)^m = x^m y^m \quad \text{(P5)}$$

1.1 Negative and fractional indices

You will already know how to work with indices when they are positive integers, but what does 3^{-2} mean?

If Equation (P2) is to work,

then $\quad 3^{-2} \times 3^2 = 3^{-2+2} = 3^0$, which equals 1

But since $\quad 3^2 = 9$,

and because $\quad 9 \times \frac{1}{9} = 1$

then $\quad 3^{-2}$ must mean $\dfrac{1}{3^2}$

Since $\dfrac{1}{x}$ is called the **reciprocal** of x, then 3^{-2} is the reciprocal of 3^2.

Now suppose that the power is a fraction. For example, what does $8^{\frac{1}{3}}$ mean? Looking back to Equation (P2),

$$8^{\frac{1}{3}} \times 8^{\frac{1}{3}} \times 8^{\frac{1}{3}} = 8^{\frac{1}{3}+\frac{1}{3}+\frac{1}{3}} = 8^1 = 8$$

Since $8^{\frac{1}{3}}$ multiplied by itself three times gives 8, then $8^{\frac{1}{3}}$ must be the **cube root** of 8

i.e. $8^{\frac{1}{3}} = \sqrt[3]{8} = 2$ (because $2 \times 2 \times 2 = 8$).

These ideas are now expanded in the following example.

Example 1.1

Try these first without using a calculator:

(i) $16^{\frac{3}{4}}$

(ii) $\left(\frac{4}{9}\right)^{-\frac{3}{2}}$

then carry out the same calculation using a calculator to check your answer.

Solution

(i) $16^{\frac{3}{4}} = 16^{\frac{1}{4}} \times 16^{\frac{1}{4}} \times 16^{\frac{1}{4}}$ or $\left(16^{\frac{1}{4}}\right)^3$ (Using Equation (P3).)

$= \sqrt[4]{16} \times \sqrt[4]{16} \times \sqrt[4]{16}$

$= 2 \times 2 \times 2 = 8$

(**Note**: $16^{\frac{1}{4}}$ means the **fourth root of 16**)

(ii) Remove the negative power first.

So $\left(\frac{4}{9}\right)^{-\frac{3}{2}} = \dfrac{1}{\left(\frac{4}{9}\right)^{\frac{3}{2}}}$

But $\left(\frac{4}{9}\right)^{\frac{3}{2}} = \left(\left(\frac{4}{9}\right)^{\frac{1}{2}}\right)^3 = \left(\frac{2}{3}\right)^3$

$= \frac{2^3}{3^3} = \frac{8}{27}$ (Using Equation (P3).)

So $\left(\frac{4}{9}\right)^{-\frac{3}{2}} = 1 \div \frac{8}{27} = \frac{27}{8}$

$= 3\frac{3}{8}$

> **MEMORY JOGGER**
>
> Always try to change negative powers to expressions with positive powers. Use the idea of reciprocal to do this.

We shall now look at how to evaluate these on a calculator.

Make sure you read the instruction manual on your calculator before using buttons mentioned here.

You will need the power button [x^y] and a fraction button [$a^b/_c$].

Hence:

(i) [16] → [x^y] → [3] → [$a^b/_c$] → [4] → [=] (16 to the power 3/4)

Display

[8.]

(ii) [4] → [$a^b/_c$] → [9] → [x^y] → [3] → [$a^b/_c$] → [2] → [+/−] → [=]

($\frac{4}{9}$ to the power $-\frac{3}{2}$)

Display

[3.375]

Here you can see one weakness of a calculator. It will not give the answer as a simple fraction. You need to confirm that

$\frac{27}{8} = 3.375$

These examples lead to the following equations:

$$a^{\frac{m}{n}} = \left(a^{\frac{1}{n}}\right)^m = (a^m)^{\frac{1}{n}} \qquad \text{(P6)}$$

$$a^{\frac{1}{n}} = \sqrt[n]{a} \qquad \text{(P7)}$$

$$a^{-n} = \frac{1}{a^n} \qquad \text{(P8)}$$

The previous example dealt with numbers only. The following example shows how to use these equations in algebraic simplification.

Example 1.2

Simplify:

(i) $(2x^2)^3 \div 4x^2$ (ii) $(4xy^2)^2 \times 3xy^3$

(iii) $\sqrt{4x^2y^6}$ (iv) $\sqrt[3]{(8x^6y^3)^2}$

Solution

(i) $(2x^2)^3$ means $2x^2 \times 2x^2 \times 2x^2$

$= 8x^6$

So $(2x^2)^3 \div 4x^2 = 8x^6 \div 4x^2$

$$= \frac{8x^6}{4x^2} = 2x^{6-2}$$

$$= 2x^4$$

(ii) $(4xy^2)^2$ means $4xy^2 \times 4xy^2 = 16x^2y^4$

$(4xy^2)^2 \times 3xy^3 = 16x^2y^4 \times 3xy^3 = 48x^3y^7$

(iii) $\sqrt{4x^2y^6} = (4x^2y^6)^{\frac{1}{2}} = (4)^{\frac{1}{2}}(x^2)^{\frac{1}{2}}(y^6)^{\frac{1}{2}}$ (Using Equation (P5).)

$= 2xy^3$

(iv) $\sqrt[3]{(8x^6y^3)^2} = \sqrt[3]{(8)^2(x^6)^2(y^3)^2}$

$= \sqrt[3]{64x^{12}y^6} = (64)^{\frac{1}{3}}(x^{12})^{\frac{1}{3}}(y^6)^{\frac{1}{3}}$

$= 4x^4y^2$

Very often, in order to simplify an expression, you have to express each part as the power of the same number (or numbers) called the **base**. This technique is quite difficult to learn, and the following examples give you an idea of how to look for that number (or numbers).

Example 1.3

Express 4^{2+x}, 8^{3-2x}, 16^{3+2x} in the form 2^y. Hence, solve the equation:

$$\frac{4^{2+x}}{16^{3+2x}} = \frac{2^{6x}}{8^{3-2x}}$$

Solution

Here we are told to express the numbers as powers of 2.

(i) $4^{2+x} = (2^2)^{2+x} = 2^{4+2x}$

(ii) $8^{3-2x} = (2^3)^{3-2x} = 2^{9-6x}$

(iii) $16^{3+2x} = (2^4)^{3+2x} = 2^{12+8x}$

(Note we are using Equation (P3) here.)

The equation can now be written:

$$\frac{2^{4+2x}}{2^{12+8x}} = \frac{2^{6x}}{2^{9-6x}}$$

$$\therefore \quad 2^{(4+2x)-(12+8x)} = 2^{6x-(9-6x)} \qquad \text{(Using Equation (P2).)}$$

This simplifies to:

$$2^{-8-6x} = 2^{12x-9}$$

Since both sides are powers of 2, the powers on each side must be equal. Hence:

$$-8 - 6x = 12x - 9$$

$$\therefore \quad 1 = 18x,$$

giving $x = \frac{1}{18}$

Example 1.4

Simplify without using a calculator:

(i) $(9^2 \times 4^3) \div (3^{-2} \times 8^2)$

(ii) $\left(6^{\frac{1}{2}} \times 12^2\right) \div \left(4^2 \times 24^{\frac{1}{2}}\right)$

Solution

(i) We are not told this time what numbers to use. Looking at the numbers involved, we have 9, 4, 3 and 8. Each of these numbers can be expressed as a power of either 2 or 3.

So: $9^2 = (3^2)^2 = 3^4$

$4^3 = (2^2)^3 = 2^6$

$8^2 = (2^3)^2 = 2^6$

Hence we can write the expression to be evaluated as:

$$\frac{3^4 \times 2^6}{3^{-2} \times 2^6} = \left(\frac{3^4}{3^{-2}}\right) \times \left(\frac{2^6}{2^6}\right) = 3^6 \times 1 = 3^6$$

(ii) The numbers involved in this example, 6, 12, 4 and 24, initially do not appear to be powers of one or two numbers.

However, $6 = 3 \times 2$

$12 = 3 \times 2^2$

$4 = 2^2$

and $24 = 3 \times 2^3$

Hence,

$$6^{\frac{1}{2}} = (3 \times 2)^{\frac{1}{2}} = 3^{\frac{1}{2}} \times 2^{\frac{1}{2}}$$

(Using Equation (P5).)

$$12^2 = (3 \times 2^2)^2 = 3^2 \times 2^4$$

$$24^{\frac{1}{2}} = (3 \times 2^3)^{\frac{1}{2}} = 3^{\frac{1}{2}} \times 2^{\frac{3}{2}}$$

$$4^2 = (2^2)^2 = 2^4$$

Hence the expression we are evaluating is:

$$\frac{6^{\frac{1}{2}} \times 12^2}{4^2 \times 24^{\frac{1}{2}}} = \frac{3^{\frac{1}{2}} \times 2^{\frac{1}{2}} \times 3^2 \times 2^4}{2^4 \times 3^{\frac{1}{2}} \times 2^{\frac{3}{2}}}$$

$$= \frac{2^{\frac{9}{2}} \times 3^{\frac{5}{2}}}{2^{\frac{11}{2}} \times 3^{\frac{1}{2}}} = 2^{-1} \times 3^2 \text{ or } \tfrac{9}{2}$$

Example 1.5

In a sports competition, the formula $P = 0.1186\,(253 - t)^{1.94}$ is used to calculate the number of points awarded in the 800m event, where t is the time in seconds taken by the athlete for the race. If an athlete scores 953 points for the race, what was the athlete's time?

Formulae like these are used in the heptathlon and similar events.

Solution

Since $p = 953$

$953 = 0.1186\,(253 - t)^{1.94}$

We need to rearrange this to make t the subject.

$$\therefore \quad \frac{953}{0.1186} = (253 - t)^{1.94}$$

$$\therefore \quad 253 - t = \left(\frac{953}{0.1186}\right)^{\frac{1}{1.94}}$$

$$\therefore \quad t = 253 - \left(\frac{953}{0.1186}\right)^{\frac{1}{1.94}} = 150 \quad \text{(to the nearest second).}$$

The time is 150 seconds.

Exercise 1(a)

1 Evaluate the following, first giving your answer exactly without the help of a calculator, and then check your answers with a calculator.

(i) $\left(\frac{1}{3}\right)^{-2}$ (ii) $\dfrac{1}{4^{-2}}$ (iii) $16^{-\frac{1}{4}}$

(iv) $\left(\frac{8}{27}\right)^{\frac{1}{3}}$ (v) $\left(\frac{2}{3}\right)^{0}$ (vi) $\left(\frac{25}{9}\right)^{-\frac{3}{2}}$

(vii) $\left(-\frac{1}{4}\right)^{-3}$ (viii) $\left(2\frac{1}{4}\right)^{-\frac{1}{2}}$ (ix) $(0.001)^{\frac{2}{3}}$

(x) $\left(\frac{100}{9}\right)^{-1.5}$ (xi) $\left(\frac{4}{9}\right)^{-\frac{1}{2}}$ (xii) $\left(\frac{1}{125}\right)^{-\frac{1}{3}}$

2 Simplify the following:

(i) $(2x)^3 \div (3x)^2$ (ii) $(4x^3)^{\frac{1}{2}} \times \left(3x^{\frac{1}{2}}\right)^3$

(iii) $(4^2 \times 8^3) \div \left(2^{-2} \times 4^{\frac{1}{2}}\right)$ (iv) $(9^2 \times 5^3) \div \left(25^{\frac{1}{2}} \times 3^4\right)$

(v) $\left(x^{\frac{1}{4}} \times x^{\frac{2}{3}}\right) \div x^{\frac{1}{2}}$ (vi) $8^{\frac{1}{2}} \times 2^{\frac{1}{2}}$ (vii) $\sqrt{x} \times \sqrt{x^3}$

(viii) $\sqrt{4x} \div 2x^2$ (ix) $(\sqrt{x})^2 \div \sqrt{x^4}$ (x) $\left(4^{\frac{1}{2}} \times 8^{\frac{1}{2}}\right) \div 16^{-\frac{1}{2}}$

1.2 Surds

Very often in a calculation, you may be working with a square root such as $\sqrt{3}$ which cannot be written exactly as a decimal. It may be that further on in the calculation, this needs to be squared again producing an exact answer of 3.

Hence the $\sqrt{3}$ need not have been evaluated. For this reason, numbers such as $\sqrt{3}$ are often left written in that way with the square root sign remaining. Numbers written like this are called **surds**.

The following example shows how surds can be manipulated.

Example 1.6

Write as simply as possible in surd form:

(i) $\sqrt{18}$ (ii) $\sqrt{50} + \sqrt{18}$ (iii) $(\sqrt{8})^3$

Solution

(i) $\sqrt{18} = \sqrt{9} \times \sqrt{2} = 3\sqrt{2}$

(ii) $\sqrt{50} = \sqrt{25 \times 2} = 5\sqrt{2}$

so $\sqrt{50} + \sqrt{18} = 5\sqrt{2} + 3\sqrt{2} = 8\sqrt{2}$

(iii) $\sqrt{8} = \sqrt{4 \times 2} = 2\sqrt{2}$

$$\text{so} \quad \left(\sqrt{8}\right)^3 = 2\sqrt{2} \times 2\sqrt{2} \times 2\sqrt{2} = 8\left(\sqrt{2}\right)^3$$

$$= 8\left(2\sqrt{2}\right) = 16\sqrt{2}$$

When a surd appears in the denominator of a fraction, there are a couple of techniques worth learning to help simplify expressions. These methods are known as **rationalisation of the denominator**.

For example,

(i) $\dfrac{2}{\sqrt{3}}$

Multiply the numerator and denominator by $\sqrt{3}$ $\quad \dfrac{2}{\sqrt{3}} = \dfrac{2}{\sqrt{3}} \times \dfrac{\sqrt{3}}{\sqrt{3}} = \dfrac{2\sqrt{3}}{3}$

The denominator no longer contains a surd.

(ii) $\dfrac{4}{\sqrt{6} - \sqrt{3}}$

Multiply top and bottom by $\sqrt{6} + \sqrt{3}$

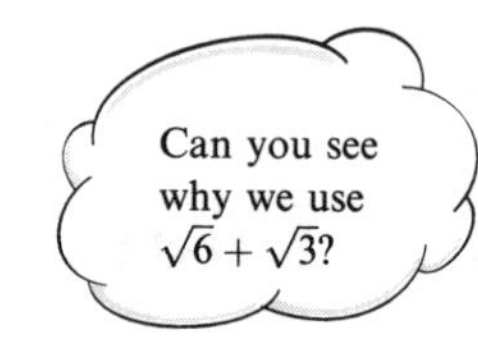

$$\text{So} \quad \frac{4}{\sqrt{6} - \sqrt{3}} = \frac{4 \times (\sqrt{6} + \sqrt{3})}{(\sqrt{6} - \sqrt{3})(\sqrt{6} + \sqrt{3})}$$

$$= \frac{4(\sqrt{6} + \sqrt{3})}{6 + \sqrt{18} - \sqrt{18} - 3} = \frac{4(\sqrt{6} + \sqrt{3})}{3}$$

MEMORY JOGGER

Never leave a surd as part of a denominator if it can be removed.

Exercise 1(b)

1 Simplify when possible, but still leave surds in your answer.

(i) $\sqrt{72}$ (ii) $\sqrt{32} + \sqrt{24}$

(iii) $\sqrt{32} \times \sqrt{50}$ (iv) $\sqrt{2}(1 + \sqrt{2})$

(v) $\sqrt{392}$ (vi) $(1 + \sqrt{2})(1 - \sqrt{2})$

(vii) $(3\sqrt{2} + 1)^2$ (viii) $(\sqrt{3} + \sqrt{2})(\sqrt{2} + 3)$

2 Rationalise the denominator in the following fractions and simplify your answer if possible.

(i) $\dfrac{4}{\sqrt{2}}$ (ii) $\dfrac{1}{\sqrt{8}}$ (iii) $\dfrac{1}{1+\sqrt{2}}$ (iv) $\dfrac{1}{\sqrt{3}-\sqrt{2}}$

(v) $\dfrac{1}{1+\sqrt{2}}+\dfrac{1}{1-\sqrt{2}}$ (vi) $\dfrac{1}{3\sqrt{2}+4\sqrt{3}}$

(vii) $\dfrac{1}{(1+\sqrt{3})^2}$ (viii) $\dfrac{1+\sqrt{2}}{1-\sqrt{2}}$

1.3 Logarithms

Logarithms have been used since 1614 to help the processes of multiplication and division.

Although this aspect of their use has now been replaced by the calculator, the logarithm function is still essential in scientific calculations. As many readers of this book will not have met logarithm tables before, we will begin with a simple introduction of how they work.

The logarithm tables need to work from a number called the **base**. Suppose we work with base 2.

Listing some of the powers of two:

$2^{-2} = 0.25$
$2^{-1} = 0.5$
$2^{0} = 1$
$2^{1} = 2$
$2^{2} = 4$
$2^{3} = 8$
$2^{4} = 16$
$2^{5} = 32$

It is easy to see that $8 \times 0.5 = 4$, but this calculation could be done with the **powers** of two.

Since 8 equals 2 to the power 3
and 0.5 equals 2 to the power -1,
then if you **add** the powers $3 + -1 = 2$, and 2 to the power $2 = 4$
So $8 \times 0.5 = 4$

In other words, the multiplication problem has been replaced by an addition problem. The power is referred to as the **logarithm to base 2**, (written $\log_2$) of the given number.

Hence:

$$\log_2 8 = 3 \qquad [\text{log to base 2 of } 8 = 3]$$

and

$$\begin{aligned} \log_2 0.5 &= -1 \\ \log_2 8 + \log_2 0.5 &= 3 + (-1) \\ &= 2 = \log_2 4 \end{aligned}$$

Hence: $\quad 8 \times 0.5 = 4$

The logarithm can be seen as a sort of code number. Hence 8 codes to 3, 0.5 codes to -1. You add the codes to give 2, and 2 decodes to give 4. The process of decoding can be called *finding an antilogarithm*. Old-fashioned tables would contain antilogarithms as well as logarithms.

In our original list, there are only a few numbers, and you might ask how these can help in more complicated problems. The list can be extended by using a graphical approach. We need to look at the graph of $y = 2^x$ shown in Figure 1.1. Suppose we wanted to find:

3.65×0.85

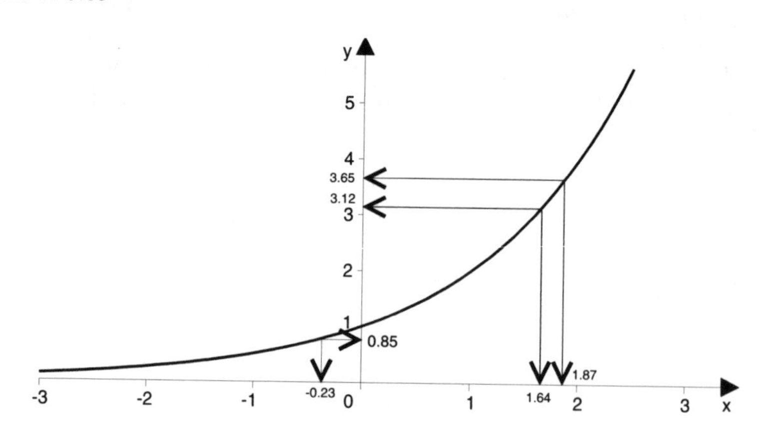

Figure 1.1

To find the power of 2 corresponding to each number, read from the graph as shown.

Hence:

$\log_2 3.65 \approx 1.87$

$\log_2 0.85 \approx -0.23$

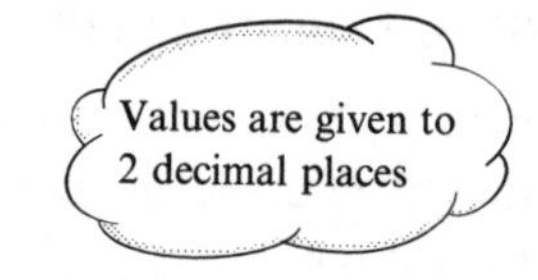

(**Note**: $\approx$ means approximately equal to)

Hence: $\log_2(3.65 \times 0.85) \approx 1.87 + -0.23$
≈ 1.64

Reading from the graph, 1.64 gives the value of 3.12.

Hence: $3.65 \times 0.85 \approx 3.12$

(The correct answer is, in fact, 3.1025.)

Division is a similar process, except that the logarithms are subtracted.

Obviously, you would not be able to draw and read a graph precisely enough to get an accurate answer.

The number 10 is also a much better number to use as the base than 2. To avoid the graphical approach, logarithms were written in tables; these first appeared in 1614, calculated by John Napier. Without these tables, scientific developments would have been held back by hundreds of years. Today, all these tables are stored in the calculator. The button [log] means logarithms to base 10 (the ten is omitted).

We can summarise this by saying that:

If $a^x = y$, then $x = \log_a y$ (L1)

The rules of using logarithms are, then,

$$\log M + \log N = \log MN \quad \text{(L2)}$$

$$\log M - \log N = \log \frac{M}{N} \quad \text{(L3)}$$

If n is a positive integer, then

$$\log M^n = \underbrace{\log M + \log M + \ldots\ldots\ldots + \log M}_{n \text{ terms}}$$

$$= n \log M$$

Conveniently, however, this rule is also true for any value of n, that is,

$$\log M^n = n \log M \quad \text{(for all values of } n\text{)} \quad \text{(L4)}$$

Another important result, often forgotten, follows from the fact that since $a^0 = 1$ for all values of a, then whatever base you use, the logarithm of 1 will be zero.

Hence $\log_a 1 = 0$ (L5)

Since $a^1 = a$ for all a, then

$$\log_a a = 1 \quad \text{(L6)}$$

Also, $\log \frac{1}{N} = \log 1 - \log N = 0 - \log N$

Hence, $\log \frac{1}{N} = -\log N$ (L7)

Also, **any** number N can be written as

$$N = a^{\log_a N} \tag{L8}$$

This result is again true for all positive values of a.

You may find formulae (L5) to (L8) quite difficult to understand at a first reading. They are included here for completeness and you may need to keep returning to them.

The following examples illustrate ways in which the logarithm notation is used.

Example 1.7

Write in logarithmic form

(i) $4^2 = 16$; (ii) $10^{-3} = 0.001$; (iii) $x^y = t$

Solution

Here we are being asked to introduced logarithms into an expression. We are using Equation (L1).

MEMORY JOGGER

Remember a logarithm is a power

(i) Here the power is 2, so the logarithm equals 2.
4 is the base,
therefore $\log_4 16 = 2$

(ii) Here the power is -3, so the logarithm equals -3.
10 is the base.
$\log_{10} 0.0001 = -3$

(iii) Here the power is y, so the logarithm equals y.
The base is x.
$\log_x t = y$.

Example 1.8

Express the following in index form (using powers):

(i) $\log_4 16 = 2$ (ii) $\log_{10} 0.1 = -1$; (iii) $\log_e N = t$

Solution

Once again, we are using Equation (L1):

(i) 2 is the power, and 4 is the base, so:

$$16 = 4^2$$

(ii) -1 is the power and 10 is the base, hence:

$$10^{-1} = 0.1$$

(iii) Here t is the power, and e the base:

Hence $e^t = N$

You should try to feel comfortable changing quickly between logarithmic form and index form.

Example 1.9

Evaluate:

(i) $\log_3 81$ (ii) $\log_{16} 4$ (iii) $\log_2 0.25$ (iv) $4^{\log_4 16}$

Solution

(i) To find $\log_3 8$ you need to answer the question: What power of the base gives the number 81?
Since $3^4 = 81$, then the power is 4.

i.e. $\log_3 81 = 4$

(ii) To find $\log_{16} 4$, what power of 16 gives the value 4. Now 4 is the square root of 16, and so the power is $\frac{1}{2}$.

$\therefore \quad \log_{16} 4 = 0.5$

(iii) $0.25 = \dfrac{1}{4} = \dfrac{1}{2^2} = 2^{-2}$

Hence the power of the base 2 is -2.

$\therefore \log_2 0.25 = -2$

(iv) Now $\log_4 16 = 4^2 = 16$

This illustrates the truth of equation L8.

The following examples illustrate ways in which the logarithm notation can be used to simplify expressions.

Example 1.10

Simplify and then evaluate:

(i) $3\log 2 + 2\log 5$ (ii) $\frac{1}{2}\log 16 - \frac{1}{3}\log 27$.

Solution

(i) $3\log 2 = \log 2^3$
$= \log 8$ (Using Equation (L4).)

and $2\log 5 = \log 5^2 = \log 25$

So $\log 8 + \log 25 = \log 8 \times 25 = \log 200$ (Using Equation (L2).)

Using the calculator,

Display

AC → 200 → log → 2.301029996 (AC clears the screen)

Hence $3\log 2 + 2\log 5 = 2.30$ (3 sig. figs).

(ii) $\frac{1}{2}\log 16 = \log 16^{\frac{1}{2}}$ using formula (L4)

$= \log 4$ (**Remember**: power $\frac{1}{2}$ is $\sqrt{\ }$)

$\frac{1}{3}\log 27 = \log 27^{\frac{1}{3}}$

$= \log 3$ (power $\frac{1}{3}$ is $\sqrt[3]{\ }$)

So $\frac{1}{2}\log 16 - \frac{1}{3}\log 3 = \log\frac{4}{3}$

$= 0.125$ (3 sig. figs).

Example 1.11

Write as a single logarithm:

(i) $2\log x + 5\log x^2$ (ii) $m\log n - n\log m$

Solution

(i) $2\log x = \log x^2$

and $5\log x^2 = \log(x^2)^5 = \log x^{10}$

So $2\log x + 5\log x^2$

$= \log x^2 + \log x^{10} = \log x^2 \times x^{10}$

$= \log x^{12}$

(ii) $m\log n - n\log m$

$$= \log n^m - \log m^n = \log\frac{n^m}{m^n}$$

This cannot be simplified any further.

Example 1.12

Express the following, in terms of $\log x$ and $\log y$.

(i) $\log 5x^2y$ (ii) $\log\sqrt{\frac{2x}{y^2}}$

Solution

In this example, we see how to expand out a logarithm into simple parts. This is often quite a useful technique.

(i) $\log 5x^2y = \log 5 + \log x^2 + \log y$ (Using Equation (L2).)

$= \log 5 + 2\log x + \log y$ (Using Equation (L4).)

(ii) $$\log\sqrt{\frac{2x}{y^2}} = \log\left(\frac{2x}{y}\right)^{\frac{1}{2}} = \tfrac{1}{2}\log\frac{2x}{y^2} \qquad \text{(Using Equation (L1).)}$$

$$= \tfrac{1}{2}\left[\log 2x - \log y^2\right] \qquad \text{(Using Equation (L3).)}$$

$$= \tfrac{1}{2}\left[\log 2 + \log x - 2\log y\right] \qquad \text{(Using Equation (L2).)}$$

Logarithms are very important in the solution of equations similar to the type $2^x = 7$. In other words, the unknown x appears in the power. Study the next example carefully.

Example 1.13

Solve the equations:

(i) $3^x = 7$ (ii) $4^{2x+1} = 5^{x-2}$

Solution

(i) $3^x = 7$

Take logs of both sides,

i.e. $\log 3^x = \log 7$

Use formula (L4) on the L.H.S.

$$x\log 3 = \log 7$$

Hence $$x = \frac{\log 7}{\log 3}$$

Note: This cannot be simplified in logarithmic form.

It can now be worked out:

$$x = \frac{0.84509804}{0.477121254} = 1.77$$

This can be carried on in one series of operations using a calculator:

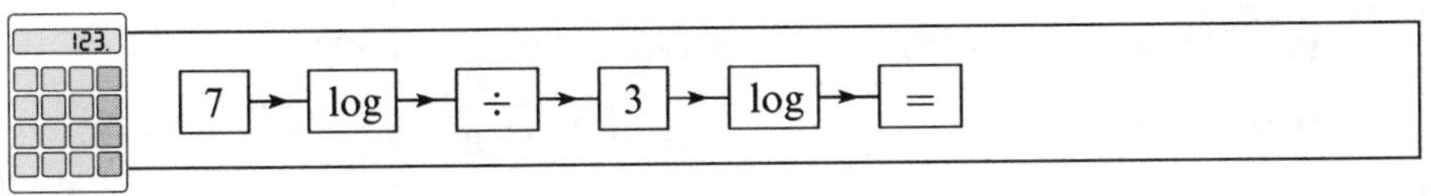

(ii) Once again, take logs of both sides

$$\log 4^{2x+1} = \log 5^{x-2}$$

$$\therefore \quad (2x+1)\log 4 = (x-2)\log 5 \qquad \text{(Make sure you use brackets here.)}$$

$$\therefore \quad 2x\log 4 + \log 4 = x\log 5 - 2\log 5$$

$$\therefore \quad 2x\log 4 - x\log 5 = -\log 4 - 2\log 5$$

$$x[2\log 4 - \log 5] = -[\log 4 + \log 5^2]$$

$$x[\log 4^2 - \log 5] = -\log 4 \times 25$$

$$\therefore \quad x = -\frac{\log 100}{\log\left(\frac{16}{5}\right)} = -3.96$$

Exercise 1(c)

1 Express the following using logarithms:

(i) $2^4 = 16$ (ii) $5^2 = 25$ (iii) $729 = 9^3$

(iv) $8^{\frac{2}{3}} = 4$ (v) $10^{-4} = 0.0001$ (vi) $16^{-\frac{1}{2}} = 0.25$

(vii) $1 = 7^0$ (viii) $t = x^3$ (ix) $t^x = q$

(x) $p^{\frac{1}{x}} = y$

2 Write these using index notation:

(i) $\log_4 64 = 3$ (ii) $\log_{10} 10000 = 4$

(iii) $\log_x y = t$ (iv) $\log_4 q = t$

(v) $\log_x 4 = y$ (vi) $\log_4 \frac{1}{16} = -2$

(vii) $z = \log_x y$ (viii) $\log_e 2 = t$

(ix) $\log_t 4q = 6$

3 Evaluate:

(i) $\log_2 8$ (ii) $\log_{10} 100$ (iii) $\log_4 64$

(iv) $\log_5 0.2$ (v) $\log_x x^4$ (vi) $\log_{10} 0.01$

(vii) $\log_{0.5} 8$ (viii) $3^{\log_3 9}$

4 Simplify:

(i) $\log 2 + \log 3$ (ii) $\log 12 - \log 4$

(iii) $2\log 4$ (iv) $2\log 3 + 3\log 2$

(v) $\log x + \log y + \log z$ (vi) $\frac{1}{2}\log 16 + \log 3$

(vii) $2\log x + \log y$ (viii) $1 + \log x$

(ix) $\log(x+1) - \log(x-1)$ (x) $\dfrac{\log 64}{\log 4}$

5 Expand

(i) $\log xy$ (ii) $\log \dfrac{p}{2q}$ (iii) $\log \dfrac{xy}{t}$

(iv) $\log \dfrac{x^2}{y}$ (v) $\log \sqrt{\dfrac{p}{q}}$ (vi) $\log \dfrac{a^3}{100}$

(vii) $\log \dfrac{1}{10x}$ (viii) $\log \sqrt{\dfrac{4x^2}{y^3}}$

6 Solve the following equations:

(i) $2^x = 5$ (ii) $3^t = 4$ (iii) $4^{x+2} = 9$

(iv) $2^{3t-1} = 2^{1+t}$ (v) $3^{2x-1} = 2^{x-1}$

(vi) $4^{x-1}.3^{2x-1} = 8$

1.4 The number e and other bases

There is a particular number in mathematics denoted by $e = 2.7182818$ (8 sig. figs), which assumes a particular importance in logarithms. The origins of this number will be investigated later on in the book. If e is used as the base of logarithms, i.e. $\log_e$, then usually this is abbreviated to 'ln' on the calculator and also in the written form,

that is, $\log_e 12$ will be written $\ln 12$

It is important that you use the correct logarithm button in a calculation, although sometimes it doesn't matter. For instance, in Example 1.12, when we solved $3^x = 7$, this could just as easily have been written:

$$x = \frac{\ln 7}{\ln 3} = \frac{1.945910}{1.098612} = 1.77 \text{ (3 sig. figs)}$$

Sometimes it may be necessary to change the base of a logarithm to other than 10 or e. To do this you use the following formula:

$$\log_a N = \frac{\log_b N}{\log_b a} \qquad \text{(L9)}$$

Hence if you wanted to find the logarithm to base 2 of 6, then:

$$\log_2 6 = \frac{\log_{10} 6}{\log_{10} 2} = 2.58. \quad \text{(3 sig. figs)}$$

In Equation (L9), if N happens to be the same as b, then

$$\log_a b = \frac{\log_b b}{\log_b a} = \frac{1}{\log_b a} \qquad \text{(Using formula (L9).)}$$

$$\text{Hence} \quad \log_a b = \frac{1}{\log_b a} \qquad \text{(L10)}$$

The following examples illustrate more difficult problems with equations and logarithms.

Example 1.14

Solve the equations:

(i) $\log_4 x = 7$ (ii) $\log(x+2) + \log(x-3) = \log 2$
(iii) $\log x + \log_2 x = 1$.

Solution

(i) $x = 4^7 = 16384$ using (L1)
(ii) The L.H.S. $= \log(x+2)(x-3)$

$$= \log(x^2 - x - 6)$$

So we are solving $\log(x^2 - x - 6) = \log 2$

$$\therefore \quad x^2 - x - 6 = 2$$

$$x^2 - x - 8 = 0$$

$$x = \frac{1 \pm \sqrt{1 - -32}}{2} = 3.37 \text{ or } -2.37$$

It is tempting to say that these are the two answers. However, if you try and substitute -2.37 into the original equation, you get:

$$\log(-2.37+2) + \log(-2.37-3)$$
$$= \log(-0.37) + \log(-5.37)$$

The logarithm of a negative number does not exist, hence -2.37 is not a solution.

MEMORY JOGGER

Always check your solutions back into the equation when it contains a logarithm of a function of x.

(iii) This is a very difficult problem. The bases are mixed, and you need to change the base 2 part into base 10, using formula L9.

$$\text{Hence} \quad \log x + \frac{\log x}{\log 2} = 1$$

$$\therefore \quad \log x + 3.322 \log x = 1$$

$$4.322 \log x = 1$$

$$\therefore \quad \log x = \frac{1}{4.322} = 0.2314$$

Hence $x = 10^{0.2314} = 1.70$

Exercise 1(d)

Solve the following equations:

1 $\ln x = 3$

2 $\ln x(x+1) = \ln 6$

3 $\log_2 x = 6$

4 $\log(x+3) + \log(x-2) = \log 50$

5 $\log_4 x + \log_x 4 = 2$

6 Solve the simultaneous equations:

$$\log_2 x = \log_4 y - 5$$

$$\log_4 x = \log_2 y + 3$$

(This is a more challenging question. Well done if you solve it without looking up the solution.)

Miscellaneous Examples 1

1 Simplify the following:

(i) 4^{-2} (ii) $16^{\frac{1}{2}}$ (iii) $2^3 \times 4^2$ (iv) $(10^2)^3$

(v) $64^{-\frac{1}{2}}$ (vi) $(\frac{1}{2})^{-2}$ (vii) $(\frac{1}{3})^{-2} \div (\frac{1}{4})^{-1}$

(viii) $125^{\frac{2}{3}}$ (ix) $(0.0001)^{\frac{1}{4}}$ (x) $(\frac{2}{5})^{-2} \times (\frac{1}{10})^2$

2 Write the following algebraic expressions as simply as possible:

(i) $x^2 \times 2x^4$ (ii) $(2x)^3$ (iii) $8a \times 4a^2$

(iv) $(x^3)^2 \div x^4$ (v) $(8at)^2 \div 2at^2$ (vi) $\sqrt{x^2 y^4}$

(vii) $\sqrt[3]{(8x^3 y^3)^2}$ (viii) $(4p)^2 \div (2p)^{-2}$

(ix) $\left(x + \dfrac{1}{x}\right) \div (x^2+1)^2$ (x) $(2\sqrt{x})^3 \div x^4$

3 Write the following as a single logarithm:

(i) $\log x + 2\log y$ (ii) $3\log t - 2\log x$

(iii) $\frac{1}{2}\log A - 2\log y$ (iv) $\frac{1}{2}\log 4x - \frac{1}{2}\log y^2$

(v) $4\log x + 3\log y - 2\log t$ (vi) $2 + 3\log x$

4 Expand the following expressions to give answers of the form $m\log x + n\log y$

(i) $\log x^5 y$ (ii) $\log\sqrt{\dfrac{x}{y}}$ (iii) $\log\dfrac{1}{\sqrt{x}}$

(iv) $\log\dfrac{x^3}{y^3}$ (v) $\log\sqrt[3]{\dfrac{x^2}{y^3}}$

5 Solve the equations:

(i) $4^x = 3$ (ii) $2^{2x+1} = 4$ (iii) $2^x = 4^{x+3}$
(iv) $3^{2x+1} . 3^{x+2} = 8$

6 Rationalise the denominator in the following expressions:

(i) $\dfrac{2}{\sqrt{8}}$ (ii) $\dfrac{1}{1+\sqrt{2}}$ (iii) $\dfrac{1}{\sqrt{6}-\sqrt{3}}$

(iv) $\dfrac{\sqrt{3}}{\sqrt{6}}$ (v) $\dfrac{1+\sqrt{5}}{1-\sqrt{5}}$ (vi) $\dfrac{\sqrt{3}}{1+\sqrt{2}}$

7 Solve the equations:
(i) $\log(x+1) + \log(3x+1) = \log 65$
(ii) $\log_4 x + \log_2 x = 9$
(iii) $\log(3x+2) - \log(x+1) = \log(4x+1) - \log 2$

Revision Problems 1

1 The value, V, of a particular car, can be modelled by the equation

$$V = ke^{-\lambda t}$$

where t years is the age of the car. The car's original price was £7499, and after one year it is valued at £6000. State the value of k and calculate λ, giving your answer to 2 decimal places. Obtain the value of the car when it is three years old.

(NEAB)

2 The population of Britain about a century ago is given in the table below

Year	1881	1891	1901	1911
Population (millions)	30	33	37	41

(a) Calculate the growth factors of the three decades covered in the table, giving your answers to 2 decimal places.

Explain why the population growth for this period can be modelled using an exponential function.

(b) If the population had continued to grow exponentially, with growth factor r per decade, write down a formula in terms of r for the poulation t decades after 1911.

(c) Show that the population would take T decades to double in size, where:

$$T = \frac{\ln 2}{\ln r}$$

(d) State a suitable value for r. Hence, calculate the value of T to the nearest whole number.

(NEAB)

2 Polynomials

A large part of mathematics centres around the solution of some sort of equation. You should already be able to solve simple equations including simultaneous and perhaps quadratic equations. This chapter takes you forward to other types.

DO YOU KNOW

(i) $(a+b)(c+d) = ac + bc + bd + ad$

(ii) $f(x)$, known as the **function** of x, can be used to represent any expression containing a variable, x.

(iii) $f(a)$ means replace (or substitute) $x = a$ into that expression in place of x. This is referred to as function of a.

(iv) The solution of $ax^2 + bx + c = 0$ is:

$$x = \frac{-b \pm \sqrt{b^2 - 4ac}}{2a}$$

2.1 Polynomials (manipulation)

A polynomial is an expression such as $x^4 + 3x^3 - 2x + 1$ (the sum or difference of **terms** containing powers of x usually arranged in order of decreasing powers). The highest power of x (in this example, the power 4), is called the **degree** of the polynomial. A simple polynomial of the type $3x - 5$ is often referred to as a **linear** function. It is possible to multiply and divide polynomials as shown below.

Example 2.1

(i) Find $(x^3 + 1)(x^2 + 2x - 3)$.

(ii) Divide $x^5 + 3x^3 + x^2 + 2x + 2$ by $(x^2 + 2)$

(iii) Find the remainder when $x^3 + 4x^2 + 6x + 1$ is divided by $(x - 4)$.

Solution

(i) We are going to multiply out the brackets.

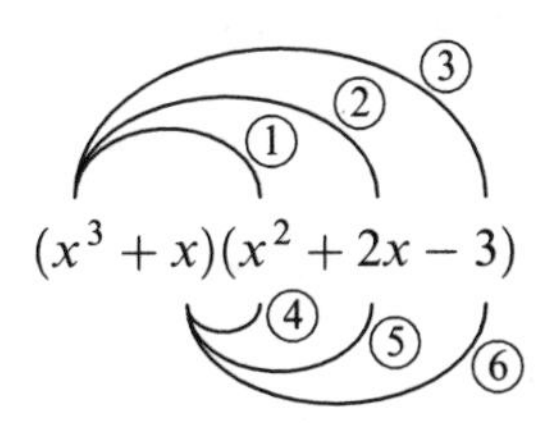

There are $2 \times 3 = 6$ possible products of one term taken from each bracket. The answer is:

$$x^5 + 2x^4 - 3x^3 + x^3 + 2x^2 - 3x$$

① ② ③ ④ ⑤ ⑥

This simplifies to:

$$x^5 + 2x^4 - 2x^3 + 2x^2 - 3x$$

(Note that ③ and ④ have been combined.)

This is called a polynomial of **degree** 5, because the highest power of x that occurs is 5.

(ii) Division follows a process similar to that of long division with ordinary numbers:

Step (i)

$(x^2 + 2)$ divides into x^5, x^3 times. (Ⓐ)

$$\begin{array}{r|l} & x^3 \quad \swarrow Ⓐ \\ \hline x^2 + 2 & x^5 + 3x^3 + x^2 + 2x + 2 \\ & \underline{x^5 + 2x^3} \end{array}$$

Now write $x^3 \times (x^2 + 2)$ under $x^5 + 3x^3 + x^2 + 2x + 2$ and subtract. (Ⓑ)

Step (ii)

(x^2+2) divides into x^3, x times. (Ⓒ)

$$
\begin{array}{r l}
 & x^3 + x \quad \leftarrow \text{Ⓒ} \\
x^2+2 & \overline{)x^5+3x^3+x^2+2x+2} \\
 & \underline{x^5+2x^3} \qquad \text{subtract} \leftarrow \text{Ⓑ} \\
 & x^3+x^2+2x+x
\end{array}
$$

($x^3 \times (x^2+2)$)

Write $x \times (x^2+2)$ under x^3+x^2+2x+2 and subtract. (Ⓓ)

Step (iii)

x^2+2 divides exactly in x^2+2 once, (Ⓔ)

giving a zero remainder. (Ⓕ)

$$
\begin{array}{r l}
 & x^3 + x + 1 \quad \leftarrow \text{Ⓔ} \\
x^2+2 & \overline{)x^5+3x^3+x^2+2x+2} \\
 & \underline{x^5+2x^3} \\
 & x^3+x^2+2x+2 \\
 & \underline{x^3+2x} \qquad \text{subtract} \leftarrow \text{Ⓓ} \\
 & x^2+2 \\
 & \underline{x^2+2} \\
 & 0 \quad \leftarrow \text{Ⓕ}
\end{array}
$$

($x \times (x^2+2)$)

($1 \times (x^2+2)$)

Hence: $(x^5+3x^3+x^2+2x+2) \div (x^2+2) = x^3+x+1$

(iii) As with ordinary division, algebraic division does not always give an exact answer.

$$
\begin{array}{r l}
 & x^2 + 8x + 38 \\
x-4 & \overline{)x^3+4x^2+6x+1} \\
 & \underline{x^3-4x^2} \\
 & 8x^2+6x+1 \\
 & \underline{8x^2-32x} \qquad \text{subtract} \\
 & 38x+1 \\
 & \underline{38x-152} \\
 & 153
\end{array}
$$

($x^2 \times (x-4)$)

($8x \times (x-4)$)

($38 \times (x-4)$)

The division process is followed as in Example (ii) and you can see that this time there is a remainder of 153. It should be pointed out that the remainder can be found in a much simpler way.

If you let $f(x) = x^3 + 4x^2 + 6x + 1$, then find:

$$f(4) = 4^3 + 4 \times 4^2 + 6 \times 4 + 1$$
$$= 153$$

This leads to stating the **remainder theorem** which is as follows:

> If a polynomial $f(x)$ is divided by $(x - a)$,
> the remainder will be $f(a)$ (F 1)

Example 2.2

Find the remainder when $x^3 + 2x^2 - 5$ is divided by $x + 2$.

Solution

$x + 2$ can be written $x - (-2)$, and so $a = -2$.

$$f(x) = x^3 + 2x^2 - 5$$
$$\therefore \quad f(-2) = (-2)^3 + 2(-2)^2 - 5 = -5$$

The remainder is -5.

Exercise 2(a)

1 Evaluate the following, and simplify your answer:

(i) $(2x + 1)(3x - 4)$ (ii) $(t + 1)(t^2 + 1)$

(iii) $(x^2 + x + 1)(x - 1)$ (iv) $(3x - 1)(2x + 1)(x - 2)$

(v) $(x + 2)(x^3 - x^2 + 1)$ (vi) $(2t^2 + 1)(t^2 + t + 2)$

2 Work the following:

(i) $(3x^3 + 4x^2 + 7x - 14) \div (x - 1)$

(ii) $(4x^4 - 3x^2 + 7x + 6) \div (x + 1)$

(iii) $(7x^3 - 7x - 42) \div (x - 2)$

(iv) $(8x^3 + 4x^2 - 2) \div (2x - 1)$

3 Find the remainder when:

(i) $x^2 + 7x + 8$ is divided by $x + 1$

(ii) $x^3 + 2x + 1$ is divided by $x - 2$

(iii) $x^4 - 7x^3 + 6x^2 + 5$ is divided by $x - 1$

2.2 Finding linear factors of a polynomial

In Example 2.1, you were shown how to divide a polynomial by a linear factor. Finding what a linear factor $(ax + b)$ is requires a systematic approach in the first instance. The **factor formula** can be used. It states:

> If a polynomial $f(x)$ is exactly divisible by a linear factor
>
> $ax - b$, then $f\left(\frac{b}{a}\right) = 0.$ (F2)
>
> Very often, $a = 1$, and so if $x - b$ is a factor, $f(b) = 0$

The use of this theorem is demonstrated in the following example.

Example 2.3

Find one factor of the expression

$$4x^3 - 20x^2 + 19x + 15$$

and hence factorise the expression completely.

Solution

Let $f(x) = 4x^3 - 20x^2 + 19x + 15$
Now try values of x until you find a solution of $f(x) = 0$.
Choose small values of x first:

$$\begin{aligned} f(1) &= 4 - 20 + 19 + 15 = 18 \\ f(2) &= 32 - 80 + 38 + 15 = 5 \\ f(3) &= 108 - 180 + 57 + 15 = 0 \quad \checkmark \end{aligned}$$

Since $f(3) = 0$, then $x - 3$ is a factor.
Using the division process, you will find:

$$(4x^3 - 20x^2 + 19x + 15) \div (x - 3) = 4x^2 - 8x - 5$$

$$\begin{aligned} \text{So} \quad f(x) &= (x - 3)(4x^2 - 8x - 5) \\ &= (x - 3)(2x + 1)(2x - 5) \end{aligned}$$

The polynomial has now been completely factorised.
You will need to take care if the solution is not a whole number. Work carefully through the following example:

Example 2.4

Solve the equation $2x^3 + 3x^2 + 3x + 1 = 0$.

$$
\begin{aligned}
\text{Let}\quad f(x) &= 2x^3 + 3x^2 + 3x + 1 \\
f(0) &= 1 \\
f(1) &= 2 + 3 + 3 + 1 = 9 \\
f(2) &= 16 + 12 + 6 + 1 = 35
\end{aligned}
$$

Clearly, as x increases, $f(x)$ is getting bigger and not closer to zero.
Try negative values of x.

$$
\begin{aligned}
f(-1) &= -2 + 3 - 3 + 1 = -1 \\
f(-2) &= -16 + 12 - 6 + 1 = -9 \\
f(-3) &= -54 + 27 - 9 + 1 = -35
\end{aligned}
$$

Here you can see that $f(x)$ is becoming more negative.

However, since $\quad f(0) = 1$

and $\quad f(-1) = -1$

there must be a value of x between 0 and -1 for which $f(x) = 0$.
Try $x = -\frac{1}{2}$.

$$f(-\tfrac{1}{2}) = -\tfrac{1}{4} + \tfrac{3}{4} - \tfrac{3}{2} + 1 = 0$$

Hence $(x - -\frac{1}{2}) = x + \frac{1}{2}$ is a factor.

Since all the coefficients in the polynomial are whole numbers, the factor is:

$2 \times (x + \frac{1}{2})$ or $(2x + 1)$

> **MEMORY JOGGER**
>
> You can locate a solution to $f(x) = 0$ by finding two values of x (a and b) which give a different signs for $f(a)$ and $f(b)$. The solution lies between $x = a$ and $x = b$

Now find $(2x^3 + 3x^2 + 3x + 1) \div (2x + 1)$ using the division process to give $x^2 + x + 1$

Hence the equation is $(2x + 1)(x^2 + x + 1) = 0$

Now $\quad x = -\frac{1}{2}$ or $x^2 + x + 1 = 0$

If you use the quadratic formula on this,

$$x = \frac{-1 \pm \sqrt{1 - 4}}{2} = \frac{-1 \pm \sqrt{-3}}{2}$$

Hence this part has no real solutions, as $\sqrt{-3}$ is not a real number. The final solution of the equation is just $x = -\frac{1}{2}$.

Exercise 2(b)

1 Factorise the following expressions. One factor is given:

(i) $x^3 + 3x^2 - 6x - 8$: $(x - 2)$

(ii) $4x^3 - 12x^2 - x - 3$: $(x + 3)$

(iii) $6x^3 + 5x^2 - 21x + 10$: $(x - 1)$

(iv) $2t^3 - 15t^2 + 13t + 60$: $(2t + 3)$

2 Factorise the following expressions:

(i) $5x^2 - 12x + 7$ (ii) $6t^2 - t - 15$

(iii) $t^3 - 6t^2 + 11t - 6$ (iv) $x^5 - x^4 - x + 1$

(v) $2t^3 + t^2 + 2t + 1$ (vi) $y^3 - y^2 - 14y + 24$

3 Solve the equations

(i) $x^3 - 2x^2 + x - 2 = 0$ (ii) $x^4 - 9 = 0$

(iii) $6x^3 - 5x^2 - 2x + 1 = 0$ (iv) $4y^3 + y^2 - 8y - 2 = 0$

2.3 Solving equations

DO YOU KNOW

(i) $ax = b$ has the solution $x = \dfrac{b}{a}$.

(ii) $ax + b = 0$ is called a linear equation.

(iii) $ax^2 + bx + c = 0$ is a quadratic equation.
It can be solved by factors, or using the formula:

$$x = \frac{-b \pm \sqrt{b^2 - 4ac}}{2a}$$

(iv) $\left.\begin{aligned} ax + by &= c \\ dx + ey &= f \end{aligned}\right\}$ are called linear simultaneous equations. They can be solved by eliminating either x or y first.

(v) $(a + b)^2 = a^2 + 2ab + b^2$
$(a - b)^2 = a^2 - 2ab + b^2$
$(a - b)(a + b) = a^2 - b^2$

It is likely that the only types of equation you will be able to solve with confidence at the moment are linear, quadratics, and linear simultaneous equations. The purpose of this section is to help you increase that range. Work carefully through the following equations, and see how your knowledge grows.

MEMORY JOGGER

Any equation that contains fractions should have the denominators removed, by multiplying through by the common denominator first.

Type I Solve $\frac{1}{3}(x+2) - \frac{1}{5}(2x-1) = \frac{1}{4}$.

The smallest number that 3, 4 and 5 divide into exactly is 60. Hence each term should be multiplied by 60.

$$\therefore \quad 60 \times \frac{1}{3}(x+2) - 60 \times \frac{1}{5}(2x-1) = 60 \times \frac{1}{4}$$

$$\begin{aligned}
\text{i.e.} \quad 20(x+2) - 12(2x-1) &= 15 \\
20x + 40 - 24x + 12 &= 15 \\
-4x = 15 - 12 - 40 &= -37 \\
\therefore \quad x = \frac{-37}{-4} &= 9.25
\end{aligned}$$

Type II Solve $\frac{4}{(3x-1)} = \frac{3}{(1-2x)}$

The common denominator here is $(3x-1)(1-2x)$

$$\text{So} \quad (3x-1)(1-2x) \times \frac{4}{(3x-1)} = (3x-1)(1-2x) \times \frac{3}{(1-2x)}$$

$$\begin{aligned}
\text{i.e.} \quad 4(1-2x) &= 3(3x-1) \\
\therefore \quad 4 - 8x &= 9x - 3 \\
4 + 3 &= 9x + 8x \\
7 &= 17x \\
x &= \tfrac{7}{17}
\end{aligned}$$

Note: The equation was in fact a linear one.

Type III $3x + \frac{2}{x} = 8$

The common denominator here is x.

$$\begin{aligned}
\text{So} \quad x \times 3x + x \times \frac{2}{x} &= 8 \times x \\
\therefore \quad 3x^2 + 2 &= 8x \\
3x^2 - 8x + 2 &= 0
\end{aligned}$$

Using the formula for a quadratic equation:

$$\begin{aligned}
x &= \frac{8 \pm \sqrt{(-8)^2 - 4 \times 3 \times 2}}{6} \\
&= \frac{8 \pm \sqrt{64 - 24}}{6} \\
&= 2.39 \quad \text{or} \quad 0.279
\end{aligned}$$

Type IV $t^4 - 5t^2 - 36 = 0$

At first sight, this looks like a **quartic** (degree 4) equation, but in fact it is not. If you let $x = t^2$, the equation becomes:

$$x^2 - 5x - 36 = 0$$
$$(x - 9)(x + 4) = 0$$
$$\therefore \quad x = 9 \quad \text{or} \quad x = -4$$

But, since $x = t^2$, this means:

$$t^2 = 9 \quad \text{or} \quad t^2 = -4$$

Now t^2 cannot be negative, and so $t^2 = -4$ is impossible. If $t^2 = 9$, then $t = \pm 3$.

The complete solution is $t = -3, 3$.

Type V $\sqrt{2x + 3} + \sqrt{6x - 2} = 7$

This is quite a difficult equation to solve. For the moment, let $A = \sqrt{2x + 3}$ and $B = \sqrt{6x - 2}$.

So $\quad A + B = 7$

Now square both sides:

$$(A + B)^2 = 7^2 = 49$$
$$\text{i.e.} \quad A^2 + B^2 + 2AB = 49$$

Now substitute back:

$$\text{So} \quad (2x + 3) + (6x - 2) + 2\sqrt{2x + 3}\sqrt{6x - 2} = 49$$
$$\therefore \quad 8x + 1 + 2\sqrt{2x + 3}\sqrt{6x - 2} = 49$$
$$\therefore \quad 2\sqrt{2x + 3}\sqrt{6x - 2} = 49 - 8x - 1 = 48 - 8x$$

Divide by 2

$$\therefore \quad \sqrt{2x + 3}\sqrt{6x - 2} = 24 - 4x$$

You will notice that the $\sqrt{\ }$ has not yet been removed, and so both sides must be squared again.

$$\text{So} \quad (2x + 3)(6x - 2) = (24 - 4x)^2$$
$$\therefore \quad 12x^2 + 18x - 6 - 4x = 576 + 16x^2 - 192x$$
$$\therefore \quad 0 = 4x^2 - 206x + 582$$
$$0 = 2x^2 - 103x + 291$$
$$0 = (2x - 97)(x - 3)$$

(You can always use the quadratic formula if you are not able to factorise it.)

$$\therefore \quad x = 3 \quad \text{or} \quad 48.5$$

It would appear that there are two answers, but you have squared the original equation twice, and this may produce problems.

Checking $x = 3$ $\quad \sqrt{2 \times 3 + 3} + \sqrt{6 \times 3 - 2} = 7$

$3 + 4 = 7$ ✔

For $x = 48.5$ $\quad \sqrt{2 \times 48.5 + 3} + \sqrt{6 \times 48.5 - 2} = 7\,?$

But $10 + 17 \neq 7$

Hence $x = 3$ is the only solution.

Exercise 2(c)

Solve the following equations:

1 $4(x+1) = 7(3-2x)$

2 $\dfrac{5}{x+1} = \dfrac{4}{x-3}$

3 $\dfrac{1}{4}(x+1) - \dfrac{2}{3} = \dfrac{1}{5}(x+2)$

4 $4x^2 - 7x - 3 = 0$

5 $5 + \dfrac{2}{x} = 6x$

6 $\sqrt{7x+1} - \sqrt{2x-1} = 3$

7 $t^4 + 8t^2 - 6 = 0$

8 $\sqrt{2x-1} = 3\sqrt{3x+5}$

9 $\dfrac{1}{x} + \dfrac{3}{x^2} = 4$

10 $\dfrac{x+1}{x-3} = \dfrac{2x-5}{2x+11}$

2.4 The quadratic polynomial

The expression $ax^2 + bx + c$ is a polynomial of degree 2. However, it is usually referred to as a quadratic expression (or quadratic function). It is used extensively in many aspects of mathematics and deserves special consideration. You might like to try the following activity before continuing.

ACTIVITY 1

The quadratic function $y = ax^2 + bx + c$.

There are many aspects of this function which depend on relationships between the coefficients a, b and c. Here are some suggestions for you to look at. You can of course look at your own.

(i) Where is the line of symmetry of the curve?

(ii) What does the sign of a tell you about the graph?

(iii) What happens if $b^2 < 4ac$?

(i) Completing the square

A quadratic expression, such as $x^2 + 6x + 11$, can often be manipulated more easily if it is changed into a different form, obtained by a process called **completing the square**.

Now $(x+3)(x+3) = x^2 + 6x + 9$, so if you write $x^2 + 6x + 11$ as $x^2 + 6x + 9 + 2$, then this becomes $(x+3)(x+3) + 2$ or $(x+3)^2 + 2$.

This means that x only appears once in the expression. It can be represented by a flow chart.

$$\boxed{x} \rightarrow \boxed{+3} \rightarrow \boxed{\text{square}} \rightarrow \boxed{+2} \rightarrow \boxed{y}$$

That is, $y = (x+3)^2 + 2$.

If you reverse this flow chart, you get:

$$\boxed{x} \leftarrow \boxed{-3} \leftarrow \boxed{\pm\sqrt{\ }} \leftarrow \boxed{-2} \leftarrow \boxed{y}$$

Hence $x = -3 \pm \sqrt{(y-2)}$

You would find it impossible to rearrange the original expression $y^2 + 6x + 9$ to make x the subject without this technique. The technique is also particularly useful when trying to find inverse functions.

Try another example:

$$x^2 - 5x - 3$$

$$\left(x - \tfrac{5}{2}\right)^2 = x^2 - 5x + \tfrac{25}{4} = x^2 - 5x + 6\tfrac{1}{4}$$

$$\therefore \quad x^2 - 5x = \left(x - \tfrac{5}{2}\right)^2 - 6\tfrac{1}{4}$$

$$\therefore \quad x^2 - 5x - 3 = \left(x - \tfrac{5}{2}\right)^2 - 6\tfrac{1}{4} - 3$$

$$= \left(x - \tfrac{5}{2}\right)^2 - 9\tfrac{1}{4}$$

This result can be summarised as follows:

If $y = x^2 + bx + c$

Completing the square gives

$$y = \left(x + \frac{b}{2}\right)^2 + c - \frac{b^2}{4} \qquad \text{(Q1)}$$

If the coefficient of x^2 is not 1, proceed as follows:

For example, If $y = 4x^2 + 8x + 5$

then
$$y = 4\left[x^2 + 2x + \tfrac{5}{4}\right]$$
$$= 4\left(x+1)^2 + \tfrac{5}{4} - 1\right]$$
$$= 4\left[(x+1)^2 + \tfrac{1}{4}\right]$$
$$= 4(x+1)^2 + 1$$

> **MEMORY JOGGER**
>
> If trying to complete the square on $ax^2 + bx + c$, always take out a as a factor first.

You can use completing the square to find the greatest or least value of a quadratic expression, as the following example shows:

Example 2.5

Find the greatest or least values of the following expressions, stating at what value of x they occur:

(i) $x^2 + 8x + 2$ (ii) $4 - 3x - 2x^2$

Solution

(i) $x^2 + 8x + 2 = (x+4)^2 + 2 - 16$
$\qquad = (x+4)^2 - 14$

Now $(x+4)^2$ is always > 0 unless $x = -4$. So if $x = -4$, the least value of the expression $x^2 + 8x + 2$ is -14 (because $(x+4)^2 = 0$ at $x = -4$).

(ii) Here, $a = -2$

$$\begin{aligned}\therefore\quad 4 - 3x - 2x^2 &= -2[x^2 + \tfrac{3}{2}x - 2]\\ &= -2\left[\left(x + \tfrac{3}{4}\right)^2 - 2 - \tfrac{9}{16}\right]\\ &= -2\left[\left(x + \tfrac{3}{4}\right)^2 - \tfrac{41}{16}\right]\\ &= \tfrac{41}{8} - 2\left(x + \tfrac{3}{4}\right)^2\end{aligned}$$

The greatest value that can be obtained is $\frac{41}{8}$ and this occurs when $x = -\frac{3}{4}$

(ii) Properties of roots of the equation

If the roots (or solutions) of a quadratic equation are denoted by α and β, then it means the equation will have factorised as:

$$(x - \alpha)(x - \beta) = 0$$

That is, $x^2 - (\alpha + \beta)x + \alpha\beta = 0.$ (i)

If the original equation was $ax^2 + bx + c = 0$, then:

Dividing by a $\quad x^2 + \dfrac{b}{a}x + \dfrac{c}{a} = 0$ (ii)

Since (i) and (ii) represent the same equation, it follows that:

the **sum** of the roots $\alpha + \beta = -\dfrac{b}{a}$ (Q2)

the **product** of the roots $\alpha\beta = \dfrac{c}{a}$ (Q3)

Equations (Q2) and (Q3) state the symmetry properties of the roots. They can be used if you are only interested in changing the roots of an equation without finding the individual values.

Example 2.6

The equation $4x^2 - 3x + 1 = 0$ has roots α and β. Without evaluating α and β, find:

(i) $\alpha + \beta$ (ii) $\alpha\beta$ (iii) $\frac{1}{\alpha} + \frac{1}{\beta}$

(iv) $\alpha^2 + \beta^2$ (v) $\alpha^3 + \beta^3$ (vi) $\alpha^4 + \beta^4$

Solution

Dividing the equation by 4, we get

$$x^2 - \tfrac{3}{4}x + \tfrac{1}{4} = 0$$

Hence $\alpha + \beta = - -\tfrac{3}{4} = \tfrac{3}{4}$

$$\alpha\beta = \tfrac{1}{4}$$

We have the answers to (i) $\frac{3}{4}$ and (ii) $\frac{1}{4}$

(iii) $\frac{1}{\alpha} + \frac{1}{\beta} = \frac{\beta + \alpha}{\alpha\beta} = \frac{3}{4} \div \frac{1}{4} = 3$

(iv) $\alpha^2 + \beta^2$ cannot be evaluated immediately.

However $(\alpha + \beta)^2 = \alpha^2 + 2\alpha\beta + \beta^2$

Rearranging, $\alpha^2 + \beta^2 = (\alpha + \beta)^2 - 2\alpha\beta$

$$= \left(\tfrac{3}{4}\right)^2 - 2 \times \tfrac{1}{4}$$

$$= \tfrac{1}{16}$$

(v) $\alpha^3 + \beta^3$ suggests working out $(\alpha + \beta)^3$

$$(\alpha + \beta)^3 = \alpha^3 + 3\alpha^2\beta + 3\alpha\beta^2 + \beta^3$$

$$= \alpha^3 + \beta^3 + 3\alpha\beta(\alpha + \beta)$$

Hence $\alpha^3 + \beta^3 = (\alpha + \beta)^3 - 3\alpha\beta(\alpha + \beta)$

$$= \left(\tfrac{3}{4}\right)^3 - 3 \times \tfrac{1}{4} \times \tfrac{3}{4}$$

$$= -\tfrac{9}{64}$$

(vi) $\alpha^4 + \beta^4 = (\alpha^2)^2 + (\beta^2)^2 = (\alpha^2 + \beta^2)^2 - 2\alpha^2\beta^2$

$$= \left(\tfrac{1}{16}\right)^2 - 2 \times \left(\tfrac{1}{4}\right)^2 = -\tfrac{31}{256}$$

It is possible, by a technique of transforming the equation, to find a new equation where the roots are related to the roots of a given equation. Look at the following example.

Example 2.7

If α and β are the roots of $3x^2 + 2x + 5 = 0$, find new equations with roots

(i) $3\alpha, 3\beta$ (ii) $\frac{1}{\alpha}, \frac{1}{\beta}$ (iii) $\frac{\alpha}{\beta}, \frac{\beta}{\alpha}$

Solution

(i) If the new equation has solutions X_1 and X_2, then

$$\left.\begin{array}{l} X_1 = 3\alpha = 3x_1 \\ X_2 = 3\beta = 3x_2 \end{array}\right\} \quad \text{where } x_1 \text{ and } x_2 \text{ are the original roots.}$$

Hence: $X = 3x$ or $x = \frac{X}{3}$

Substitute this in the original equation:

$$3\left(\frac{X}{3}\right)^2 + 2\left(\frac{X}{3}\right) + 5 = 0$$

$$\frac{X^2}{3} + \frac{2X}{3} + 5 = 0$$

or $X^2 + 2X + 15 = 0$ is the new equation.

(ii) Here $X = \frac{1}{x}$ so $x = \frac{1}{X}$

$$\therefore \quad 3\left(\frac{1}{X}\right)^2 + 2\left(\frac{1}{X}\right) + 5 = 0$$

Hence: $3 + 2X + 5X^2 = 0$

(iii) This is not quite so easy because α and β appear in each new root.

However, $\alpha + \beta = -\frac{2}{3}$, so $\alpha = -\frac{2}{3} - \beta$

or $\beta = -\frac{2}{3} - \alpha$

Hence: $$\frac{\alpha}{\beta} = \frac{1}{\beta}\left(-\frac{2}{3} - \beta\right) = -\frac{2}{3\beta} - 1$$

and $$\frac{\beta}{\alpha} = \frac{1}{\alpha}\left(-\frac{2}{3} - \alpha\right) = -\frac{2}{3\alpha} - 1$$

that is, $$X = -\frac{2}{3x} - 1$$

$$\therefore \quad X + 1 = -\frac{2}{3x}, \quad x = -\frac{2}{3(X+1)}$$

$\therefore \qquad 3 \times \dfrac{4}{9(X+1)^2} - \dfrac{4}{3(X+1)} + 5 = 0$

$\therefore \qquad 12 - 12(X+1) + 45(X+1)^2 = 0$

giving $\qquad 45X^2 + 78X + 45 = 0$

or $\qquad 15X^2 + 26X + 15 = 0$

Exercise 2(d)

1 Complete the square for the following quadratic equations. In each case, state whether the function will have a greatest or least value, and the value of x where it occurs.

(i) $x^2 + 6x + 12$ (ii) $x^2 - 3x + 1$

(iii) $1 - x - x^2$ (iv) $4x^2 + 8x + 15$

(v) $9x^2 - 6x - 11$ (vi) $1 - 3x + 2x^2$

(vii) $-2 - 3x - 6x^2$ (viii) $7x^2 - 11x - 8$

2 If α and β are the roots of the equation $x^2 + 4x + 5 = 0$, evaluate

(i) $\alpha^2 + \beta^2$ (ii) $\dfrac{1}{\alpha} + \dfrac{1}{\beta}$

(iii) $\dfrac{\alpha}{\beta} + \dfrac{\beta}{\alpha}$ (iv) $\dfrac{1}{\alpha^2} + \dfrac{1}{\beta^2}$

3 If α and β are the roots of the equation $2x^2+x+2 = 0$. Form new equations with roots

(i) $2\alpha, 2\beta$ (ii) $\dfrac{1}{\alpha^2}, \dfrac{1}{\beta^2}$

(iii) $\dfrac{\alpha}{\beta}, \dfrac{\beta}{\alpha}$ (iv) $\dfrac{1}{\alpha^3}, \dfrac{1}{\beta^3}$

2.5 Non-linear simultaneous equations

You should be able to solve simultaneous equations of the linear type, for example: $2x + 3y = 6$ and $5x - 3y = 5$.

Now consider the following situation:

$$x^2 + 2xy = 5 \qquad \text{(i)}$$

$$x + y = 3 \qquad \text{(ii)}$$

Equation (ii) is linear, but equation (i) is not, because it contains an x^2 and an xy term. Rearrange the linear equation to make x or y the subject.

So $y = 3 - x$ (iii)

Now replace y in equation (i) by using y given in equation (iii)

$$\therefore \quad x^2 + 2x(3 - x) = 5$$
$$x^2 + 6x - 2x^2 = 5$$
$$\therefore \quad 0 = x^2 - 6x + 5$$
$$0 = (x - 5)(x - 1)$$
$$\therefore \quad x = 1 \quad \text{or} \quad x = 5$$

If $x = 1$ equation (i) gives $y = 2$
If $x = 5$ equation (i) gives $y = -2$

Hence, there are two solutions: $x = 1, y = 2$
and $x = 5, y = -2$

If both equations are non-linear, then very often an exact solution cannot be found, and numerical methods must be used. The following example shows how sometimes it is possible to solve them exactly.

$$x^2 + y^2 = 10 \qquad \text{(i)}$$
$$x^2 + 2xy = 3 \qquad \text{(ii)}$$

From (ii) $y = \dfrac{3 - x^2}{2x}$

Substitute this into equation (i):

$$x^2 + \frac{(3 - x^2)^2}{(2x)^2} = 10$$
$$\therefore \quad x^2 + \frac{(9 + x^4 - 6x^2)}{4x^2} = 10$$

Multiply through by $4x^2$

$$4x^4 + (9 + x^4 - 6x^2) = 40x^2$$
$$\therefore 5x^4 - 46x^2 + 9 = 0$$
$$(5x^2 - 1)(x^2 - 9) = 0, \text{ hence } x^2 = 9 \text{ or } 0.2$$
$$\therefore \quad x = 3, \ -3, \ 0.447, \ -0.447$$

These four values must now be substituted into (ii) to find the y values from $y = \dfrac{3 - x^2}{2x}$

Hence, if $x = 3, \quad y = -1$
$x = -3, \quad y = 1$
$x = 0.447, \quad y = 3.13$
$x = -0.447, \quad y = -3.13$ } (3 sig. figs)

Exercise 2(e)

Solve the following pairs of simultaneous equations:

1 $x + y = 6; \quad x^2 + y^2 = 26$
2 $3x - 2y = 1; \quad x^2 + xy = 2$
3 $y^2 = 4x; \quad 5x + 7y = 19$
4 $x^2 - y^2 = 15; \quad x^2 + y^2 = 17$
5 $5x - 3y = 12; \quad 5x^2 - 7xy = 24$

Miscellaneous Examples 2

1 Expand and simplify:

(i) $(x + 1)(x^2 + x + 2)$
(ii) $(3x - 1)(2x^2 - x - 3)$
(iii) $(x + 2y)(3x - 4y)$
(iv) $(x^2 + 4)^2$
(v) $(t + 4)(t^2 - 3t - 2)$
(vi) $(t^2 + 2t + 3)(t^2 - 3t - 1)$

2 In the following examples, one factor of the expression is given. Use this to factorise the polynomial expressions completely.

(i) $x^3 + x^2 - 10x + 8$; $(x - 2)$ is a factor.
(ii) $2x^3 + 9x^2 - 2x - 24$; $(x + 4)$ is a factor.
(iii) $2x^3 + 25x^2 + 71x - 42$; $(2x - 1)$ is a factor.

3 Factorise the following expressions:

(i) $t^3 + 2t^2 - 23t - 60$
(ii) $x^3 - 3x^2 - 54x + 112$

4 Find the remainder when $x^4 + 2x^2 + x + 1$ is divided by $x + 3$.

5 Given that $x + 2$ is a factor of $x^3 + kx^2 + 7x + 2$, find the value of k.

6 Solve completely the equations:

(i) $x(x - 1) = 12$
(ii) $\dfrac{2}{x + 1} = \dfrac{3}{x + 2}$
(iii) $x(x - 2) + 2x(x + 3) = 4$
(iv) $x^4 = 9$
(v) $\sqrt{3x + 5} - \sqrt{2x - 3} = 2$
(vi) $x^4 - 13x^2 + 36 = 0$

7 Rewrite the following expressions in the form $a(x + b)^2 + c$.

(i) $4x^2 + 8x + 11$
(ii) $x^2 - 7x - 5$
(iii) $1 - 6x - 2x^2$
(iv) $9 + 8x - 3x^2$

8 If α and β are the roots of the equation $2x^2 + 3x + 5 = 0$, find new equations with roots:

(i) $\alpha - 2, \beta - 2$
(ii) $\frac{1}{2}\alpha, \frac{1}{2}\beta$
(iii) $\alpha^2 + 1, \beta^2 + 1$
(iv) $\dfrac{\alpha}{\beta}, \dfrac{\beta}{\alpha}$
(v) $2\alpha^3, 2\beta^3$

9 Solve the simultaneous equations:

(i) $x + 2y = 3$ $\quad x^2 + y^2 = 2$

(ii) $x + y = 8$ $\quad y^2 + xy = 48$

(iii) $x^2 + y^2 = 10$ $\quad x^2 + 2y^2 = 11$

(iv) $y^2 = 8x + 8$ $\quad 2x + 3y = 14$

Revision Problems 2

1 The polynomial $x^5 - 3x^4 + 2x^3 - 2x^2 + 3x + 1$ is denoted by $f(x)$.

(i) Show that neither $(x - 1)$ nor $(x + 1)$ is a factor of $f(x)$.

(ii) By substituting $x = 1$ and $x = -1$ in the identity

$$f(x) \equiv (x^2 - 1)q(x) + ax + b$$

where $q(x)$ is a polynomial and a and b are constants, or otherwise, find the remainder when $f(x)$ is divided by $(x^2 - 1)$.

(iii) Show, by carrying out the division, or otherwise, that when $f(x)$ is divided by $(x^2 + 1)$, the remainder is $2x$.

(iv) Find all the real roots of the equation $f(x) = 2x$

(UCLES)

2 Show that $x + 3$ is a factor of $2x^3 + 15x^2 + 22x - 15$.
Find the set of values for x for which:

$$2x^3 + 15x^2 + 22x - 15 > 0$$

3 An enclosure $PQRS$ is to be made, as shown in Figure 2.1. PQ and QR are fences of total length 300 m. The other two sides are hedges. The angles at Q and R are right angles and the angle at S is 135°. The length of QR is x metres.

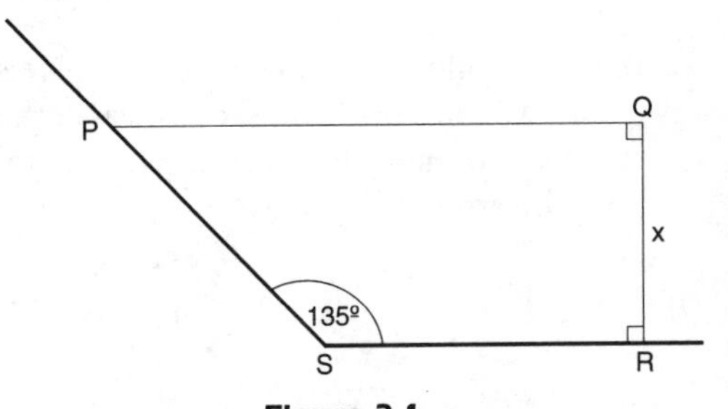

Figure 2.1

(a) Show that the area, $A\,\text{m}^2$, of the enclosure is given by

$$A = 300x - \frac{3x^2}{2}$$

(b) Show that A can be written as $-\frac{3}{2}\left((x - a)^2 - b\right)$ where a and b are constants whose values you should determine.

Hence show that A cannot exceed 15 000.

(AEB)

3 Series

Sequences and series have fascinated mathematicians for hundreds of years. There are many famous sequences that are beyond the scope of this book, but are worth researching if you continue with the study of mathematics. The following activity might give you some idea of the possibilities.

ACTIVITY 2

Look at the following series which all give approximations to π. By taking a few terms of each of them, see how accurate they are. If you can program a computer, you might be able to generate the sequence and sum it for a large number of terms.

(i) $\pi = 4\left(1 - \frac{1}{3} + \frac{1}{5} - \frac{1}{7} + \ldots\right)$

(ii) $\pi = 4\left(\frac{2 \times 4 \times 4 \times 6 \times 6 \times 8 \times 8 \times \ldots}{3 \times 3 \times 5 \times 5 \times 7 \times 7 \times \ldots}\right)$

(iii) $\pi = 6\sqrt{\tfrac{1}{3}}\left(1 - \frac{1}{3 \times 3} + \frac{1}{3^2 \times 5} - \frac{1}{3^3 \times 7} + \frac{1}{3^4 \times 9} - \ldots\right)$

You might also like to read about the history of calculating π. It is a fascinating subject.

To study a series systematically, basic skills are required. These will be developed in this chapter.

3.1 The summation notation ($\sum$)

There are many occasions in mathematics when you want to add up a series of numbers that have a pattern to them. If you wanted to add up the integers from 1 to 100, you could write this:

$$1 + 2 + 3 + \ldots\ldots\ldots\ldots\ldots + 100$$

Most people would assume that $+\ldots\ldots$ meant 'carry on increasing the numbers by one each time'. Mathematically, however, this is not precise enough, because there are many number sequences that start $1, 2, 3, \ldots$ The following notation is used to avoid the confusion. The greek letter Σ (Sigma) is used to denote 'find the sum'.

$$\text{Hence} \quad 1 + 2 + 3 + \ldots + 100 = \sum_{r=1}^{100} r$$

The expression r represents the rth term (T_r) in the series, and the instruction above and below the sigma sign means you start at $r = 1$, increase r by one each time, substitute into the expression for the rth term, and add up the results until r reaches 100. In the above example, T_1 is the first term, that is, $T_1 = 1$. T_2 is the second term, that is, $T_2 = 2$, and so on.

Let us look at another example:

$$\sum_{r=3}^{8} (2r + 1)$$

Hence the rth term is $T_r = 2r + 1$

r starts at 3, $\quad T_3 = 2 \times 3 + 1 = 7$

$\quad T_4 = 2 \times 4 + 1 = 9$

$$\text{Hence} \quad \sum_{r=3}^{8} (2r + 1) = 7 + 9 + 11 + 13 + 15 + 17$$

This notation can be used quite easily to express a series where the terms alternate in signs.

$$\text{Consider:} \quad \sum_{r=0}^{8} \frac{(-r)^r}{r + 1}$$

Here the rth term is $T_r = \dfrac{(-1)^r}{r + 1}$

r starts at 0, $\quad T_0 = \dfrac{(-1)^0}{0 + 1} = 1$

$\quad T_1 = \dfrac{(-1)^1}{1 + 1} = -\frac{1}{2}$

$$\text{Hence} \quad \sum_{r=0}^{8} \frac{(-1)^r}{r + 1} = 1 - \tfrac{1}{2} + \tfrac{1}{3} - \tfrac{1}{4} + \tfrac{1}{5} - \tfrac{1}{6} + \tfrac{1}{7} - \tfrac{1}{8} + \tfrac{1}{9}$$

The notation can also be used for a series that carries on indefinitely. (We say goes on to infinity (∞)).

$$\text{So} \quad \sum_{r=1}^{\infty} = \frac{1}{2^r} = \tfrac{1}{2} + \tfrac{1}{4} + \tfrac{1}{8} + \dots$$

The reverse process is not quite so easy. If we are given the series, how can we express it using the sigma notation?

Consider the series $1 + 4 + 7 + 10 + \dots + 28$. We assume that the terms increase by 3 each time, hence they are related to the multiples of 3 $(3, 6, 9, \dots)$. They are all 2 smaller than the corresponding multiple of 3. We can write $T_r = 3r - 2$.

$$\therefore \quad 1 + 4 + 7 + 10 + \dots + 28 = \sum_{r=1}^{10} (3r - 2)$$

Now work through the following exercise:

Exercise 3(a)

1 Write out the series represented by the following:

(i) $\sum_{r=1}^{6} (3r + 1)$ (ii) $\sum_{k=1}^{4} k^2$ (iii) $\sum_{r=2}^{6} \frac{r}{r+1}$

(iv) $\sum_{t=1}^{4} (-1)^t t^3$ (v) $\sum_{r=1}^{5} \frac{1}{2^r}$

2 Write out the following series using the $\sum$ notation:

(i) $1 + 3 + 5 + 7 + 9$ (ii) $4 + 7 + 10 + 13$

(iii) $\frac{1}{5} + \frac{2}{6} + \frac{3}{7} + \frac{4}{8} + \frac{5}{9}$ (iv) $1 - 4 + 9 - 16 + 25$

(v) $40 + 35 + 30 + 25 + 20$ (vi) $1 + \frac{1}{3} + \frac{1}{9} + \frac{1}{27}$

(vii) $n + (n + k) + (n + 2k) + (n + 3k) + \dots + (n + (n - 1)k)$

3.2 Arithmetic series

There is a very well-known story that in 1787 the famous mathematician, Carl Friedrich Gauss, was kept in at school and was asked to find the sum of a sequence similar to $1 + 2 + \dots + 1000$. He was able to come up with the answer of 500 500 almost immediately. What is it about this number sequence that makes it easy to work with?

If you reverse the numbers in the sequence, you have the sequence written twice:

$$1 + \quad 2 + \quad 3 + \ldots + 999 + 1000$$
$$1000 + 999 + 998 + \ldots + \quad 2 + \quad 1$$

If you add together each number to that directly below it, you get:

$$1001 + 1001 + 1001 + \ldots + 1001 + 1001$$

but there are 1000 of these 1001s, and so the sum is

$$1001 \times 1000 = 1001000$$

However, we have written down the sequence twice, so this total needs to be halved, giving:

$$1001000 \div 2 = 500500 \quad \text{(simple if you know how)}$$

A sequence where the numbers increase by the same amount each time is called an **arithmetic sequence** (or **arithmetic progression**). A formula for the sum of an arithmetic series can be found in exactly the same way as with Gauss's solution.

If the first term of the sequence is a, and the difference between the terms (called the common difference) is d, then the series will be:

$$S_n = a + (a + d) + (a + 2d) + \ldots + (a + (n - 1)d)$$

(S_n is another way of writing the sum of n terms.)

The nth term is clearly

$$T_n = a + (n - 1)d \qquad \text{(AS1)}$$

Writing down this series in reverse, you get:

$$S_n = (a + (n - 1)d) + (a + (n - 2)d) + \ldots + (a + d) + a$$

If you add each term to the term above it, you always get $2a + (n - 1)d$. Hence:

$$2S_n = n \times (2a + (n - 1)d)$$

$$\therefore \quad \div 2 \text{ gives} \quad S_n = \frac{n}{2}(2a + (n - 1)d) \qquad \text{(AS2)}$$

Although this formula is straightforward in its use, it is sometimes worth rewriting in the following way:

$$S_n = \frac{n}{2}(a + a + (n - 1)d)$$
$$= \frac{n}{2}(a + T_n)$$

Since a is the first term, and T_n the last term, we have:

$$S_n = n\left(\frac{\text{first term} + \text{last term}}{2}\right)$$

$$= n \times \text{(the average of the first and last terms)} \qquad \text{(AS2a)}$$

Example 3.1

Find the number of terms and the sum of the following series:

(i) $2 + 5 + 8 \ldots + 299$

(ii) $205 + 199 + 193 + \ldots - 11$

(iii) $(2n + 1) + (2n + 3) + (2n + 5) + \ldots + (4n + 9)$

Solution

(i) $a = 2$ and $d = 3$

$T_n = a + (n - 1)d$

So $2 + 3(n - 1) = 299$

$\therefore \quad 3(n - 1) = 297$

$n - 1 = 99$

hence $n = 100$

$\therefore \quad S_{100} = \dfrac{100}{2}(4 + 99 \times 3) = 15050$

(ii) $a = 205$, $d = -6$

So $205 + (n - 1) \times -6 = -11$

i.e. $(n - 1) \times -6 = -216$

$\therefore \quad n - 1 = 36$

So $n = 37$

Hence: $S_{37} = \dfrac{37}{2}(410 + 36 \times -6)$

$= 3589$

(iii) This type of question often leads to confusion in the algebra. You will notice that n appears in the series, but n also appears in the formulae that you are going to use. When this happens, alter the letter in the formula.

For example, let $T_N = a + (N - 1)d$

and $S_N = \dfrac{N}{2}(2a + (N - 1)d)$

in this case, $a = 2n + 1$ and $d = 2$

So $T_N = 2n + 1 + (N - 1)2 = 4n + 9$ (the last term given)

i.e. $2(N - 1) = 4n + 9 - 2n - 1 = 2n + 8$

So $N - 1 = n + 4$

$\therefore \quad N = n + 5$

Hence the number of terms is $n + 5$

$$\therefore \quad S_{n+5} = \frac{n+5}{2}[2(2n+1) + (n+5-1)2]$$

$$= \frac{(n+5)}{2}(4n + 2 + 2n + 8)$$

$$= \frac{(n+5)}{2}(6n + 10)$$

$$= (n+5)(3n+5)$$

Example 3.2

The twenty-fifth term of an arithmetic progression is 43, and the fifty-ninth term is 26. Find the sum of the first one hundred terms.

Solution

The formula for the nth term of an arithmetic series is:

$$T_n = a + (n-1)d$$

So $\quad T_{25} = a + 24d = 43 \qquad$ (i)

and $\quad T_{59} = a + 58d = 26 \qquad$ (ii)

Try not to use trial and error methods

(ii) – (i) $\quad 34d = -17$

$$\therefore \quad d = -0.5$$

Substitute into (i)

$$a + 24 \times -0.5 = 43$$

$$\therefore \quad a = 55$$

The sum of n terms is

$$S_n = \frac{n}{2}(2a + (n-1)d)$$

$$\therefore \quad S_{100} = \frac{100}{2}(2 \times 55 + 99 \times -0.5) = 3025$$

Arithmetic series are often used in financial calculations. The following example shows how to calculate total income on a pay scale with a regular increment.

Example 3.3

John is just starting a new job. His salary is £15 000 per annum, with an annual increment of £500. If this did not change, how much would he earn in total over a period of 12 years?

Solution

This situation is an arithmetic progression.

The total earnings $= £15\,000 + (15\,000 + 500) + (15\,000 + 1\,000) + \ldots$

Using the sum $= \frac{n}{2}(2a + (n-1)d)$

$a = 15\,000$, $n = 12$, $d = 500$

$$\text{Total} = £\frac{12}{2}(2 \times 15\,000 + 11 \times 500)$$

$$= £213\,000$$

The following example shows how the formula for an arithmetic series can easily generate a quadratic equation.

Example 3.4

Find the largest value of n, such that:

$$1 + 2 + 3 + \ldots + n < 10\,000$$

Solution

You could, of course, simply start adding the positive integers until you reach the value nearest to 10 000. This method, however, is not recommended.

For an arithmetic series, $S_n = \frac{n}{2}(2a + (n-1)d)$

Here $a = 1$, $d = 1$

$$\text{Hence} \quad S_n = \frac{n}{2}(2 + (n-1)) = \frac{n(n+1)}{2}$$

Suppose S_n can be exactly 10 000

$$\text{Then} \quad \frac{n(n+1)}{2} = 10\,000$$

$$\text{So} \quad n^2 + n = 20\,000$$

$$n^2 + n - 20000 = 0$$

This quadratic equation should be solved using the formula,

$$\text{So} \quad n = \frac{-1 \pm \sqrt{1 + 80\,000}}{2}$$

n cannot be negative, and so $n = 140.9$

But n must be a whole number, and so this value must be rounded down, if the total is to be less than 10 000.

Hence $n = 140$

Exercise 3(b)

1 Find the sum of 40 terms of the arithmetic series

$$3+4\tfrac{1}{2}+6+\ldots$$

2 How many terms are there in the following arithmetic series?
(i) $1+3+5+7+\ldots+221$
(ii) $3+7+11+15+\ldots+91$
(iii) $200+192+184+\ldots-120$
(iv) $4n+(4n-3)+(4n-6)+\ldots+(3-5n)$

3 Find the sum of the following arithmetic progressions
(i) $4+9+14+19+\ldots+189$
(ii) $2+5+8+\ldots+122$
(iii) $1+1.5+2+\ldots+18.5$
(iv) $100+97+94+91+\ldots-20$

4 How many integers are there between 200 and 350 which are exactly divisible by 9? What is their sum?

5 The second term of an arithmetic progression is 4 and the sixth term is 12. Find the sum of the first 30 terms.

6 The sum of the first n terms of an arithmetic progression is $2n^2+n$. Find the first four terms of the series.

7 How many terms of the AP $2+5+8+\ldots$ are required to have a sum greater than 2 000?

3.3 Geometric series

Consider the series:

$$S_n=1+2+4+8+\ldots+2^{n-2}+2^{n-1}$$

Each term is obtained from the previous term by multiplication by 2 (the common ratio). The rth term T_r is given by $T_r=2^{r-1}$.

Multiply the series S_n by 2:

$$\begin{aligned} 2S_n &= 2+4+8+\ldots 2^{n-1}+2^n \\ 2S_n &= (S_n-1)+2^n \\ \therefore\quad 2S_n-S_n &= 2^n-1 \\ S_n &= 2^n-1 \end{aligned}$$

A series where each term is multiplied by a constant ratio to give the next term is a **geometric series (progression)** or GP for short.

The general series, with first term a and common ratio r, can be written:

$$S_n = a + ar + ar^2 + \ldots + ar^{n-2} + ar^{n-1} \qquad \text{(i)}$$

Multiply each term by r:

$$\begin{aligned} rS_n &= ar + ar^2 + ar^3 + \ldots + ar^{n-1} + ar^n \qquad \text{(ii)} \\ &= (S_n - a) + ar^n \end{aligned}$$

Hence

$$rS_n - S_n = ar^n - a$$

So $S_n(r-1) = a(r^n - 1)$

We have the formula for the sum of a geometric series:

$$S_n = \frac{a(r^n - 1)}{(r-1)} \qquad \text{(GS1a)}$$

Use this version if $|r| > 1$.

This can also be written:

$$S_n = \frac{a(1 - r^n)}{(1-r)} \qquad \text{(GS1b)}$$

Use this version if $|r| < 1$.

The nth term $T_n = ar^{n-1}$ (GS2)

Now study the following examples to see how these formulae are used.

Example 3.5

Find the tenth term, and the sum of the first ten terms of the geometric series $2 + 3 + \ldots$

Solution

If you were not told this was a geometric series, you could deduce nothing from being given just the first two terms.

However, here $a = 2$ (i)

$ar = 3$ (ii)

(ii) ÷ (i) $\dfrac{ar}{a} = \dfrac{3}{2}$ (cancel by a)

i.e. $r = 1.5$

The tenth term,

$$\begin{aligned} T_{10} &= ar^9 = 2 \times 1.5^9 \\ &= 76.89 \quad \text{(4 sig. figs)} \end{aligned}$$

The sum of the first ten terms is:

$$S_{10} = \frac{a(r^{10} - 1)}{(r - 1)} = \frac{2(1.5^{10} - 1)}{(1.5 - 1)}$$
$$= 226.7 \quad \text{(4 sig. figs)}$$

Example 3.6

The second term of a geometric series is 8, and the fifth term is 64. Find the first term and the common ratio.

Solution

$$\text{Here} \qquad ar = 8 \qquad \text{(i)}$$
$$\text{and} \qquad ar^4 = 64 \qquad \text{(ii)}$$
$$\text{(ii)} \div \text{(i)} \qquad \frac{ar^4}{ar} = \frac{64}{8} \quad \text{(cancel by } a\text{)}$$
$$\text{Hence} \qquad r^3 = 8$$
$$\therefore \qquad r = 2$$

Substitution into (i) gives $a = 4$

The first term is 4, the common ratio is 2.

Example 3.7

How many terms are there in the geometric series

$$\frac{1}{4} + \frac{38}{8} + \frac{9}{16} + \ldots + \frac{3^{24}}{3^{26}}?$$

Solution

$$\text{Here} \quad a = \tfrac{1}{4}$$
$$\text{and} \quad T_n = ar^{n-1} = \frac{3^{24}}{2^{26}}$$
$$\text{Now} \quad ar = \tfrac{3}{8},$$
$$\text{so} \quad r = \tfrac{3}{8} \div \tfrac{1}{4} = \tfrac{3}{2}$$
$$\therefore \quad T_n \text{ becomes } \tfrac{1}{4} \times \left(\tfrac{3}{2}\right)^{n-1} = \frac{3^{24}}{2^{26}}$$
$$\therefore \quad \left(\tfrac{3}{2}\right)^{n-1} = \frac{3^{24}}{2^{24}} = \left(\tfrac{3}{2}\right)^{24}$$
$$\therefore \quad n - 1 = 24, \quad n = 25$$

There are twenty-five terms in the series.

Example 3.8

If $(x-2)$, $(x+2)$ and $(5x-2)$ are consecutive terms of a geometric sequence, what can you say about x?

Solution

The simplest way of tackling this type of problem is to remember that, if a, b, c are consecutive terms in a geometric sequence, then:

$$\frac{b}{a}=\frac{c}{b}=\text{the common ratio.}$$

Hence:

$$\begin{aligned}
\frac{x+2}{x-2} &= \frac{5x-2}{x+2} \\
\therefore \quad (x+2)(x+2) &= (x-2)(5x-2) \\
\therefore \quad x^2+4x+4 &= 5x^2-12x+4 \\
\text{i.e.} \quad 0 &= 4x^2-16x \\
0 &= 4x(x-4)
\end{aligned}$$

Hence: $x=0$ or 4

If $x=4$, the numbers are 2, 6 and 18 with a common ratio of 3.

If $x=0$, the numbers are -2, 2 and -2 with a common ratio of -1.

Exercise 3(c)

1 Find the sum of the first twenty terms of the GP $2+3+4.5+\ldots$

2 How many terms are there in the GPs

(i) $1+2+4+8+\ldots+1024$

(ii) $2+6+18+\ldots+28697814$

(iii) $4-2+1-\frac{1}{2}+\ldots-\frac{1}{128}$

3 The third term of a GP is 6 and the eighth term is 192. Find the first term of the series and the common ratio.

4 If a, b, c are the first three terms of a GP, and $a+b=7$ and $b+c=9\frac{1}{3}$, find the numbers.

3.4 Infinite geometric series

The idea of a series that goes on for ever, is rather a difficult one to imagine. If you look at the sequence $1, 2, 4, 8, 16, \ldots$ the numbers are getting larger and larger, and so if you try to add them up: $1+2+4+8+16+\ldots$, the total will get larger and larger, and we say it becomes infinite. This series is said to **diverge**. However, look at the sequence $1, \frac{1}{2}, \frac{1}{4}, \frac{1}{8}, \ldots$. Here the numbers are getting smaller and smaller, and it might be interesting to look at this geometric series:

$$S_1 = 1$$
$$S_2 = 1 + \tfrac{1}{2} = 1.5$$
$$S_3 = 1 + \tfrac{1}{2} + \tfrac{1}{4} = 1.75$$
$$S_4 = 1 + \tfrac{1}{2} + \tfrac{1}{4} + \tfrac{1}{8} = 1.875$$

If we now use the formula, we can show that:

$$S_{10} = 1.998047$$
$$S_{100} = 2 \quad \text{(7 sig. figs)}$$

However many terms we take, the sum never becomes greater than 2. We say that the sum to infinity (written S_∞) = 2.

This becomes clear if we look at the formula:

$$S_n = \frac{a(1-r^n)}{(1-r)} = \frac{(1-0.5^n)}{(1-0.5)} = 2(1-0.5^n)$$

Now as $n \to \infty$ (n approaches infinity), $0.5^n \to 0$ (gets closer and closer to zero)

Hence: $S_n \to 2$

This type of series is called **convergent**, and **all** geometric sequences with $-1 < r < 1$ are convergent.

Since $r^n \to 0$ as $n \to \infty$ and $-1 < r < 1$, we have the result that, for an infinite geometric series:

$$S_\infty = \frac{a}{1-r} \quad \text{if } |r| < 1 \quad [\text{note } -1 < r < 1 \text{ can be written } |r| < 1] \qquad \text{(GS3)}$$

($|r|$ is the modulus or numerical value of r. So $|2| = 2$, $|-3| = 3$.)

ACTIVITY 3

Recurring decimals

You may well have come across infinite geometric series before, without realising it.

Consider the recurring decimal $0.\dot{3}$, that is, 0.3333333... What does this decimal mean in terms of fractions?

Well, $0.\dot{3}$ is really $\frac{3}{10}$ and 0.03 is $\frac{3}{100}$ and so on.

Hence $0.\dot{3} = \frac{3}{10} + \frac{3}{100} + \frac{3}{1000} + \ldots$

You can see that this is just a geometric series, with $a = \frac{3}{10}$ and $r = \frac{1}{10}$.

$$\text{So:} \quad S_\infty = \frac{\frac{3}{10}}{1-\frac{1}{10}} = \frac{3}{10} \div \frac{9}{10} = \frac{1}{3}$$

Now you knew from the beginning that $\frac{1}{3} = 0.\dot{3}$, but now you can explore a whole range of recurring decimals and see what they are equal to as a fraction.

Try 0.12121212..., and then some of your own.

Example 3.9

Find the sum to infinity of the geometric series $2 + \frac{2}{3} + \frac{2}{9} + \ldots$. How many terms of this series do you need to get a sum within 0.05% of that sum to infinity?

Solution

This is a very common type of question.

Now $a = 2$ and $r = \frac{1}{3}$

Hence: $S_\infty = \dfrac{2}{1 - \frac{1}{3}} = 3$

To be within 0.05% of this, means you want 99.95% of S_∞.

$$\begin{aligned}
\text{That is,} \quad S_n &= 3 \times \frac{99.95}{100} = 2.9985 \\
\therefore \quad \frac{a(1 - r^n)}{(1 - r)} &= 2.9985 \\
\text{Hence:} \quad \frac{2}{(1 - \frac{1}{3})}\left(1 - \left(\tfrac{1}{3}\right)^n\right) &= 2.9985 \\
\therefore \quad 3\left(1 - \left(\tfrac{1}{3}\right)^n\right) &= 2.9985 \\
\text{That is,} \quad 1 - \left(\tfrac{1}{3}\right)^n &= 0.9995 \\
\text{So:} \quad 0.0005 &= \left(\tfrac{1}{3}\right)^n \\
\text{taking logs:} \quad \log 0.0005 &= n \log \tfrac{1}{3} \\
\therefore \quad n &= \log 0.0005 \div \log \tfrac{1}{3} = 6.92
\end{aligned}$$

Hence: Seven terms are required to be **within** 0.05%.
(Six terms would not be enough.)

Example 3.10

The sum to infinity of a geometrical progression is 8. The sum to infinity of the odd terms of the series is 6. Find the common ratio of the series.

Solution

The full series can be represented by:

$$a + ar + ar^2 + ar^3 + \ldots$$

where r is the common ratio and, of course, $-1 < r < 1$.

The sum to infinity of this series S_1 is given by:

$$S_1 = \frac{a}{1 - r} = 8 \quad \therefore \quad a = 8(1 - r)$$

The odd terms of the series are:

$$a + ar^2 + ar^4 + \ldots$$

This is another infinite geometric series, with common ratio r^2. The sum to infinity of this series S_2 will be given by:

$$S_2 = \frac{a}{1 - r^2} = 6 \qquad \therefore \quad a = 6(1 - r^2)$$

We now have two expressions for a, which must be equal.

So: $8(1 - r) = 6(1 - r^2) = 6(1 - r)(1 + r)$

Since $r \neq 1$, ($\neq$ means not equal to), $1 - r$ is not zero, and we can cancel $(1 - r)$ from both sides.

Hence: $8 = 6(1 + r)$

$\therefore \quad r = \frac{1}{3}$

Exercise 3(d)

1 Find the sum to infinity of the following GPs.

(i) $\frac{1}{2} + \frac{1}{4} + \frac{1}{8} + \ldots$

(ii) $1 - \frac{1}{3} + \frac{1}{9} - \frac{1}{27} + \ldots$

(iii) $4 + \frac{4}{3} + \frac{4}{9} + \frac{4}{27} + \ldots$

(iv) $0.2 - 0.1 + 0.05 - \ldots$

2 Find how many terms of the following series are needed to have a sum within 0.01% of the sum to infinity.

(i) $1 + \frac{1}{4} + \frac{1}{16} + \ldots$

(ii) $2 - 1 + 0.5 - 0.25 + \ldots$

(iii) $1 + \frac{1}{3} + \frac{1}{9} + \ldots$

(iv) $1 + \frac{1}{5} + \frac{1}{25} + \ldots$

3.5 The binomial expansion

Before moving on to another type of infinite series, you need to learn about a method for expanding powers of brackets.

Look at the following:

$$(a + b)^0 = 1$$

$$(a + b)^1 = a + b$$

$$(a + b)^2 = a^2 + 2ab + b^2$$

To find $(a+b)^3$, multiply $(a+b)^2$ by $(a+b)$.

So: $$(a+b)(a^2+2ab+b^2) = a^3+2a^2b+ab^2+ba^2+2ab^2+b^3$$
$$= a^3+3a^2b+3ab^2+b^3$$

Similarly, $$(a+b)^4 = (a+b)(a^3+3a^2b+3ab^2+b^3)$$
$$= a^4+4a^3b+6a^2b^2+4ab^3+b^4$$

If we write these out again underneath each other, we get:

$$1$$
$$a+b$$
$$a^2+2ab+b^2$$
$$a^3+3a^2b+3ab^2+b^3$$
$$a^4+4a^3b+6a^2b^2+4ab^3+b^4$$

If you just write out the coefficients contained in this table, they are:

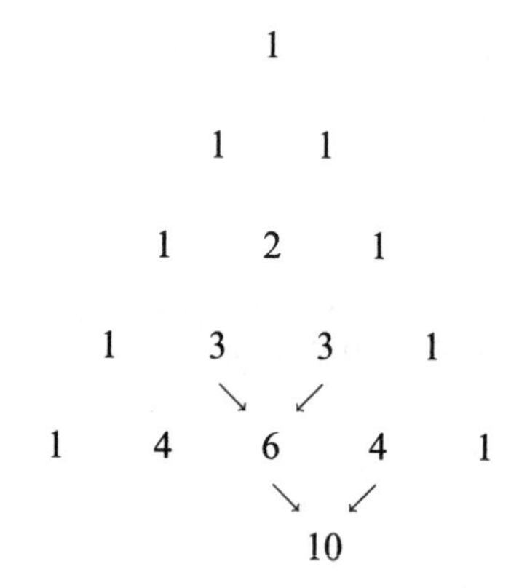

This famous number pattern was investigated by the French mathematician Blaise Pascal in about 1650, and is known as Pascal's triangle. However, it first appeared in print in China in 1303. Each number in the triangle is the sum of the two numbers directly above it. Hence the next row in the triangle will be:

1 5 10 10 5 1

In other words

$$(a+b)^5 = a^5+5a^4b+10a^3b^2+10a^2b^3+5ab^4+b^5$$

Notes:

(i) Each line of the triangle is symmetrical about a vertical line through the centre.

(ii) The powers of a decrease by one in each term, the powers of b increase by one in each term.

Clearly, you can extend this as much as you like, but eventually it gets unwieldy, and there is another way of calculating the numbers in the triangle. This involves a mathematical quantity labelled on your calculator as nCr or ${}^nC_r$ or $\binom{n}{r}$. It is actually the number of ways of choosing r objects selected from n, and is defined as:

$$nCr = \frac{n!}{r!(n-r)!}$$

where $n!$ (n factorial) means $n \times (n-1) \times (n-2) \times \ldots \times 2 \times 1$

So $8! = 8 \times 7 \times 6 \times 5 \times 4 \times 3 \times 2 \times 1 = 40320$

There is a factorial button $\boxed{x!}$ on your calculator.

You will need to note that 0! is defined as 1.

So, to find, say, $_{10}C_8$, you proceed as follows:

Display

$\boxed{10} \rightarrow \boxed{nCr} \rightarrow \boxed{8} \rightarrow \boxed{45.}$

These ideas are explained much more fully in Chapter 13, but for the moment, as long as you can use the button on your calculator, this will cause you no problems.

To use it to find the entries in Pascal's triangle, proceed as follows:

If you wanted to expand $(a+b)^{10}$, then $n = 10$.

Now work out $_{10}C_0$, $_{10}C_1$, $_{10}C_2$ and so on. (**Note:** You do not start at $_{10}C_1$). The numbers obtained are:

1, 10, 45, 120, 210, 252, 210, 120, 45, 10, 1

$$\text{Hence } (a+b)^{10} = a^{10} + 10a^9b + 45a^8b^2 + 120a^7b^3 + 210a^6b^4 + 252a^5b^5 + 210a^4b^6 + 120a^3b^7 + 45a^2b^8 + 10ab^9 + b^{1}0$$

We can summarise this in the **binomial theorem** as follows:

$$(a+b)^n = \binom{n}{0}a^n + \binom{n}{1}a^{n-1}b + \binom{n}{2}a^{n-2}b^2 + \ldots \binom{n}{n}b^n \qquad \text{(B1)}$$

The notation $\binom{n}{r}$ tends to be more commonly used.

If you want to select the rth term in this expansion, you need to be careful. This term does not contain $\binom{n}{r}$, but $\binom{n}{r-1}$. You can see this if you realise that $\binom{n}{0}$ is the **first** term, $\binom{n}{1}$ is the **second** term etc.

Hence the rth term $T_r = \binom{n}{r-1} a^{n-r+1} b^{r-1}$ (B2)

The term containing b^r is $\binom{n}{r} a^{n-r} b^r$ (B3)

Example 3.11

(i) Find the seventh term in the expansion of $(3 - 2x)^{12}$.
(ii) Find the term containing x^6 in the expansion of $(2 + 3x)^{15}$.

Solution

(i) Here $n = 12$, $r = 7$, $a = 3$, $b = -2x$

$$\begin{aligned} \therefore \quad T_7 &= \binom{12}{6} a^6 b^6 \\ &= \binom{12}{6} (3)^6 (-2x)^6 \\ &= 43110144x^6 \end{aligned}$$

(ii) Here $b = 3x$ and so the term in x^6 is the term in b^6, this is:

$$\binom{15}{6} (2)^9 (3x)^6 = 1868106240x^6$$

The following example shows how the binomial expansion is used.

Example 3.12

Expand the following:

(i) $(1 + 2x)^3$ (ii) $(4 - 3x)^4$ (iii) $\left(x + \frac{2}{x}\right)^5$

Solution

(i) Here $a = 1$ and $b = 3x$

$$\begin{aligned} (a+b)^3 &= a^3 + 3a^2b + 3ab^2 + b^3 \\ \text{Hence} \quad (1+2x)^3 &= (1)^3 + 3(1)^2(2x) + 3(1)(2x)^2 + (2x)^3 \\ &= 1 + 6x + 12x^2 + 8x^3 \end{aligned}$$

MEMORY JOGGER

Always enclose each term with a bracket when using the binomial expansion – it avoids many mistakes

(ii) Here $a = 4$, $b = -3x$

$$(a+b)^4 = a^4 + 4a^3b + 6a^2b^2 + 4ab^3 + b^4$$

$$\therefore \quad (4-3x)^4 = (4)^4 + 4(4)^3(-3x) + 6(4)^2(-3x)^2 + 4(4)(-3x)^3 + (-3x)^4$$

$$= 256 - 768x + 864x^2 - 432x^3 + 81x^4$$

(**Notice** that the signs alternate +, −.)

(iii) Here $a = x$ and $b = \dfrac{2}{x}$

$$(a+b)^5 = a^5 + 5a^4b + 10a^3b^2 + 10a^2b^3 + 5ab^4 + b^5$$

$$\left(x+\frac{2}{x}\right)^5 = (x)^5 + 5(x)^4\left(\frac{2}{x}\right) + 10(x)^3\left(\frac{2}{x}\right)^2$$

$$+ 10(x)^2\left(\frac{2}{x}\right)^3 + 5(x)\left(\frac{2}{x}\right)^4 + \left(\frac{2}{x}\right)^5$$

$$= x^5 + 10x^3 + 40x + \frac{80}{x} + \frac{80}{x^3} + \frac{32}{x^5}$$

Example 3.13

Find the term independent of x in the expansion of $\left(2x + \dfrac{1}{x^2}\right)^{12}$

Solution

Let us look first at what this question is asking for. If you were to expand this expression fully using the binomial theorem, then the coefficients needed would be:

1, 12, 66, 220, 495, 792, 924, 792, 495, 220, 66, 12, 1

Hence: $$\left(2x+\frac{1}{x^2}\right)^{12} = (2x)^{12} + 12(2x)^{11}\left(\frac{1}{x^2}\right) + 66(2x)^{10}\left(\frac{1}{x^2}\right)^2$$

$$+ 220(2x)^9\left(\frac{1}{x^2}\right)^3 + 495(2x)^8\left(\frac{1}{x^2}\right)^4 + 792(2x)^7\left(\frac{1}{x^2}\right)^5$$

$$+ 924(2x)^6\left(\frac{1}{x^2}\right)^6 + 792(2x)^5\left(\frac{1}{x^2}\right)^7 + 495(2x)^4\left(\frac{1}{x^2}\right)^8$$

$$+ 220(2x)^3\left(\frac{1}{x^2}\right)^9 + 66(2x)^2\left(\frac{1}{x^2}\right)^{10} + 12(2x)\left(\frac{1}{x^2}\right)^{11} + \left(\frac{1}{x^2}\right)^{12}$$

This simplifies to give:

$$\left(2x+\frac{1}{x^2}\right)^{12} = 4096x^{12} + 24576x^9 + 67584x^6 + 112640x^3 + 126720$$

$$+ \frac{101376}{x^3} + \frac{59136}{x^6} + \frac{25344}{x^9} + \frac{7920}{x^{12}} + \frac{1760}{x^{15}} + \frac{264}{x^{18}} + \frac{24}{x^{21}} + \frac{1}{x^{24}}$$

Clearly, the whole process takes some time. However, you will notice that one of the terms, namely 126720, does **not** contain x. This, then, is the term **independent** of x. Hence the solution to the question is 126720.

Can this answer be found in a simpler way? The secret is to factorise $2x + \dfrac{1}{2x^3}$ first, by making the first term **one**.

So $$2x + \frac{1}{x^2} = 2x\left(1 + \frac{1}{2x^3}\right)$$

Then $$\left(2x + \frac{1}{x^2}\right)^{12} = (2x)^{12}\left(1 + \frac{1}{2x^3}\right)^{12}$$

When expanding you only need the term that contains the same reciprocal power of $(2x)^{12}$ (i.e. x^{-12}).

This is: $$\left(\frac{1}{2x^3}\right)^4$$

Now in the expansion $(1+b)^{12}$, the term in b^4 is ${}_{12}C_4 b^4$.

The term required, therefore, is:

$$\begin{aligned}(2x)^{12}\,{}_{12}C_4 &= 4096x^{12} \times 495 \times \frac{1}{16x^{12}}\\ &= 126720\end{aligned}$$

Exercise 3(e)

1 Expand and simplify:

(i) $(a+2b)^3$ (ii) $(x-2y)^4$ (iii) $(2a+3t)^4$

(iv) $\left(x + \dfrac{2}{x}\right)^4$ (v) $\left(3x - \dfrac{1}{x^2}\right)^3$

2 Find the fourth term in the following expansions:

(i) $(a+2b)^{20}$ (ii) $(2x-3y)^{10}$ (iii) $\left(x + \dfrac{2}{x}\right)^8$

3 Find the term containing x^5 in the following expansions:

(i) $(1+2x)^{10}$ (ii) $(3-4x)^8$ (iii) $(2-3x)^9$

4 Find the term independent of x in the expansion of $\left(2x + \dfrac{1}{x^2}\right)^{18}$

3.6 The binomial expansion when n is not a positive integer

Consider the series $1 + x + x^2 + x^3 + \dots$

It is a geometric series, and if $|x| < 1$, it has a sum to infinity given by:

$$S_\infty = \frac{a}{1-r} = \frac{1}{1-x} = (1-x)^{-1}$$

Hence $(1-x)^{-1} = 1 + x + x^2 + x^3 + \dots$

It could be argued that this is the binomial theorem with $n = -1$.

Does the formula B1 work? The first problem is that ${}_{-1}C_r$ does not have any meaning, but if you start to expand ${}_nC_0$, ${}_nC_1$, ${}_nC_2$ in full, then the situation looks more hopeful.

${}_nC_0 = 1$ always

$${}_nC_1 = \frac{n!}{(n-1)!} = n$$

$${}_nC_2 = \frac{n!}{(n-2)!2!} = \frac{n(n-1)}{2!}$$

$${}_nC_3 = \frac{n!}{(n-3)!3!} = \frac{n(n-1)(n-2)}{3!} \text{ and so on.}$$

The other requirement (the proof is beyond the scope of this book) is that $a = 1$, and $|b| < 1$. If these conditions are satisfied, then B1 becomes:

$$(1+b)^n = 1 + nb + \frac{n(n-1)b^2}{2!} + \frac{n(n-1)(n-2)b^3}{3!} + \dots \qquad \text{(B4)}$$

This is the general form of the binomial theorem.

Now study the following examples.

Example 3.14

Expand the following functions as far as the term in x^3. Simplify your answers where possible, and state the values of x for which the expansion is valid.

(i) $(1+2x)^{-2}$ (ii) $(1-3x)$ (iii) $(4+x)^{\frac{1}{2}}$

Solution

(i) Here $b = 2x$ and $n = -2$

$$\therefore \ (1+2x)^{-2} = 1 + (-2)(2x) + \frac{(-2)(-3)(2x)^2}{2!} + \frac{(-2)(-3)(-4)(2x)^3}{3!} + \dots$$

$$= 1 - 4x + 12x^2 - 32x^3 \quad \text{(as far as the term in } x^3\text{)}$$

The series is valid if $|b| < 1$, i.e. $|2x| < 1$

$\therefore \quad -\frac{1}{2} < x < \frac{1}{2}$

(ii) Here $b = -3x$ and $n = \frac{1}{3}$

$$\therefore \quad (1-3x)^{\frac{1}{3}} = 1 + \left(\tfrac{1}{3}\right)(-3x) + \frac{\left(\frac{1}{3}\right)\left(-\frac{2}{3}\right)(-3x)^2}{2!} + \frac{\left(\frac{1}{3}\right)\left(-\frac{2}{3}\right)\left(-\frac{5}{3}\right)(-3x)^3}{3!}$$

$$= 1 - x - x^2 - \frac{5}{3}x^3$$

The series is valid if $|3x| < 1$, that is, $-\frac{1}{3} < x < \frac{1}{3}$.

(iii) In this example, $a \neq 1$.

> **MEMORY JOGGER**
>
> If $a \neq 1$, then remove a common factor a from the expression.

Here, $a = 4$, and so first you must rewrite $(4 + x) = 4\left(1 + \frac{x}{4}\right)$

$$\therefore \quad (4+x)^{\frac{1}{2}} = 4^{\frac{1}{2}}\left[1 + \frac{x}{4}\right]^{\frac{1}{2}}$$

$$= 2\left[1 + \left(\tfrac{1}{2}\right)\left(\tfrac{x}{4}\right) + \frac{\left(\frac{1}{2}\right)\left(-\frac{1}{2}\right)\left(\frac{x}{4}\right)^2}{2!} + \frac{\left(\frac{1}{2}\right)\left(-\frac{1}{2}\right)\left(-\frac{3}{2}\right)\left(\frac{x}{4}\right)^3}{3!}\right]$$

$$= 2 + \frac{1}{4}x + \frac{1}{64}x^2 + \frac{1}{512}x^3$$

The binomial expansion can be used to find approximations to a wide variety of rational (fractions) and irrational (surds) numbers by choosing suitable values of x. Look at the following example.

Example 3.15

Expand $\dfrac{1}{\sqrt{1+2x}}$ as a series of ascending powers of x up to the term including x^2. State the range of values of x for which the series is valid.

By using a suitable value for x, find a fraction which is approximately equal to $\sqrt{11}$. Find the percentage error in this fraction from the exact value, giving your answer correct to 3 significant figures.

Solution

$$\frac{1}{\sqrt{1+2x}} = (1+2x)^{-\frac{1}{2}}$$

$$= 1 + \left(-\tfrac{1}{2}\right)(2x) + \frac{\left(-\frac{1}{2}\right)\left(-\frac{3}{2}\right)(2x)^2}{2!} + \dots$$

$$= 1 - x + \tfrac{3}{2}x^2$$

The series is valid if $\quad -1 < 2x < 1$

$$\therefore \quad -\tfrac{1}{2} < x < \tfrac{1}{2}$$

To get a value of 11 inside the square root, you cannot use $x = 5$, because although $1 + 2x$ would be 11, $x = 5$ is not in the allowed range.

However, if you put $x = \frac{1}{9}$,

$$\text{then} \quad (1+2x)^{-\frac{1}{2}} = \left(1+\tfrac{2}{9}\right)^{-\frac{1}{2}} = \left(\tfrac{11}{9}\right)^{-\frac{1}{2}}$$

$$= \sqrt{\frac{9}{11}} = \frac{3}{\sqrt{11}}$$

Also, if you put $x = \frac{1}{9}$ in the expansion, you get:

$$\frac{3}{\sqrt{11}} = 1 - \tfrac{1}{9} + \tfrac{3}{2} \times \tfrac{1}{81} + \ldots \approx \tfrac{49}{54}$$

$$\therefore \quad \frac{3}{\sqrt{11}} \approx \tfrac{49}{54} \quad \text{which gives} \quad \sqrt{11} \approx \tfrac{162}{49}$$

The percentage error in using this value is given by:

$$\left(\frac{\sqrt{11} - \frac{162}{49}}{\sqrt{11}}\right) \times 100 = 0.317\%$$

Expansions can be combined, as the following example shows.

Example 3.16

Expand the rational function $\dfrac{(1-x)^2}{\sqrt{4+x}}$ as far as the term in x^2. State the range of values of x for which the expansion is valid.

Solution

$$\frac{(1-x)^2}{\sqrt{4+x}} = (1-x)^2(4+x)^{-\frac{1}{2}}$$

$$= \left(1 - 2x + x^2\right)(4+x)^{-\frac{1}{2}}$$

$$= \left(1 - 2x + x^2\right)4^{-\frac{1}{2}}\left(1 + \frac{x}{4}\right)^{-\frac{1}{2}} \qquad \left[\textbf{Note: } 4^{-\frac{1}{2}} = \tfrac{1}{2}\right]$$

$$= \tfrac{1}{2}(1 - 2x + x^2)\left(1 + \left(-\tfrac{1}{2}\right)\left(\tfrac{x}{4}\right) + \frac{\left(-\frac{1}{2}\right)\left(-\frac{3}{2}\right)\left(\frac{x}{4}\right)^2}{2!} + \ldots\right)$$

$$= \tfrac{1}{2}\left(1 - 2x + x^2\right)\left(1 - \tfrac{1}{8}x + \tfrac{3}{128}x^2\right)$$

$$= \tfrac{1}{2}\left(1 - \tfrac{1}{8}x + \tfrac{3}{128}x^2 - 2x + \tfrac{1}{4}x^2 + x^2 + \ldots\right)$$

$$= \tfrac{1}{2} - \tfrac{17}{16}x + \tfrac{163}{256}x^2$$

The expansion is valid if $-1 < \frac{x}{4} < 1$

that is, $\quad -4 < x < 4$

Exercise 3(f)

1 Expand the following functions as far as the term in x^3. Simplify the coefficients, and state the range of values of x for which the expansion is valid.

(i) $(1+2x)^{-2}$ (ii) $(1-3x)^{\frac{1}{3}}$ (iii) $\sqrt{16-4x}$

(iv) $\dfrac{1}{\sqrt[3]{1+3x}}$ (v) $\dfrac{1-x}{(1+x)^2}$ (vi) $\dfrac{\sqrt{1+x}}{1-x}$

2 Expand $(1-2x)^{\frac{1}{2}}$ as far as the term in x^2. By using $x = \frac{1}{9}$, show that $\sqrt{7} = \frac{143}{54}$. Find the percentage error in using this as an approximation.

3 Expand $(1-5x)^{\frac{1}{3}}$ as far as the term in x^2. By choosing a suitable value f o r

x, find a rational approximation to $\sqrt[3]{2}$.

Miscellaneous Examples 3

1 Write in full:

(i) $\sum_{k=1}^{5} 2k^2 - 1$ (ii) $\sum_{r=1}^{5} (-1)^r \frac{r}{r+1}$

2 Express the following series using the $\sum$ notation

(i) $1+3+5+7+9$ (ii) $1^2+2^2+3^3+\ldots+8^2$

(iii) $\frac{1}{2}+\frac{2}{3}+\frac{3}{4}+\frac{4}{5}$ (iv) $n+(n+1)+(n+2)+(n+3)$

3 Find the number of terms in the following arithmetic series:

(i) $8+11+\ldots+456$ (ii) $2+6+10+\ldots+146$

(iii) $87+84+\ldots-30$ (iv) $1+1.5+2+\ldots+101$

4 Find the number of terms in the following geometric series:

(i) $2+4+8+\ldots+4096$ (ii) $3+4.5+\ldots+3\times 1.5^8$

(iii) $1+\frac{1}{2}+\frac{1}{4}+\ldots+\frac{1}{4096}$ (iv) $2-\frac{2}{3}+\frac{2}{9}\ldots+\frac{2}{6561}$

5 Find the sum of the first ten terms of the series whose nth terms are as follows:

(i) $(2n+1)$ (ii) $(5-3n)$

(iii) $(7n-2)$ (iv) $\dfrac{1}{2^n}$

(v) $\dfrac{3}{4^n}$ (vi) $\dfrac{4}{5^{n-1}}$

6 Expand

(i) $(3+4x)^4$ (ii) $(3x-2y)^5$ (iii) $\left(2x-\frac{3}{x}\right)^4$

7 Find the term independent of t in the expansion of $\left(2t-\frac{5}{t^2}\right)^{24}$

8 Expand the following as a series in ascending powers of x as far as the term in x^2. State the range of values of x for which it is valid.

(i) $(1+2x)^6$ (ii) $(3-2x)^{-2}$ (iii) $(1-\frac{1}{2}x)^{-\frac{1}{2}}$

(iv) $\dfrac{(1+x)}{(1-x)}$ (v) $(4+x)^{\frac{1}{2}}$ (vi) $(1-3x)^{-4}$

(vii) $(1-x)\sqrt[3]{1+4x}$ (viii) $(9-4x)^{-\frac{1}{2}}$

9 Expand $(1-4x)^{-\frac{1}{2}}$ as far as the term in x^2. Choose a suitable value of x and hence find a rational approximation to $\sqrt{6}$.

Revision Problems 3

1 At the beginning of a month, a customer owes a credit card company £1000. In the middle of the month, the customer pays £A to the company, where $A < 1000$, and at the end of the month, the company adds interest at a rate of 3% of the amount still owing. This process continues, with the customer paying £A in the middle of each month, and the company adding 3% of the amount outstanding at the end of each month.

(i) Find the value of A for which the customer still owes £1000 at the start of every month.

(ii) Find the value of A for which the whole amount owing is exactly paid off after the second payment.

(iii) Assuming that the debt has not been paid off after four payments, show that the amount still owing at the beginning of the fifth month can be expressed as

$$£1000R^4 - A(R^4 + R^3 + R^2 + R), \text{ where } R = 1.03$$

(iv) Show that the value of A for which the whole amount owing is exactly paid off after the nth payment is given by:

$$A = \frac{1000R^{n-1}(R-1)}{R^n - 1}$$

(UCLES)

2 A collector buys a set of stamps with no two of the same value. The lowest value is 10¢ and the highest is 70¢. The number of ¢ in the values form an arithmetic progression. The cost of the set is \$10. Some values involve $\frac{1}{2}$¢.

(i) Write down a formula giving the sum of an arithmetic progression involving its first term a, its last term l, and the number of terms n.
(ii) How many stamps are there in the set?
(iii) Give the values of the three cheapest stamps. The collector buys a different set of stamps, again with no two of the same value. The lowest value in this set is 6¢, and the values in ¢ form an arithmetic progression with common difference 2. The total cost of this set is \$5.
(iv) How many stamps are there in this set?
(v) What is the most expensive stamp in this set?

(Oxford & Cambridge)

3 An athlete plans a training schedule which involves running 20 km in the first week of training; in each subsequent week the distance is to be increased by 10% over the previous week. Write down an expression for the distance to be covered in the nth week according to this schedule, and find in which week the athlete would first cover more than 100 km.

(UCLES)

4

A rubber ball is dropped from a height of 2 m on to a stone floor. Every time it bounces back to 75% of the height it has just fallen.

(i) Write down and simplify an expression for the total distance the ball has travelled as it just touches the floor for the $(n+1)$th time.
(ii) How far does the ball travel in total before it stops bouncing?

4 Coordinates

The idea of locating the position of a point by means of coordinates is a very common one. Most people are familiar with the idea of a map reference, which uses a **rectangular** coordinate system (the axes are at right angles to one another).The development of coordinates resulted from the work of René Descartes, the French mathematician. His work has been extended a long way since then, and there are now many different types of coordinate system. In this chapter, however, we will just look at the algebraic developments in the Cartesian coordinate system in two and three dimensions.

DO YOU KNOW

The position of a point can be located in two or three dimensions by means of rectangular Cartesian coordinates. In two dimensions the position of a point is located at the corner of a rectangle. In three dimensions, the point is located at the corner of a cuboid. The conventions for labelling the axes are shown in Figures 4.1a and 4.1b. To understand the three-dimensional case, imagine the z-axis coming out of the paper towards you if you are looking at the two dimensional axes.

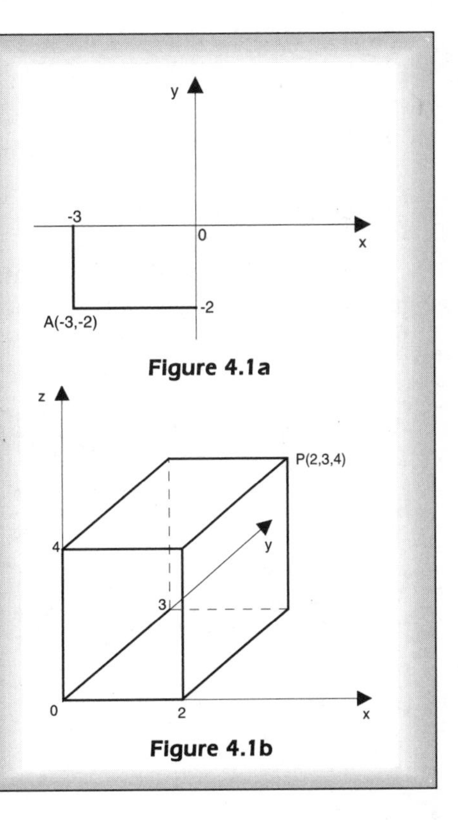

Figure 4.1a

Figure 4.1b

4.1 Distance between two points

In order to find the distance between two points with known coordinates, we can make use of Pythagoras's theorem.

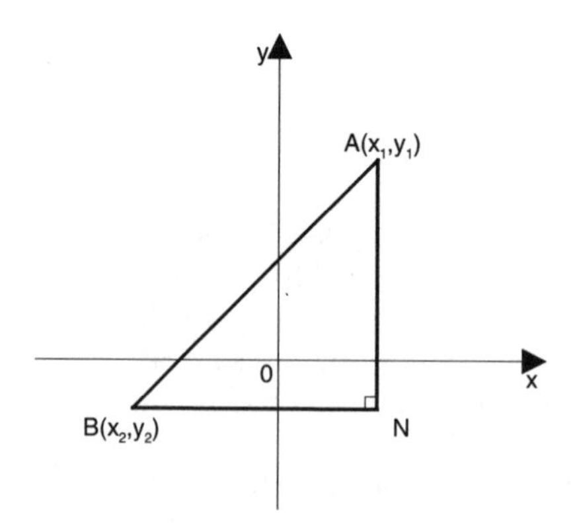

Figure 4.2

In Figure 4.2:

$$\begin{aligned} AB^2 &= AN^2 + BN^2 \\ &= (y_1 - y_2)^2 + (x_1 - x_2)^2 \end{aligned}$$

So the distance between two points is given by the expression:

$$\text{Distance} = \sqrt{(y_1 - y_2)^2 + (x_1 - x_2)^2} \qquad \text{(C1)}$$

Pythagoras's theorem extends quite naturally to three dimensions. Look at Figure 4.3:

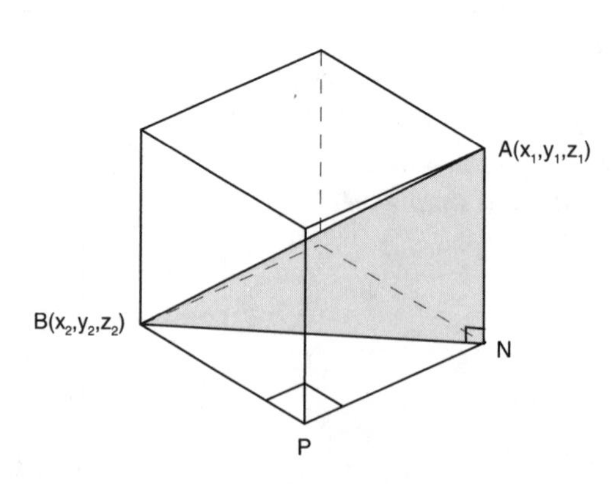

Figure 4.3

$$
\begin{aligned}
& BN^2 = BP^2 + PN^2 \\
\text{and} \quad & AB^2 = AN^2 + BN^2 \\
\text{So} \quad & AB^2 = AN^2 + BP^2 + PN^2 \\
& \quad = (z_1 - z_2)^2 + (x_1 - x_2)^2 + (y_1 - y_2)^2
\end{aligned}
$$

Hence the distance between two points is given by the expression:

$$\text{Distance} = \sqrt{(x_1 - x_2)^2 + y_1 - y_2)^2 + (z_1 - z_2)^2} \qquad \text{(C2)}$$

Example 4.1

Find the distance between the following pairs of points:

(i) $A(4, -1)$ and $B(-2, 6)$ (ii) $C(1, 3, -2)$ and $D(-2, 0, 4)$

(iii) $E(2t, t + 1)$ and $F(1 - t, t - 2)$

Solution

(i) Using the distance formula in Equation (C1), choose either point as (x_1, y_1). So if $(x_1, y_1) = (-2, 6)$ and $(x_2, y_2) = (4, -1)$,

$$
\begin{aligned}
\text{then} \quad AB &= \sqrt{(6 - -1)^2 + (-2 - 4)^2} = \sqrt{49 + 36} \\
&= 9.22 \quad \text{(3 sig. figs)}
\end{aligned}
$$

(ii) Here, let $(x_1, y_1, z_1) = (1, 3, -2)$ and $(x_2, y_2, z_2) = (-2, 0, 4)$,

$$
\begin{aligned}
\text{then} \quad CD &= \sqrt{(1 - -2)^2 + (3 - 0)^2 + (-2 - 4)^2} \\
&= \sqrt{9 + 9 + 36} = 7.35 \quad \text{(3 sig. figs)}
\end{aligned}
$$

(iii) This part is algebraic. Problems of this type occur quite frequently in questions dealing with *parameters*. The important thing is not to try and take short cuts.

$$
\begin{aligned}
\text{So if} \quad & (x_1, y_1) = (2t, t + 1) \\
\text{and} \quad & (x_2, y_2) = (1 - t, t - 2) \\
\text{Then} \quad & x_1 - x_2 = 2t - (1 - t) = 3t - 1 \\
\text{and} \quad & y_1 - y_2 = (t + 1) - (t - 2) = 3 \\
\text{So} \quad & EF = \sqrt{(3t - 1)^2 + 3^2} \\
& \quad = \sqrt{9t^2 - 6t + 1 + 9} \\
& \quad = \sqrt{9t^2 - 6t + 10}
\end{aligned}
$$

Note: This expression cannot be simplified any further.

4.2 Gradient of a line

Referring back to Figure 4.2, the gradient of the line joining the points A and B is given by the expression:

$$\text{gradient of } AB = \frac{AN}{BN} = \frac{y_1 - y_2}{x_1 - x_2} \tag{C3}$$

Note: There is no equivalent simple way of measuring gradient in three dimensions. You can only specify gradient in a two-dimensional plane.

Example 4.2

Find the distance between the points $A(1, -3)$ and $B(-2, 4)$. Also find the gradient of AB.

Solution

When using either the distance and gradient formulae, it doesn't matter which point you take as (x_1, y_1) and which point (x_2, y_2). Hence in this case we would answer the question without a diagram. However, in most problems a diagram is likely to be necessary.

If $(x_1, y_1) = (1, -3)$ and $(x_2, y_2) = (-2, 4)$

$$AB = \sqrt{(1 - -2)^2 + (-3 - 4)^2} \quad \text{(Using equation C1.)}$$

$$= \sqrt{9 + 49} = \sqrt{58}$$

$$\text{The gradient of } AB = \frac{-3 - 4}{1 - -2} = -\frac{7}{3} \quad \text{(Using equation C3.)}$$

You will notice that the gradient is **negative**.

At this point it is worth looking at Figure 4.4. The line slopes **downwards** (top left towards bottom right) in the direction of the positive x axis. Such lines have a negative gradient by convention.

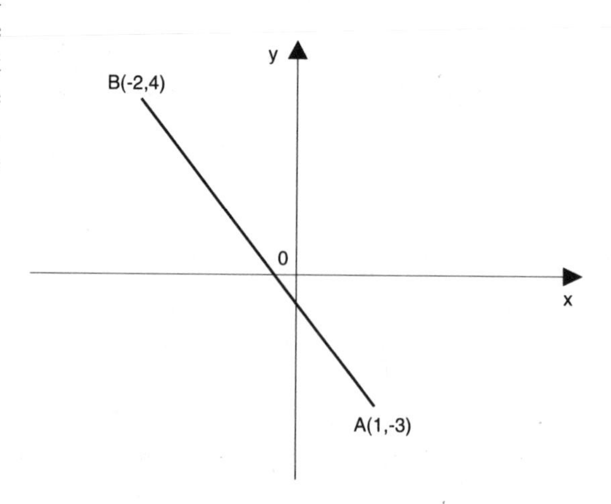

Figure 4.4

Example 4.3

The distance between the points $A(1, t, 2)$ and $B(3, 3, t)$ is 6. Find the value or values of t.

Solution

$$AB = \sqrt{(1-3)^2 + (t-3)^2 + (2-t)^2}$$

But $AB = 6$, hence

$$6 = \sqrt{4 + (t-3)^2 + (2-t)^2}$$

Square both sides and expand the brackets:

$$36 = 4 + t^2 - 6t + 9 + 4 - 4t + t^2$$

Collecting like terms:

$$0 = 2t^2 - 10t - 19$$

Using the formula for solving a quadratic equation,

$$t = \frac{10 \pm \sqrt{(-10)^2 - 4 \times 2 \times -19}}{4}$$

$$\therefore \quad t = 6.47 \quad \text{or} \quad -1.47 \quad \text{(3 sig. figs)}$$

Hence there are, in fact, two sets of points that satisfy the conditions; they are $(1, 6.47, 2)$ and $(3, 3, 6.47)$, or $(1, -1.47, 2)$ and $(3, 3, -1.47)$.

4.3 Other formulae

Although not always specifically mentioned on advanced level syllabuses, there are other formulae, considered here, that you may find useful.

(i) Mid point of a line

If M is half way between P and Q (see Figure 4.5), then the coordinates are:

$$M\left(\frac{x_1 + x_2}{2}, \frac{y_1 + y_2}{2}\right) \qquad \text{(C4a)}$$

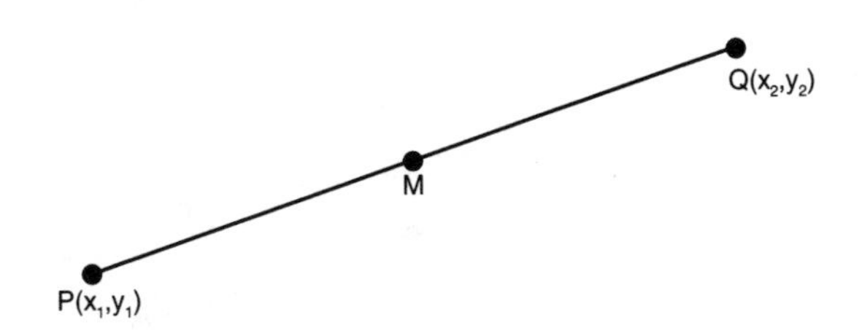

Figure 4.5

In three dimensions, this becomes:

$$M\left(\frac{x_1 + x_2}{2}, \frac{y_1 + y_2}{2}, \frac{z_1 + z_2}{2}\right) \qquad \text{(C4b)}$$

> **MEMORY JOGGER**
>
> You can remember this formula by realising you are averaging the x coordinates, averaging the y coordinates and averaging the z coordinates.

(ii) Dividing a line in any ratio

If the point T divides PQ in the ratio $m : n$ (see Figure 4.6), then the coordinates of T are:

$$T\left(\frac{mx_2 + nx_1}{m + n}, \frac{my_2 + ny_1}{m + n}\right) \qquad \text{(C5a)}$$

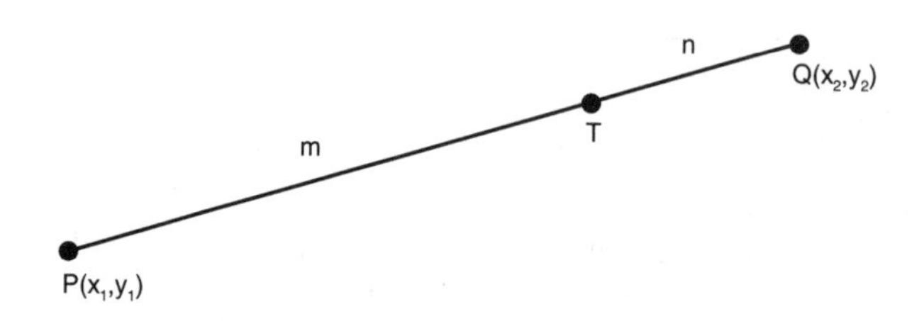

Figure 4.6

In three dimensions, this just becomes:

$$T\left(\frac{mx_2 + nx_1}{m + n}, \frac{my_2 + ny_1}{m + n}, \frac{mz_2 + nz_1}{m + n}\right) \qquad \text{(C5b)}$$

Example 4.4

A point, X, lies on the line joining $P(4, -1)$ and $Q(3, 7)$ (see Figure 4.7). If $\frac{PX}{XQ} = \frac{3}{2}$, find the coordinates of X.

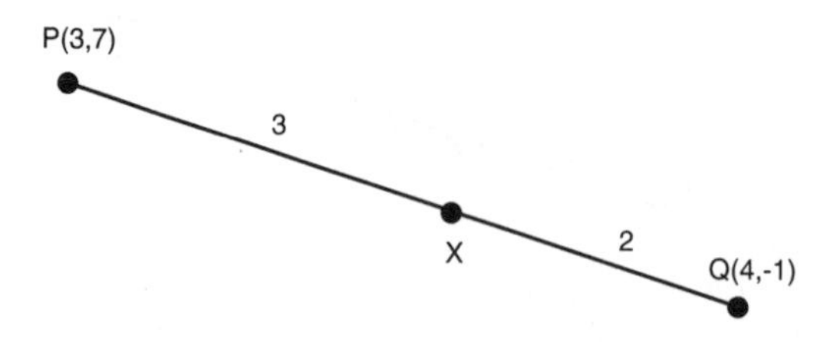

Figure 4.7

Solution

A ratio is often expressed as a fraction.

So $PX : XQ = 3 : 2$

Using formula (C5a) with $(x_2, y_2) = (4, -1)$ and $(x_1, y_1) = (3, 7)$ then $m = 3$ and $n = 2$.

$$\text{Hence } X \text{ is the point } \left(\frac{3x_2 + 2x_1}{5}, \frac{3y_2 + 2y_1}{5}\right)$$

$$= \left(\frac{3 \times 4 + 2 \times 3}{5}, \frac{3 \times -1 + 2 \times 7}{5}\right)$$

that is, X is $\left(3\frac{3}{5}, 2\frac{1}{5}\right)$

Example 4.5

The line joining $A(3, 1, -2)$ and $B(2, 2, 1)$ is extended to a point C, and $AC : BC = 5 : 2$. Find the coordinates of C.

Solution

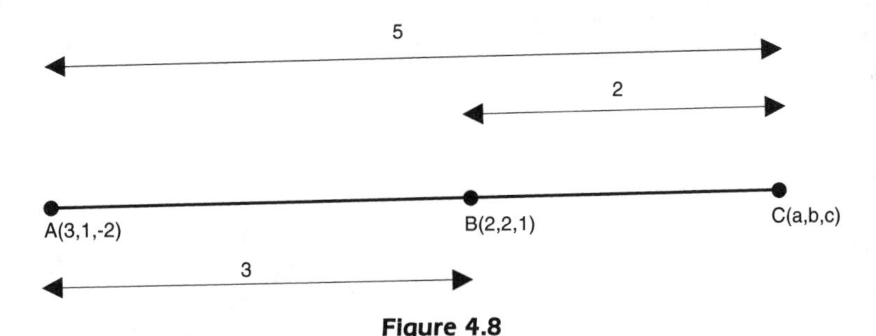

Figure 4.8

You will need to be very careful here before using equation C5b. Because the line is **extended**, B now lies between A and C. Hence $(x_1, y_1, z_1) = (3, 1, -2)$ and $(x_2, y_2, z_2) = (a, b, c)$ is the unknown point. Also $m = 5 - 2 = 3$ and $n = 2$.

Using the x coordinates:

$$2 = \frac{3a + 2 \times 3}{3 + 2}$$

$$\therefore \quad 10 = 3a + 6 \quad \text{and} \quad a = \tfrac{4}{3}$$

Using the y coordinates:

$$2 = \frac{3b + 2 \times 1}{3 + 2}$$

$$\therefore \quad 10 = 3b + 2 \quad \text{and} \quad b = \tfrac{8}{3}$$

The z coordinates give:

$$1 = \frac{3c + 2 \times -2}{3+2}$$

$$\therefore \quad 5 = 3c - 4 \quad \text{and} \quad c = 3$$

Hence C is $\left(\frac{4}{3}, \frac{8}{3}, 3\right)$.

(iii) Area of a triangle

The area of a triangle is given by the formula:

$$\text{area} = \left|\tfrac{1}{2}[x_1(y_2 - y_3) + x_2(y_3 - y_1) + x_3(y_1 - y_2)]\right| \qquad \text{(C6)}$$

You can take a closer look at this formula in the following activity.

ACTIVITY 4

There is a very useful formula using coordinates, which will give you the area of a triangle in two dimensions. It is:

$$\text{area} = \left|\tfrac{1}{2}[x_1(y_2 - y_3) + x_2(y_3 - y_1) + x_3(y_1 - y_2)]\right|$$

The modulus sign ensures that you get a positive answer.

See if you can prove this formula, by using the trapezia $AA'C'C$, $CC'B'B$ and $AA'B'B$ as shown in Figure 4.9.

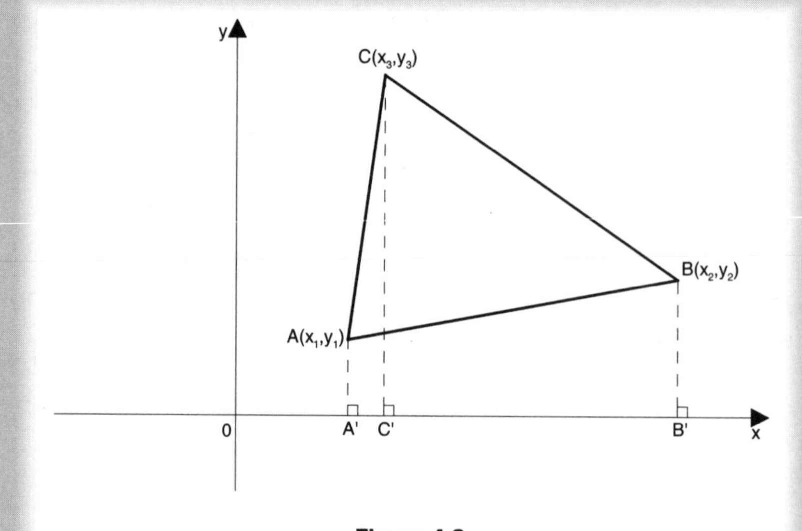

Figure 4.9

You will now be able to find the area of any shape with straight sides by dividing it into triangles. Try a few for yourself.

Do you think there is a similar formula for three dimensions?

Example 4.6

Find the area of the triangle formed by the points $(1, 3)$, $(2, -1)$ and $(4, 6)$.

Solution

It doesn't matter in which order you letter the points.

So if $(x_1 y_1) = (1, 3)$; $(x_2, y_2) = (2, -1)$; $(x_3, y_3) = (4, 6)$, then the area:

$$= \tfrac{1}{2}[1 \times (-1 - 6) + 2 \times (6 - 3) + 4 \times (3 - -1)]$$

$$= \tfrac{1}{2}[-7 + 6 + 16] = 7.5 \text{ square units}$$

(iv) The perpendicular distance of a point from a line

If the equation of a straight line is written in the form $ax + by + c = 0$, then the perpendicular distance of a point $P(h, k)$ (see Figure 4.10) from this line is given by:

Perpendicular distance

$$= \left| \frac{ah + bk + c}{\sqrt{a^2 + b^2}} \right| \qquad \text{(C7)}$$

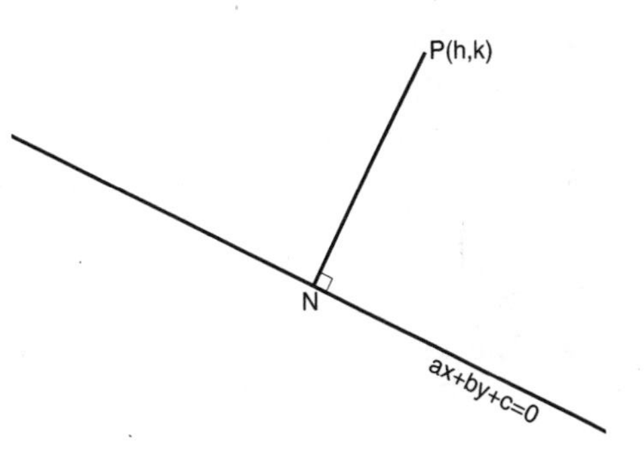

Figure 4.10

Example 4.7

Find the perpendicular distance from the point $P(2, -5)$ to the line $4y = 3x + 7$.

Solution

The straight line must be rewritten:

$$4y - 3x - 7 = 0$$

i.e. $$-3x + 4y - 7 = 0$$

then $h = 2$, $k = -5$ in Equation (C7).

Hence: the distance $= \left| \dfrac{(-3 \times 2) + (4 \times -5) - 7}{\sqrt{4^2 + (-3)^2}} \right|$

$$= 6.6$$

Exercise 4(a)

1 For the following pairs of points, find the distance between them, and the gradient of the line joining them.

(i) $(2,1),(3,-2)$ (ii) $(6,3),(1,0)$

(iii) $(2,-3),(4,1)$ (iv) $(t,2t),(3t,-t)$

(v) $(p,p+q),(q,p)$ (vi) $(2.6,-1),(1.3,4)$

(vii) $(\frac{1}{2},\frac{1}{4}),(-\frac{1}{2},0)$ viii) $(a+b,0),(0,a+b)$

2. Find the distance between the following pairs of points, simplifying where possible but leaving surds in your answers.

(i) $(1,0,2),(3,1,1)$ (ii) $(1,1,0),(-1,3,-2)$

(iii) $(4,-1,0),(3,3,2)$ (iv) $(t,4t),(3t,-t)$

(v) $(\frac{1}{2},2),(\frac{1}{4},-1)$ (vi) $(\frac{2}{3},\frac{1}{3},\frac{3}{5}),(\frac{1}{4},\frac{1}{2},\frac{3}{4})$

(vii) $(y,2y,3y),(-y,-y,2y)$

(viii) $(1-t,1+t,3),(4-t,3+t,1)$

3. Find the coordinates of the mid points of the lines joining the pairs of points given in Question 2.

4. P lies on the straight line joining $A(1,1,2)$ and $B(2,1,3)$ and $AP:PB=3:2$. Find the coordinates of P.

5. Find the area of the triangle formed by the points $P(1,-1)$, $Q(2,2)$ and $R(3,7)$.

6. The line joining points $A(1,0,2)$ and $B(1,-1,-2)$ is extended to a point C such that $AB:BC=2:5$. D is the point $(1,1,1)$. Find (i) the coordinates of C; (ii) the area of the triangle BCD; (iii) the perpendicular distance of B from CD.

7. Find the perpendicular distance from the given point to the given line:

(i) $(2,0)\ :\ 3x-2y+1=0$

(ii) $(1,-2)\ :\ 5x+12y-7=0$

(iii) $(-2,5)\ :\ y=7x+11$

4.4 The equation of a straight line

DO YOU KNOW

(i) How to plot an equation such as $y=3x+2$?

(ii) The equation $y=mx+c$ represents a straight line, m is the gradient and c is the value of y where the line cuts the y-axis (called the y-intercept).

You should already have come across the fact that $y = mx + c$ is a straight line. In A-level work, it is necessary to reach this equation from a number of situations.

(i) Suppose we know one point, $P(x_1, y_1)$ on the line, and the gradient of the line m (see Figure 4.11). The line cuts the y axis at $A(0, c)$.

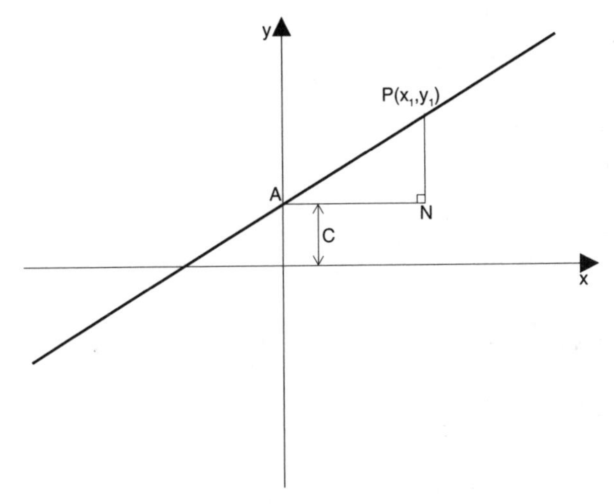

Figure 4.11

Now $\dfrac{PN}{AN} = m$, so $\dfrac{(y_1 - c)}{x_1} = m$

$\therefore \quad y_1 - c = mx$, hence: $c = y_1 - mx_1$

Now, using the equation $y = mx + c$, we can replace c by $y_1 - mx_1$

$\therefore \quad y \quad = mx + y_1 - mx_1$

or $\quad y - y_1 = m(x - x_1)$ (SL1)

(ii) Suppose we know two points $P(x_1, y_1)$ and $Q(x_2, y_2)$ on the line.

Then the gradient $\quad m = \dfrac{y_2 - y_1}{x_2 - x_1}$

Hence Equation (SL1) can be written:

$$(y - y_1) = \left(\frac{y_2 - y_1}{x_2 - x_1}\right)(x - x_1)$$

or $\quad \dfrac{y - y_1}{y_2 - y_1} = \dfrac{x - x_1}{x_2 - x_1}$ (SL2)

The following example shows you how to use these formulae.

Example 4.8

(i) Find the equation of the line that passes through the point $(4, -2)$ with gradient $-\frac{1}{2}$.
(ii) Find the equation of the line that joins the points $(3, -1)$ and $(2, 2)$.
(iii) Find the point of intersection of the lines found in parts (i) and (ii).

Solution

(i) Here, $(x_1, y_1) = (4, -2)$ and $m = -\frac{1}{2}$.

(SL1) gives $y - -2 = -\frac{1}{2}(x - 4)$

$y + 2 = -\frac{1}{2}x + 2$

$\therefore \quad y = -\frac{1}{2}x$

(ii) Here, let $(x_1, y_1) = (3, -1)$ and $(x_2, y_2) = (2, 2)$
Note: It doesn't matter which order you number them.

(SL2) gives $\dfrac{y - -1}{2 - -1} = \dfrac{(x - 3)}{2 - 3}$

$\therefore \quad \dfrac{y + 1}{3} = \dfrac{x - 3}{-1}$

$\therefore \quad -y - 1 = 3x - 9$

$\therefore \quad y + 3x = 8$

(iii) The point of intersection of two straight lines can be found by solving simultaneous equations.

So: $y + 3x = 8$ (i)

$y = -\frac{1}{2}x$ (ii)

The easiest way of solving these is to substitute in (i) for y from (ii),

So $-\frac{1}{2}x + 3x = 8$

$2\frac{1}{2}x = 8$

So $x = 3\frac{1}{5} \quad (8 \div 2\frac{1}{2})$

from (ii) $y = -\frac{1}{2} \times 3\frac{1}{5} = -1\frac{3}{5}$

Hence the point of intersection is: $(3\frac{1}{5}, -1\frac{3}{5})$

MEMORY JOGGER

The equation $y = mx + c$ often appears in the forms $ax + by + c = 0$, $ax + by = c$, or $\dfrac{x}{a} + \dfrac{y}{b} = 1$

Example 4.9

For the following equations, find the gradient of the line, and draw a sketch to indicate the position of the line:

(i) $4x + 3y = 12$ (ii) $7x - 3y - 5 = 0$

(iii) $\frac{x}{6} + \frac{y}{4} = 1$

Solution

(i) Rearranging this formula to make y the subject:

$4x + 3y = 12$ $\therefore \quad 3y = -4x + 12$

Divide the equation by 3 $y = -\frac{4}{3}x + 4$

Hence the gradient is $-\frac{4}{3}$, and it cuts the y axis at $(0, 4)$.

Another way of doing this type of problem is to find the points where the line cuts the axes.

At $x = 0$, $0 + 3y = 12$, so $y = 4$

At $y = 0$, $4x + 0 = 12$, so $x = 3$

The line is shown in Figure 4.12.

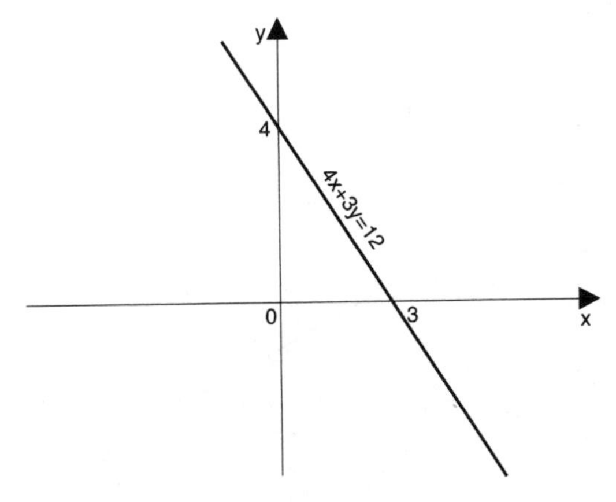

Figure 4.12

(ii) Using the second method given in part (i)

At $x = 0$, $-3y - 5 = 0$ $\therefore \quad y = -\frac{5}{3}$

At $y = 0$, $7x - 5 = 0$ $\therefore \quad x = \frac{5}{7}$

The graph can now be sketched as in Figure 4.13, without finding the gradient.

You can now look at the diagram, and, ignoring the signs, the gradient is:

$\frac{5}{3} \div \frac{5}{7} = \frac{7}{3}$

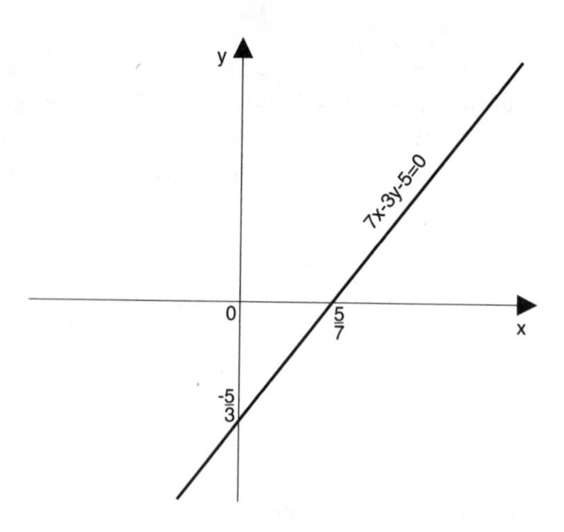

Figure 4.13

(iii) At $x = 0$, $\dfrac{y}{4} = 1$ $\therefore$ $y = 4$

At $y = 0$, $\dfrac{x}{6} = 1$ $\therefore$ $x = 6$

The sketch is shown in Figure 4.14.

The gradient $= -\frac{4}{6} = -\frac{2}{3}$

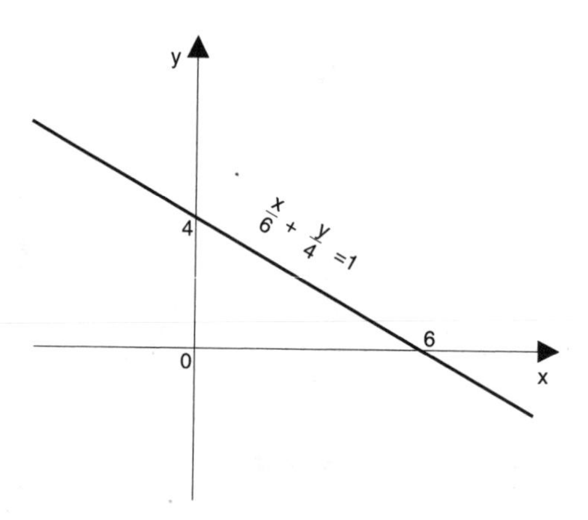

Figure 4.14

4.5 Perpendicular lines

Fig. 4.15 shows two lines that are perpendicular, namely PQ and PR. Hence, angle $QPR = 90°$. You can see this from the fact that $x° + y° = 90°$, and triangle PMQ is identical (congruent) to triangle PNR.

The gradient of $PQ = \frac{b}{a} = m_1$, say

The gradient of $PR = -\frac{a}{b} = m_2$, say

So $m_1 m_2 = \frac{b}{a} \times -\frac{a}{b} = -1$

We have the result that, if two lines are perpendicular with gradients m_1 and m_2,

Then $m_1 m_2 = -1$ (SL3)

This result is very useful.

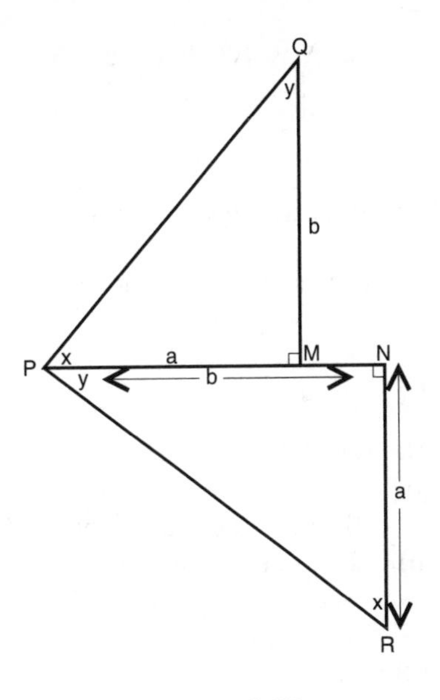

Figure 4.15

Example 4.10

Find the equation of the line through $M\ (4, 3)$ which is perpendicular to the line $x + 2y + 8 = 0$. Hence find the perpendicular distance of M from the line $x + 2y + 8 = 0$.

Solution

If $x + 2y + 8 = 0$
then $y = -\frac{1}{2}x - 4$

Hence the gradient of the line is $-\frac{1}{2}$.

Referring to Figure 4.16, if MN is perpendicular to the line, then if the gradient of MN is m,

$$m \times -\tfrac{1}{2} = -1 \quad \text{using formula (SL3)}$$

$$\therefore \quad m = 2$$

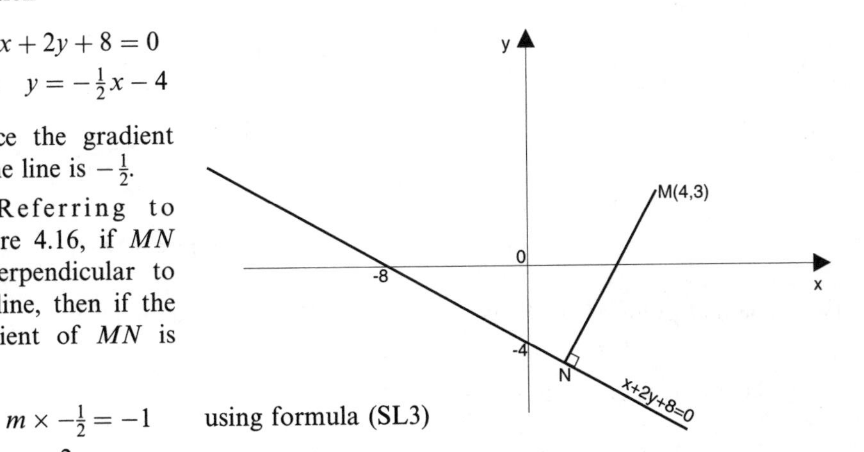

Figure 4.16

Hence the equation of MN is:

$$(y - 3) = 2(x - 4)$$

that is, $y = 2x - 5$

To find N, solve this equation with $x + 2y + 8 = 0$,

$$\text{that is,} \quad x + 2(2x - 5) + 8 = 0$$
$$\therefore \quad 5x = 2$$

Therefore $x = 0.4$, $y = -4.2$.

$$\text{The required distance} = MN = \sqrt{(4 - 0.4)^2 + (3 - -4.2)^2} = 8.05$$

Example 4.11

The points A, B, C have coordinates $(2, 5)$, $(1, -2)$ and $(6, -3)$ respectively.

(i) Find the equation of the line L which is perpendicular to AC, and passes through the mid point of AB.

(ii) The line m cuts the x-axis at P and the y-axis at Q. Find the area of the triangle POQ, where O is the origin.

Solution

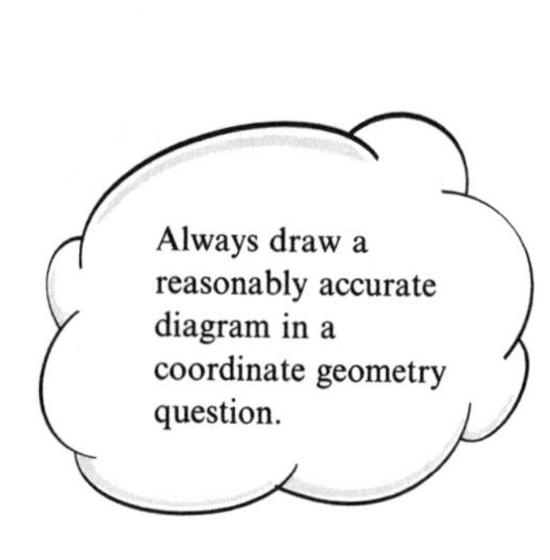

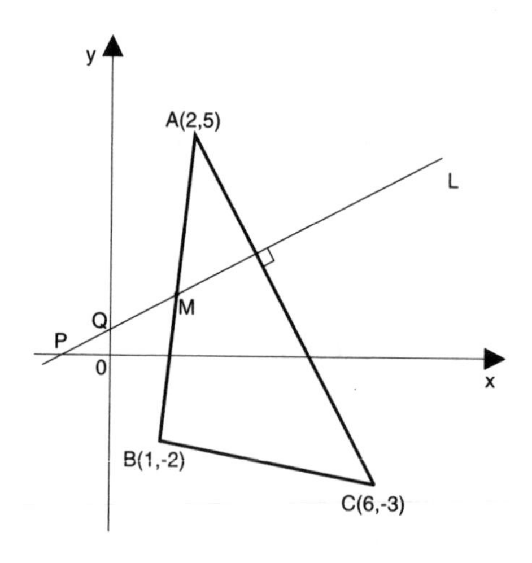

Figure 4.17

(i) The mid point M of AB is $\left(\frac{2+1}{2}, \frac{5+-2}{2}\right)$, that is, M is $\left(\frac{3}{2}, \frac{3}{2}\right)$.

The gradient of AC is $\frac{5 - -3}{2 - 6} = -2$.

The gradient of L is $\frac{1}{2}$ (using $m_1 m_2 = -1$ for perpendicular lines)
The equation of L is:

$$y - \tfrac{3}{2} = \tfrac{1}{2}\left(x - \tfrac{3}{2}\right)$$
$$\therefore \quad y - \tfrac{3}{2} = \tfrac{1}{2}x - \tfrac{3}{4}$$
$$y = \tfrac{1}{2}x + \tfrac{3}{4}$$

(ii) To find P, $y = 0$.

So $0 = \frac{1}{2}x + \frac{3}{4}$ $\quad \therefore \quad x = -\frac{3}{2}$

To find Q, $x = 0$

So $y = 0 + \frac{3}{4} = \frac{3}{4}$

The area of the triangle $= \frac{1}{2}$ base $\times$ height

$= \frac{1}{2} \times \frac{3}{4} \times \frac{3}{2} = \frac{9}{16}$ units^2

Exercise 4(b)

1 Find the equation of the straight line through the given point with the given gradient.

(i) $(2, 3)$; -2 (ii) $(1, -1)$; 4 (iii) $(1, -\frac{1}{2})$; -2

(iv) (p, q); $2p$ (v) $(-\frac{1}{3}, -\frac{1}{4})$; $-\frac{1}{2}$

2 Find the equation of the straight line joining the following pairs of points:

(i) $(1, 2)$, $(3, 1)$ (ii) $(1, 1)$, $(2, -1)$ (iii) $(\frac{1}{2}, 2)$, $(-1, \frac{1}{2})$

(iv) $\left(p, \frac{1}{p}\right)$, $\left(q, \frac{1}{q}\right)$ (v) $(2at_1, at_1{}^2)$, $(2at_2, at_2{}^2)$

3 For the following equations, find the gradient of the line and draw a sketch to indicate the position of the line.

(i) $3x + 6 = y$ (ii) $2x - 3y = 12$ (iii) $\frac{x}{4} + \frac{y}{6} = 1$

(iv) $5x + 7y + 20 = 0$

4 Find the equation of the line through $(4, -3)$, which is perpendicular to the line $4y = 3x - 5$.

4.6 Conversion of data to fit a straight line graph

There are many situations in practical work, where if the graph of related data is plotted, the resulting line is not straight. By transforming the variables in some way, the resulting graph can be made straight.

Consider the following situation.

When working with a lens of focal length f cm, the distances of the object u cm and image v cm from the lens are given by the formula $\frac{1}{u} + \frac{1}{v} = \frac{1}{f}$.

Look at the following results

u	30.8	24.6	22.0	19.6	18.2
v	28.6	37.2	44.1	61.9	71.6

If a graph of this data is plotted as it stands, the results are not on a straight line. See Figure 4.18.

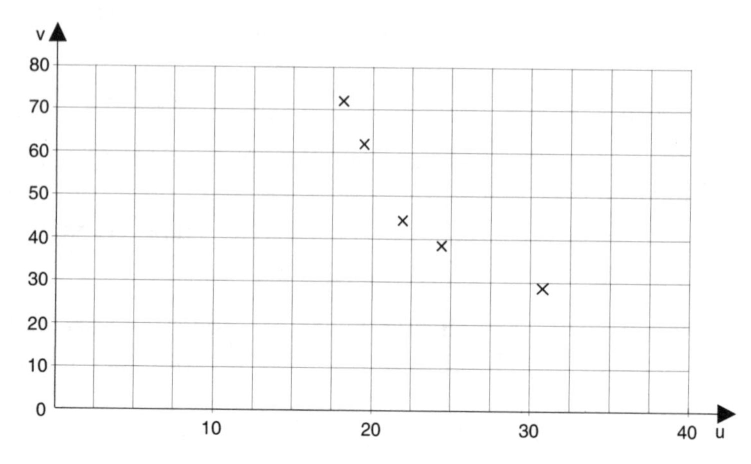

Figure 4.18

However, if you let $\frac{1}{u} = x$, and $\frac{1}{v} = y$, then the equation becomes

$$y + x = \frac{1}{f} \quad \text{or} \quad y = -x + \frac{1}{f}$$

Now tabulate $\frac{1}{u}$ and $\frac{1}{v}$.

$x = \frac{1}{u}$	0.032	0.041	0.045	0.051	0.055
$y = \frac{1}{v}$	0.035	0.027	0.023	0.016	0.014

If you plot these points, then they are almost on a straight line (see Figure 4.19). This extends back to cut the y-axis at 0.065.

Comparing $y = -x + \frac{1}{f}$ with

$$y = mx + c$$

you can see that $\frac{1}{f} = 0.065$.

Hence the focal length $f = 15.4$ cm.

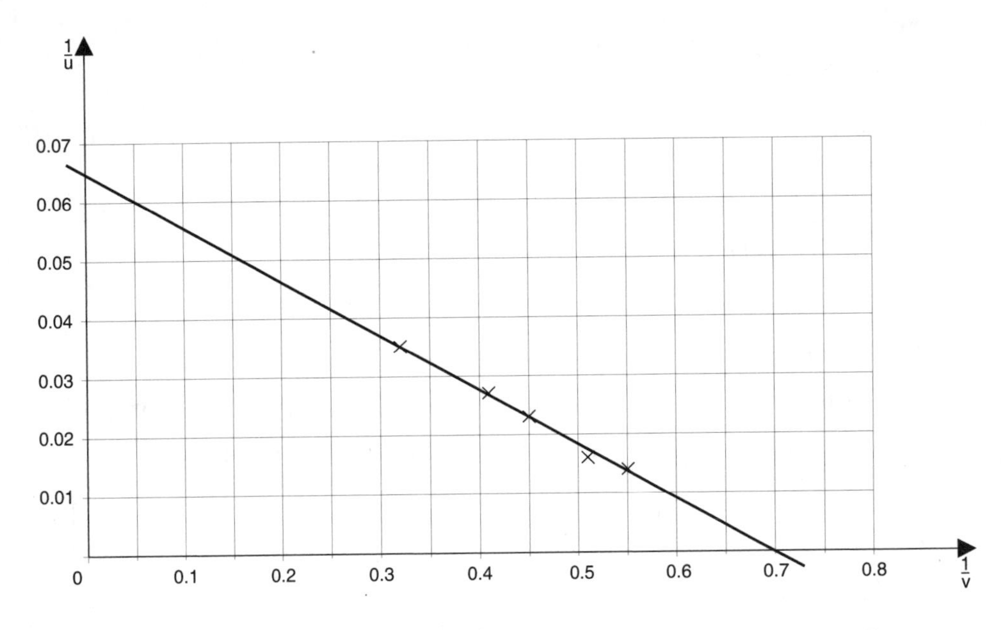

Figure 4.19

With a little practice, you can usually determine what you should choose as x and y. Look at the following example:

Example 4.12

The distance (s metres) travelled by a car as it accelerates from a speed of 5 m/s is given by the formula

$$s = 5t + \tfrac{1}{2}at^2$$

where a is the constant acceleration, and t the time, in seconds, after the start. Times are measured every 2 seconds, and the results are as follows:

s	0	10.9	24.2	40	55.4	76
t	0	2	4	6	8	10

Alter the variables in some way, so that a straight line can be obtained. Hence find a.

Solution

$$s = 5t + \tfrac{1}{2}at^2$$

We need the two variables to occur only *once* each. Here t occurs in two places:

$$\div t \quad \frac{s}{t} = 5 + \tfrac{1}{2}at$$

Let $y = \dfrac{s}{t}$ and $x = t$

So $\quad y = \tfrac{1}{2}ax + 5$

This should give us a straight line with gradient $\tfrac{1}{2}a$.

There is one small problem, because $\dfrac{s}{t}$ for the first point would be $\dfrac{0}{0}$ which cannot be evaluated. This point must be ignored. The table becomes:

$\dfrac{s}{t}$	5.45	6.05	6.67	6.93	7.6
t	2	4	6	8	10

We also know from the formula that the graph crosses the y-axis at 5. The gradient of the graph is:

$$\frac{1.93}{8} = 0.24$$

So $\quad \tfrac{1}{2}a = 0.24$

$\therefore \quad a = 0.48$

The acceleration is 0.48 m/s^2. See Figure 4.20.

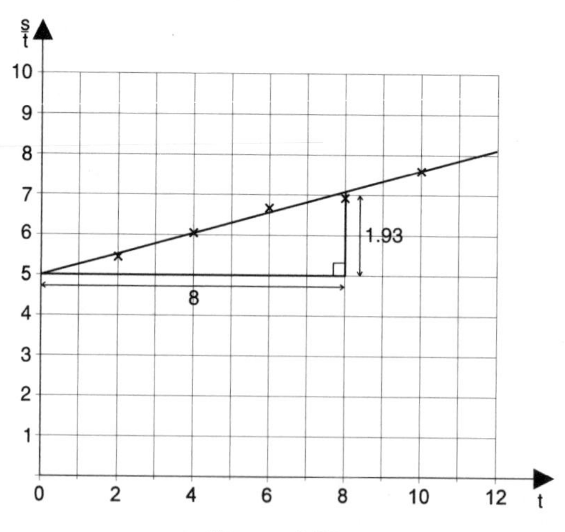

Figure 4.20

There are situations where logarithms are required to transform an equation. Work carefully through the next example.

Example 4.13

At time t minutes, the temperature of a liquid, which is cooling, exceeds room temperature by T°C. The table below shows the temperature difference at given times.

Time t (minutes)	0	5	10	15	20
Temperature difference T (°C)	23	13.5	8.0	4.4	2.5

It is believed that T and t are related by a law of the form $T = ke^{at}$ where k and a are constants.

(a) By drawing a graph of $\ln T$ against t show that this law is approximately valid.

(b) Use your graph to estimate values for k and a, giving your answers to 2 significant figures. (LONDON)

Solution

(a) If $T = ke^{at}$, taking logarithms to base e, we have:

$$\ln T = \ln ke^{at} = \ln k + \ln e^{at}$$

$$= \ln k + at \ln e$$

$$\therefore \quad \ln T = \ln k + at \quad \text{(because } \ln e = 1\text{)}$$

Let $y = \ln T$, $x = t$ and $\ln k = C$

then $\qquad y = ax + c$

This is the equation of a straight line. So if we plot a graph of $\ln T(y)$ against $t(x)$ it should be a straight line.

Using the calculator to work out the table, we have:

t	0	5	10	15	20
$\ln T$	3.14	2.60	2.08	1.48	0.92

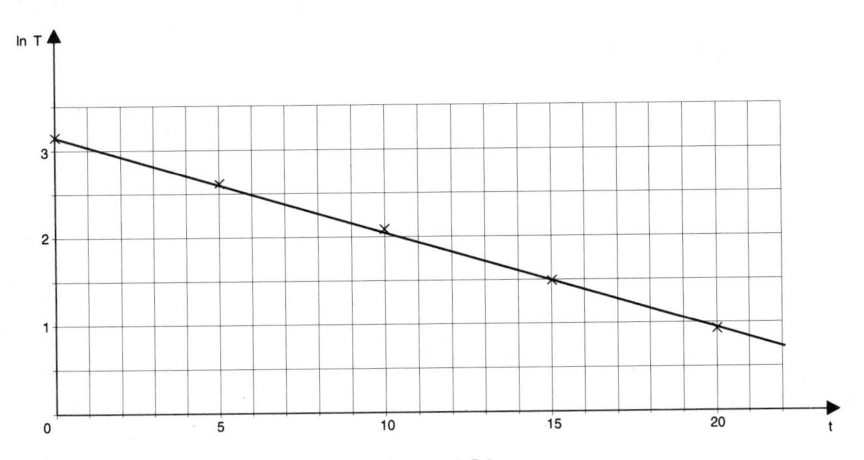

Figure 4.21

The graph is shown in Figure 4.21.

The points are more or less on a straight line, proving the assumption is probably correct.

(b) 'c' is the intercept on the y-axis, which is 3.14.

But $c = \ln k \quad \therefore \quad \ln k = 3.14$

Hence: $\quad k = e^{3.14} = 23 \quad$ (2 sig. figs)

'a' is the gradient of the line (**Note:** it is negative)

$$\therefore \quad a = -\frac{(3.14 - 0.92)}{20} = -0.11 \quad \text{(2 sig. figs)}$$

The following exercise includes enough hints for you to be able to work through a range of different types of transformation.

Exercise 4(c)

In the following questions, you are given a table of results, the equation that governs the results, and the transformations necessary to produce a straight-line graph. In each case, alter the given formula so that it becomes type $y = mx + c$. Then plot the transformed results, and hence find the unknown constants.

1

v	10	20	30	40	50
p	100	246	416	600	810

$p = kv^{\alpha}$; plot $\ln p$ against $\ln v$.

2

x	0	1	2	3	4
y	7	34	180	870	4400

$y = ka^x$; plot $\ln y$ against x.

3

x	0.1	0.2	0.3	0.4
T	6.1	7.2	8.5	9.9

$T = ke^x + q$; plot T against e^x.

4

u	10	15	20	25	30
v	−11.8	−20	−28.5	−43	−60

$\dfrac{1}{u} + \dfrac{1}{v} = \dfrac{1}{f}$; plot $\dfrac{1}{u}$ against $\dfrac{1}{v}$

Miscellaneous Examples 4

1 Find the distance between the points:

(i) $(2, 8)$ and $(-3, -2)$ (ii) $(1, 4, 6)$ and $(-2, -3, 2)$

2 Find the gradient of line joining the points:

(i) $(3, -2)$ and $(4, -7)$ (ii) $\left(cp, \frac{c}{p}\right)$ and $\left(cq, \frac{c}{q}\right)$

3 Find the mid point of the line joining the points:

(i) $(2, 3)$ and $(4, -6)$ (ii) $(\frac{1}{2}, \frac{1}{3})$ and $(-\frac{2}{3}, \frac{1}{4})$

(iii) $(1, -2, 3)$ and $(3, 1, 1)$

4 Find the area of the triangle formed by the points:
$A(1, 1, 4)$, $B(2, 1, 3)$ and $C, (1, 2, 0)$

5 Find the equation of the line joining the points $P(1, 3)$ and $Q(2, 5)$. This line cuts the line $2y + 3x = 23$ at the point R. Find the coordinates of R, and hence the area of the triangle POR.

6 Find the perpendicular distance of the point $(4, 3)$ from the line $5x + 12y = 7$.

7 $A(2, 3)$, $B(1, 6)$ and $C(p, q)$ are three corners of a square $ABCD$. Find the possible values of p and q, and the corresponding coordinates of D.

8 Two variables x and y are related by the equation $y = ka^x$. Corresponding values of the variables are given in the following table:

x	1	1.5	2	2.5
y	23.5	42	71.6	125

By plotting a suitable straight line, find the values of k and a.

Revision Problems 4

1 For certain planets, the approximate mean distance x, in millions of km from the centre of the sun, and the period of the orbit T, in Earth years, are recorded.

x	57.9	108.2	227.9	778.3
T	0.24	0.62	1.88	11.86

Assuming a law of the form $T = Ax^n$, draw a graph of $\ln T$ against $\ln x$. Estimate the values of A and n, giving your answers to two significant figures.

Use your graph to estimate the approximate mean distance in millions of km of the Earth from the Sun.

(AEB 94)

2 A researcher claims that the figures in the following table give the approximate total population (in millions) of England at the beginning of four particular years.

Year	1500	1560	1700	1800
Population	2.1	2.8	5.4	8.7

Let $N(t)$ be the population (in millions) at time t (in years after 1500). The researcher claims that during the period 1500–1800 the population approximately satisfies a rule of the form

$$N(t) = N_0 e^{kt}$$

for some constants N_0 and k.

(a) By plotting a suitable straight line graph, or otherwise, confirm that such a rule is consistent with the given data. State approximate values of N_0 and k.

(b) Estimate the size of the population in 1670, but suggest a reason against over-reliance on the above rule for predicting a population.

(AEB 94)

5 The differential calculus

The calculus was invented independently by Isaac Newton in England about 1687, and in Germany by Gottfried Leibnitz in about 1684. There has always been controversy as to whom should be given the credit, but both had in fact brought together ideas that had been developed by other mathematicians during the preceding 200 years.

DO YOU KNOW

(i) A tangent to a curve is a straight line that touches the curve in one point.

(ii) The gradient of a line$=\dfrac{\text{change in vertical distance}}{\text{change in horizontal distance}}$

5.1 The concept of differentiation

$A(x, y)$ and $B(x+h, y+k)$ are two points on the graph $y = x^2$ (see Figure 5.1).

Since B lies on the curve, then

$$(y+k) = (x+h)^2 = (x+h)(x+h)$$

that is, $\quad y + k = x^2 + 2hx + h^2$

Since $y = x^2$, we are left with

$$k = 2hx + h^2$$

divide by h: $\quad \dfrac{k}{h} = 2x + h \qquad$ (i)

The vertical distance between A and B is k.

The horizontal distance between A and B is h.

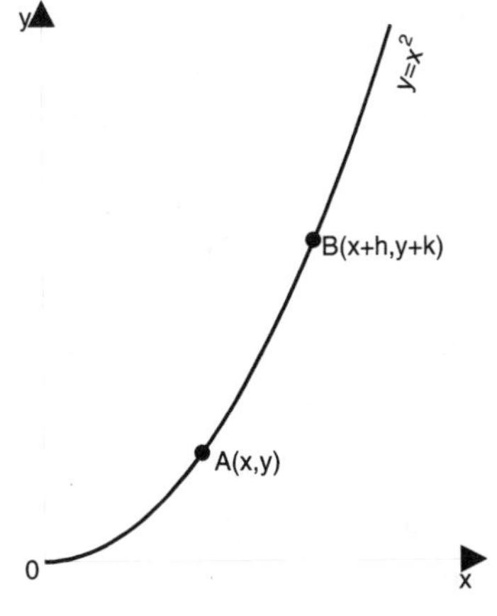

Figure 5.1

Now $\frac{k}{h}$ is the slope of AB, so we have the result;

the slope of $AB = 2x + h$

Now suppose that h is very small, almost zero.

This is written $h = \delta x$ (δx means a small change in x)
Similarly, $k = \delta y$

hence equation (i) can be written: $\frac{\delta y}{\delta x} = 2x + \delta x$.

As δx approaches zero, δy also approaches zero. $\frac{\delta y}{\delta x}$ is written $\frac{dy}{dx}$ at the point where $\delta x = 0$. (This process is called differentiation from first principles.)

$$\therefore \quad \frac{dy}{dx} = 2x$$

$\frac{dy}{dx}$ is said 'd y by d x' and it represents the gradient of the curve at the point (x, y). Note that $\frac{\delta y}{\delta x}$ is not a single symbol, but represents $\delta y \div \delta x$, since both δy and δx are not zero until point A is reached. This 'limiting' process as $\delta x \to 0$ (δx approaches zero) is written by mathematicians:

$$\lim_{\delta x \to 0} \frac{\delta y}{\delta x} = \frac{dy}{dx} \quad \text{(sometimes written } f'(x) \text{ in function notation)}$$

The process of finding is called **differentiation**. The actual answer is called the **derivative** of y with respect to x. Sometimes the symbol $\frac{d}{dx}$ is used to denote the instruction of differentiate with respect to x.

Hence: $\frac{d}{dx}(x^2)$ means 'differentiate the function x^2 with respect to x. This saves writing $y = x^2$ and then $\frac{dy}{dx}$ and so on.

$$\text{Hence:} \quad \frac{d}{dx}(x^2) = 2x$$

If you now repeat these ideas to the curve $y = x^3$, you will find:

$$\frac{\delta y}{\delta x} = 3x^2 + 3x.\delta x + (\delta x)^2$$

as $\delta x \to 0$

$$\frac{dy}{dx} = 3x^2$$

Similarly, if $y = x^4$, $\frac{dy}{dx} = 4x^3$. It would appear (as can be proved) that

$$\text{if} \quad y = x^n$$

$$\frac{dy}{dx} = nx^{n-1} \qquad \text{(D1)}$$

It is perhaps surprising that this result is true for all values of n. The result can be extended if:

$$y = kx^n, \text{ where } k \text{ is any number}$$

$$\text{then} \quad \frac{dy}{dx} = knx^{n-1} \qquad \text{(D2)}$$

Note: If $n = 0$, $y = k$ and $\frac{dy}{dx} = 0$ (D2a)

if $n = 1$, $y = kx$ and $\frac{dy}{dx} = k$ (D2B)

If we add together two (or more) functions to get $y = f(x) + g(x)$

$$\text{then:} \quad \frac{dy}{dx} = \frac{df}{dx} + \frac{dg}{dx} \qquad \text{(D3)}$$

Example 5.1

Differentiate the following functions with respect to x.

(i) $4x^6$ (ii) $3x^3 + 2x + 1$ (iii) $\sqrt{x}$

(iv) $\frac{2}{x^2}$ (v) $\sqrt[3]{8x^2}$

Solution

(i) If $y = 4x^6$

$$\frac{dy}{dx} = 4 \times 6x^5 = 24x^5$$

(ii) If $y = 3x^3 + 2x + 1$

$$\frac{dy}{dx} = 3 \times 3x^2 + 2 = 9x^2 + 2$$

(iii) If $y = \sqrt{x}$

$$y = x^{\frac{1}{2}}$$

$$\therefore \quad \frac{dy}{dx} = \frac{1}{2}x^{(\frac{1}{2}-1)} = \frac{1}{2}x^{-\frac{1}{2}} = \frac{1}{2\sqrt{x}}$$

(iv) If $y = \frac{2}{x^2}$

$$y = 2x^{-2}$$

$$\therefore \quad \frac{dy}{dx} = 2 \times -2x^{(-2-1)} = -4x^{-3}$$

$$= \frac{-4}{x^3}$$

(v) $y = \sqrt[3]{8x^2} = \sqrt[3]{8} \times (x^2)^{\frac{1}{3}}$

$$\therefore \quad y = 2x^{\frac{2}{3}}$$

Hence: $\frac{dy}{dx} = 2 \times \frac{2}{3}x^{(\frac{2}{3}-1)} = \frac{4}{3}x^{-\frac{1}{3}}$

$$= \frac{4}{3\sqrt[3]{x}}$$

MEMORY JOGGER

The letters used in the process of differentiation are not important, as the following example shows.

Example 5.2

Differentiate the following expressions with respect to t.

(i) $4t^3 + 8$

(ii) $\frac{1}{2t^3}$

(iii) $\frac{a}{t^2}$ where a is a constant.

Solution

(i) $\frac{d}{dt}(4t^3 + 8) = 12t^2$

(ii) $\frac{d}{dt}\left(\frac{1}{2t^3}\right) = \frac{d}{dt}\left(\frac{1}{2}t^{-3}\right) = -\frac{3}{2}t^{-4} = -\frac{3}{2} \times \frac{1}{t^4} = -\frac{3}{2t^4}$

(iii) Care is needed here, because a is treated like an ordinary number

so $\frac{d}{dt}\left(\frac{a}{t^2}\right) = \frac{d}{dt}(at^{-2}) = -2at^{-3} = -\frac{2a}{t^3}$

Sometimes, a function may need to be rewritten before it can be differentiated.

Example 5.3

Find the derivative with respect to x of:

(i) $\dfrac{4x+1}{x^2}$ (ii) $\dfrac{(3x-2)^2}{4x^4}$

Solution

(i) You must first divide the fraction out.

So: If $y = \dfrac{4x+1}{x^2}$

then: $y = \dfrac{4x}{x^2} + \dfrac{1}{x^2}$

$= \dfrac{4}{x} + \dfrac{1}{x^2} = 4x^{-1} + x^{-2}$

Hence: $\dfrac{dy}{dx} = -4x^{-2} - 2x^{-3} = \dfrac{-4}{x^2} - \dfrac{2}{x^3}$

(ii) Here you need to square out the top line first.

So: if $y = \dfrac{(3x-2)^2}{4x^4}$

$$y = \frac{9x^2 - 12x + 4}{4x^4} = \frac{9x^2}{4x^4} - \frac{12x}{4x^4} + \frac{4}{4x^4}$$

$$y = \frac{9}{4x^2} - \frac{3}{x^3} + \frac{1}{x^4} = \frac{9}{4}x^{-2} - 3x^{-3} + x^{-4}$$

$$\therefore \quad \frac{dy}{dx} = -\frac{9}{2}x^{-3} + 9x^{-4} - 4x^{-5}$$

$$= -\frac{9}{2x^3} + \frac{9}{x^4} - \frac{4}{x^5}$$

5.2 Differentiating a function of a function (the chain rule)

An expression such as $y = (x^2+1)^3$ is called a **function of a function**. To obtain y, first square x and add 1 (the first function). Then cube the answer (the second function). Hence you are finding a function (cubing) of a function (squaring + 1).

If you let $u = x^2 + 1$, then $y = u^3$

$\dfrac{dy}{du} = 3u^2$ (Note how the letters are not part of the process)

Now $\dfrac{du}{dx} = 2x$

The **chain rule** states:

$$\frac{dy}{dx} = \frac{dy}{du} \times \frac{du}{dx} \qquad \text{(D4)}$$

So $\dfrac{dy}{dx} = 3u^2 \times 2x = 6xu^2$

but $u = x^2 + 1$, hence

$$\frac{dy}{dx} = 6x(x^2+1)^2$$

The working can be omitted on many occasions, going straight to the answer.

Example 5.4

Use the chain rule to differentiate the following functions with respect to x.

(i) $\sqrt{2x-1}$ (ii) $\dfrac{1}{(x^2-3)^3}$ (iii) $\left(x - \dfrac{1}{x}\right)^3$

Solution

(i)
$$y = \sqrt{2x-1} = (2x-1)^{\frac{1}{2}}$$

$$u = 2x - 1 \quad \text{so} \quad \frac{du}{dx} = 2$$

$$y = u^{\frac{1}{2}} \quad \text{so} \quad \frac{dy}{du} = \frac{1}{2}u^{-\frac{1}{2}}$$

$$\therefore \quad \frac{dy}{dx} = 2 \times \frac{1}{2}u^{-\frac{1}{2}} = (2x-1)^{-\frac{1}{2}} = \frac{1}{\sqrt{2x-1}}$$

(ii)
$$y = (x^2-3)^{-3}$$

$$u = x^2 - 3 \quad \text{so} \quad \frac{du}{dx} = 2x$$

$$y = u^{-3} \quad \text{so} \quad \frac{dy}{du} = -3u^{-4}$$

$$\therefore \quad \frac{dy}{dx} = 2x \times -3u^{-4} = -\frac{6x}{u^4}$$

$$= -\frac{6x}{(x^2-3)^4}$$

(iii)
$$y = \left(x - \frac{1}{x}\right)^3$$

$$u = x - x^{-1} \quad \text{so} \quad \frac{du}{dx} = 1 + x^{-2} = 1 + \frac{1}{x^2}$$

$$y = u^3 \quad \text{so} \quad \frac{dy}{du} = 3u^2$$

$$\therefore \quad \frac{dy}{dx} = 3u^2\left(1 + \frac{1}{x^2}\right) = 3\left(x - \frac{1}{x}\right)^2\left(1 + \frac{1}{x^2}\right)$$

Example 5.5

Find the gradient of the curve $y = \sqrt{(2x - 1)}$ at the point $(5, 3)$.

Solution

Remember the gradient of the curve is simply given by $\frac{dy}{dx}$. So, using the chain rule:

$$\frac{dy}{dx} = \tfrac{1}{2}(2x - 1)^{-\frac{1}{2}} \times 2$$

$$\text{if } x = 5 \quad \frac{dy}{dx} = \tfrac{1}{2} \times 9^{-\frac{1}{2}} \times 2 = \tfrac{1}{3}$$

The gradient is $\frac{1}{3}$.

Exercise 5(a)

Differentiate the following functions with respect to x, simplifying your answers if possible.

1 $4x^3$ **2** $7x^4$ **3** $x^3 - 2x$

4 $\sqrt{x^3}$ **5** $\dfrac{1}{2x^2}$ **6** $\dfrac{3}{4\sqrt{x}}$

7 $(2x + 1)^5$ **8** $4(3x + 1)^2$ **9** $\dfrac{1}{\sqrt{1 - x^3}}$

10 $\left(\sqrt{x} + \dfrac{1}{\sqrt{x}}\right)^2$ **11** $\dfrac{1}{4(x^2 - 1)^3}$ **12** $\sqrt[3]{1 + \sqrt{x}}$

13 $\dfrac{x^3 + 1}{x}$ **14** $\dfrac{(x - 1)^2}{x^4}$ **15** $\dfrac{(2x + 1)^3}{3x^2}$

16 Find the gradient of the curve $y = 4x^6 - 3x^2$ at the point $(1, 1)$.

17 Find the value of $\dfrac{dV}{dt}$ when $t = 16$, if $V = 12\sqrt{t}$.

18 Find the values of x for which the gradient is zero, on the curve $y = 4x^3 - 3x^2 + 1$.

19 Find $\dfrac{dH}{dq}$ if $H = 4\sqrt{1 + q^3}$

20 If $p = \dfrac{24(1 + t)^2}{t^3}$, find $\dfrac{dp}{dt}$ when $t = 2$.

5.3 The product rule

If $y = uv$, where u and v are both functions of x, this is known as a **product** [multiply]. If x changes by δx, this will cause u to change by δu, v by δv and y by δy.

$$\text{Hence:}\quad y + \delta y = (u + \delta u)(v + \delta v)$$
$$= uv + v\delta u + u\delta v + \delta u.\delta v$$
$$y = uv \quad \therefore \quad \delta y = u\delta v + v\delta u + \delta u.\delta v$$

Since δu and δv are very small if δx is very small, then $\delta u.\delta v$ will be negligible.

$$\text{So:}\qquad \delta y = u\delta v + v\delta u$$
$$\therefore \qquad \frac{\delta y}{\delta x} = u\frac{\delta v}{\delta x} + v\frac{\delta u}{\delta x}, \quad \text{on dividing by } \delta x,$$
$$\text{as } \delta x \to 0 \qquad \frac{dy}{dx} = u\frac{dv}{\delta x} + v\frac{du}{dx} \qquad \text{(D5)}$$

MEMORY JOGGER

The **product rule** is often easier to remember if written in words. The derivative of a product is the sum of (the first function × the derivative of the second) and (the second function × the derivative of the first).

Note: It doesn't matter which functions you call the first and second.

Example 5.6

Use the product rule to differentiate:

(i) $(x^3 + 1)(4x^6 - x^2 + 1)$ with respect to x.

(ii) $(2t + 1)^3(t^3 + 1)$ with respect to t.

Simplify your answers where possible.

Solution

(i) Here $u = x^3 + 1$ so $\dfrac{du}{dx} = 3x^2$

$v = 4x^6 - x^2 + 1$ so $\dfrac{dv}{dx} = 24x^5 - 2x$

$$\therefore \quad \frac{dy}{dx} = (x^3 + 1)(24x^5 - 2x) + (4x^6 - x^2 + 1)3x^2$$
$$= 24x^8 - 2x^4 + 24x^5 - 2x + 12x^8 - 3x^4 + 3x^2$$
$$= 36x^8 + 24x^5 - 5x^4 + 3x^2 - 2x$$

(ii) Let $u = (2t+1)^3$.

This needs the chain rule to find $\frac{du}{dt}$.

$$\therefore \quad \frac{du}{dt} = 3(2t+1)^2 \times 2 = 6(2t+1)^2$$

If $v = t^3 + 1$

$$\frac{dv}{dt} = 3t^2$$

$$\therefore \quad \frac{dy}{dt} = (2t+1)^3 \times 3t^2 + (t^3+1) \times 6(2t+1)^2$$

$$= 3(2t+1)^2[t^2(2t+1) + 2(t^3+1)] \quad \text{after factorising}$$

$$= 3(2t+1)^2[2t^3 + t^2 + 2t^3 + 2]$$

$$= 3(2t+1)^2(4t^3 + t^2 + 2)$$

5.4 The quotient rule

If $y = \frac{x^3}{(1+x^2)}$, then this is best considered as the **quotient** of two functions, $u = x^3$ and $v = 1 + x^2$. Hence $y = \frac{u}{v}$, (quotient means divide)

$$\text{Now,} \quad y + \delta y = \frac{u + \delta u}{v + \delta v}, \quad \text{if } x \text{ changes by } \delta x, \text{ (see last section)}$$

$$\therefore \quad \delta y = \frac{u+\delta u}{v+\delta v} - y = \frac{u+\delta u}{v+\delta v} - \frac{u}{v}$$

$$= \frac{v(u+\delta u) - u(v+\delta v)}{v(v+\delta v)}$$

$$= \frac{vu + v\delta u - uv - u\delta v}{v^2 + v\delta u}$$

$$\therefore \quad \delta y = \frac{v\delta u - u\delta v}{v^2 + v\delta u}$$

$$\div \text{ by } \delta x \quad \frac{\delta y}{\delta x} = \frac{v\frac{\delta u}{\delta x} - u\frac{\delta v}{\delta x}}{v^2 + v\delta v}$$

as $\delta x \to 0 \quad v\delta v \to 0$

$$\therefore \quad \frac{dy}{dx} = \frac{v\frac{du}{dx} - u\frac{dv}{dx}}{v^2} \qquad \text{(D6)}$$

MEMORY JOGGER

To differentiate a quotient $= \dfrac{\text{top}}{\text{bottom}}$.

The derivative of a **quotient**

$$= \frac{(\text{bottom} \times \text{derivative of top}) - (\text{top} \times \text{derivative of bottom})}{(\text{bottom})^2}.$$

In the example, $u = x^3$ and $v = 1 + x^2$

So: $\dfrac{du}{dx} = 3x^2, \quad \dfrac{dv}{dx} = 2x$

Hence:
$$\frac{dy}{dx} = \frac{(1+x^2) \times 3x^2 - x^3 \times 2x}{(1+x^2)^2}$$
$$= \frac{3x^2 + 3x^4 - 2x^4}{(1+x^2)^2}$$
$$= \frac{3x^2 + x^4}{(1+x^2)^2}$$
$$= \frac{x^2(3+x^2)}{(1+x^2)^2}$$

Exercise 5(b)

Differentiate the following functions with respect to x and simplify your answers where possible.

1 $x^2(x+1)^3$ **2** $\dfrac{x}{x+1}$ **3** $(x+1)^2(x+2)^3$

4 $\dfrac{x}{(1+x)^2}$ **5** $(2x^2-1)^3(x+1)^4$ **6** $\dfrac{x}{1+x^2}$

7 $(1+x)(1+\sqrt{x})$ **8** $\dfrac{1+x}{1+\sqrt{x}}$ **9** $\left(1+\dfrac{1}{x^2}\right)(1+x)^2$

10 $\dfrac{2}{(1-\sqrt{x})^2}$ **11** $(3x+1)\sqrt{1+\dfrac{1}{x}}$ **12** $\dfrac{x^3}{(1-x^2)^2}$

13 $\dfrac{2x-1}{(x-3)^2}$ **14** $\dfrac{2x+4}{1+\sqrt{x}}$ **15** $\sqrt{\dfrac{1-x}{1+x}}$

5.5 Small increments

Since $\frac{\delta y}{\delta x} \approx \frac{dy}{dx}$, if δx and δy are small, then it follows that $\delta y \approx \frac{dy}{dx}.\delta x$. This fact can often be used to evaluate quantities close to known quantities. This method is known as the method of **small increments** (or small changes).

Example 5.7

Use the method of small increments to evaluate:

(i) $\sqrt{9.00006}$ (ii) $\sqrt[3]{124.8}$

Solution

(i) Clearly, the answer is near to $\sqrt{9} = 3$.

Let $y = \sqrt{x} = x^{\frac{1}{2}}$

$\frac{dy}{dx} = \frac{1}{2}x^{-\frac{1}{2}}$

if $x = 9$ $\frac{dy}{dx} = \frac{1}{2} \times 9^{-\frac{1}{2}} = \frac{1}{6}$

Since $9.00006 = 9 + 0.00006$, then $\delta x = 0.00006$

Using $\delta y \approx \frac{dy}{dx} \times \delta x$

$\delta y = \frac{1}{6} \times 0.00006$

$= 0.00001$

Hence: $\sqrt{9.00006} = 3 + 0.00001$

$= 3.00001$

(ii) Here the answer is near $\sqrt[3]{125} = 5$

let $y = x^{\frac{1}{3}}$

$\frac{dy}{dx} = \frac{1}{3}x^{-\frac{2}{3}}$

if $x = 125$ $\frac{dy}{dx} = \frac{1}{3} \times 125^{-\frac{2}{3}} = \frac{1}{75}$

Now $124.8 = 125 - 0.2$, so $\delta x = -0.2$ (note δx is negative)

$\therefore \quad \delta y \approx \frac{dy}{dx} \times \delta x = \frac{1}{75} \times -0.2 = -0.0027$

Hence $\sqrt[3]{124.8} = 5 - 0.0027 = 4.9973$

Example 5.8

A carton is in the shape of a square-based pyramid (see Figure 5.2). The edges of the pyramid are all 5 cm long. Calculate the volume of the container. If the sides are increased in length by 0.05 cm, use the method of small increments to find the new volume of the pyramid.

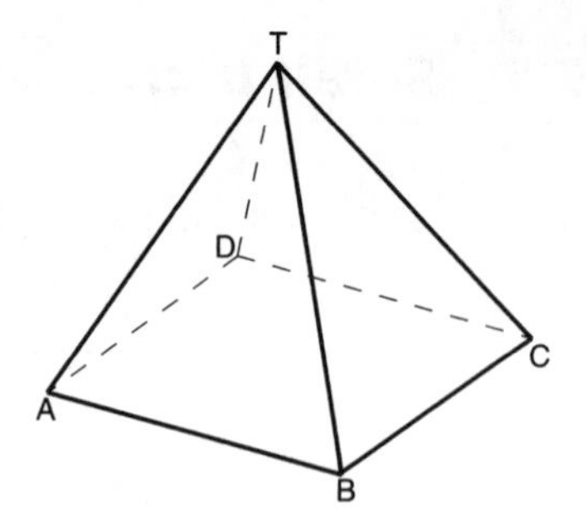

Figure 5.2

Solution

If the sides of the pyramid are of length x cm, then if the base is $ABCD$ as shown in Figure 5.3, then the diagonal AC is given by:

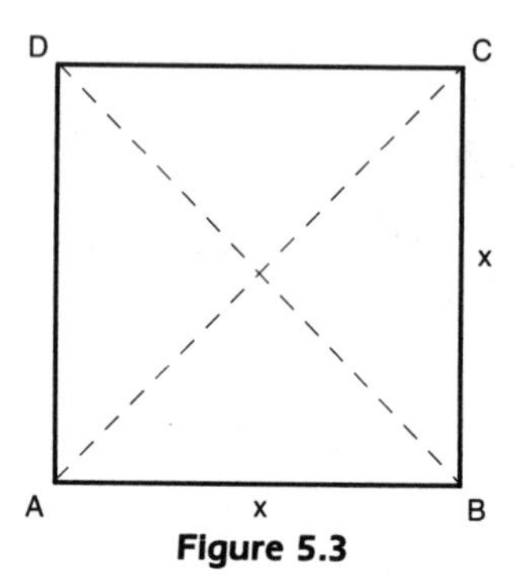

Figure 5.3

$$AC = \sqrt{x^2 + x^2} = \sqrt{2x^2}$$

A vertical cross-section TAC can be drawn as shown in Figure 5.4. If TN is the vertical height, then

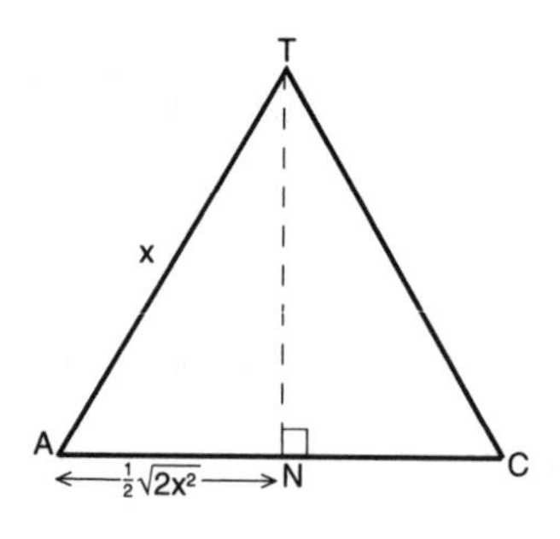

Figure 5.4

$$TN^2 = AT^2 - AN^2$$

and $\quad AN = \frac{1}{2}\sqrt{2x^2}$

$\therefore \quad TN^2 = x^2 - x^2 - \frac{1}{4} \times 2x^2 = \frac{1}{2}x^2$

Therefore, $\quad TN = \dfrac{x}{\sqrt{2}}$

The volume of the pyramid V is given by:

$$V = \tfrac{1}{3} \times \text{base area} \times \text{height}$$

$$V = \tfrac{1}{3} \times x^2 \times \frac{x}{\sqrt{2}} = \frac{x^3}{3\sqrt{2}}$$

Hence if $x = 5$, $\quad V = \dfrac{5^3}{3\sqrt{2}}$

The volume $= 29.46\ \text{cm}^3$.

$$\frac{dV}{dx} = \frac{3x^2}{3\sqrt{2}} = \frac{x^2}{\sqrt{2}}$$

$$\frac{\delta V}{\delta x} \approx \frac{dV}{dx}$$

Hence $\quad \delta V \approx \dfrac{dV}{dx}.\delta x$

$$\delta V \approx \frac{x^2}{\sqrt{2}} \times \delta x$$

If x increases from 5 cm to 5.05 cm, then $\delta x = 0.05$.

Hence $\delta V \approx \dfrac{5^2}{\sqrt{2}} \times 0.05 = 0.88 \text{ cm}^3$

The new volume $= 29.46 + 0.88$

$= 30.3 \text{ cm}^3$ (3 sig. figs)

5.6 Tangents and normals

If you are given the equation of a curve $y = f(x)$, it is possible to find the equation of the **tangent** to the curve at any point P on the curve, and also the equation of the **normal** (a line perpendicular to the tangent) at P by using differentiation. See Figure 5.5.

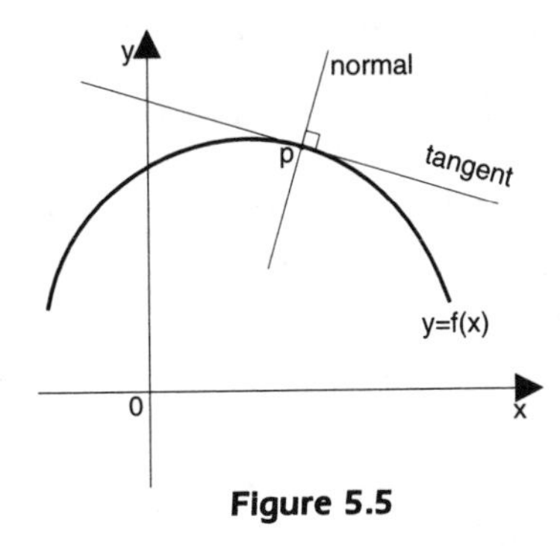

Figure 5.5

Example 5.9

Find the equation of the tangent and normal to the curve $y = 4x^3 + 3x$ at the point $(2, 38)$.

Solution

If $y = 4x^3 + 3x$

$$\frac{dy}{dx} = 12x^2 + 3$$

at $x = 2$, $\dfrac{dy}{dx} = 12 \times 2^2 + 3 = 51$

Hence the gradient of the line = 51.

Using the equation of a straight line in the form $(y - y_1) = m(x - x_1)$, we get

$$y - 38 = 51(x - 2) = 51x - 102$$

$$\therefore\ y = 51x - 64$$

Since the normal is perpendicular to the tangent, using the formula $m_1 m_2 = -1$ for perpendicular lines, the gradient of the normal is $-\dfrac{1}{51}$.

Its equation is $y - 38 = -\dfrac{1}{51}(x - 2)$

$$\therefore\ 51y - 1938 = -x + 2$$

that is, $51y + x = 1940$

Example 5.10

Find the equation of the tangent on the curve $y = 4 - x^2$, which is parallel to the line $x + y = 8$.

Solution

A diagram is essential for a question like this (Figure 5.6). You can see that the first problem is that, although you **know** the gradient of the tangent, (because it is parallel to the line $x + y = 8$), you do **not** know the point P where this tangent touches the curve.

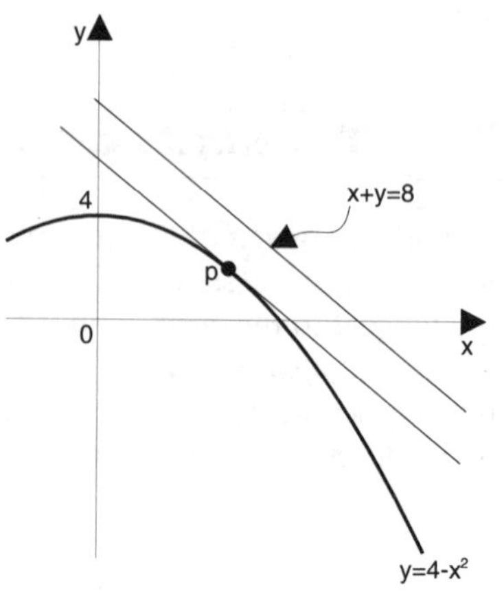

Figure 5.6

Now $\frac{dy}{dx} = -2x$ for the curve.

At P, the gradient $= -1$, hence $-2x = -1$, giving $x = \frac{1}{2}$. Substitute this value into the equation of the curve, to give

$$y = 4 - \left(\tfrac{1}{2}\right)^2 = 3\tfrac{3}{4}$$

The equation of the tangent can now be found.

$$y - 3\tfrac{3}{4} = -1\left(x - \tfrac{1}{2}\right)$$

$$\therefore \quad y - 3\tfrac{3}{4} = -x + \tfrac{1}{2}$$

$$y = -x + 4\tfrac{1}{4}$$

$$y + x = 4\tfrac{1}{4} \quad \text{or} \quad 4y + 4x = 17$$

Exercise 5(c)

1 Find the equation of the tangent at the given point, to the given curve.

(i) $(3, 28)$: $y = x^3 + 1$ (ii) $(9, 3)$: $y = \sqrt{x}$

(iii) $(4, 8)$: $y = (x - 2)^3$ (iv) $(4, 2.5)$: $y = \dfrac{x + 1}{x - 2}$

(v) $\left(1, \dfrac{1}{2}\right)$: $y = \dfrac{1}{2\sqrt{x}}$ (vi) $(-2, -27)$: $y = (1 - x)^3$

2 Find the equation of the normal at the given point, to the given curve.

(i) $(2, 10)$: $y = 3x^2 - 2$ (ii) $(1, 1)$: $y = \dfrac{1}{\sqrt[3]{x}}$

(iii) $(-2, -1)$: $y = \dfrac{1}{1 + x}$ (iv) $(-1, -1)$: $y = \dfrac{x - 2}{x + 4}$

3 Use the method of small increments to find:

(i) $\sqrt[3]{27.003}$ (ii) $\sqrt{15.998}$

4 Find the equation of the tangent to the curve $y = (1 + x)(1 - \sqrt{x})$ at the point where $x = 4$.

5.7 Higher derivatives

If $y = 3x^4 + 2x^2 + 6$, then $\frac{dy}{dx} = 12x^3 + 4x$. If you now differentiate with respect to x, you get $36x^2 + 4$. This is written $\frac{d^2y}{dx^2}$ (pronounced '*d* 2 *y* by *d x* squared'). This can also be written $f''(x)$ in function notation,

$$\text{that is,} \quad \frac{d^2y}{dx^2} = 36x^2 + 4$$

We call $\frac{d^2y}{dx^2}$ the **second derivative** of y with respect to x. This process can be continued indefinitely.

$$\text{So:} \quad \frac{d^3y}{dx^3} = 72x$$

$$\frac{d^4y}{dx^4} = 72$$

$$\text{and:} \quad \frac{d^5y}{dx^5} = 0$$

5.8 Stationary values

Referring to Figure 5.7, the curve of $y = f(x)$ has two points, A and B, at which $\frac{dy}{dx} = 0$. A is called a **maximum** point, and B a **minimum** point.

As x increases through point A, the gradient goes from positive to negative. Hence at A, the gradient is *decreasing*. The rate at which the gradient changes is given by $\frac{d^2y}{dx^2}$. Hence at a maximum point: $\frac{dy}{dx} = 0$, and $\frac{d^2y}{dx^2} < 0$.

Similarly, at B, the gradient changes from negative to positive. Hence the gradient is *increasing*. Hence, at a minimum point: $\frac{dy}{dx} = 0$, and $\frac{d^2y}{dx^2} > 0$.

Maximum and minimum points are called collectively **stationary values**.

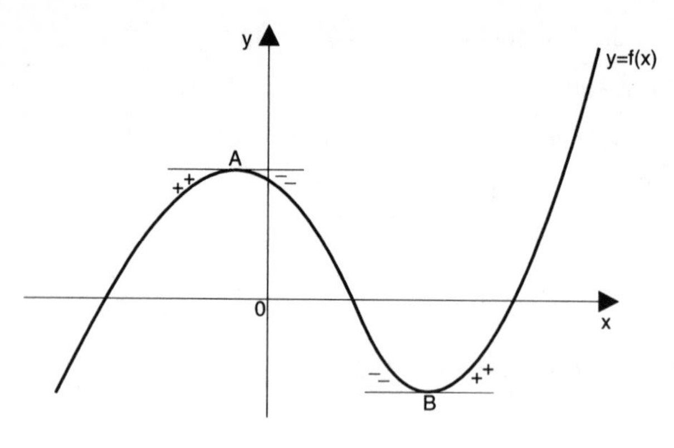

Figure 5.7

Example 5.11

Find the turning points on the graph of $y = x^3 + 3x^2 - 24x$, distinguishing between maximum and minimum values. Hence sketch the graph.

Solution

$$\frac{dy}{dx} = 3x^2 + 6x - 24; \quad \frac{d^2y}{dx^2} = 6x + 6$$

At turning points, $\dfrac{dy}{dx} = 0$.

$$\begin{aligned} \text{So:} \quad 3x^2 + 6x - 24 &= 0 \\ x^2 + 2x - 8 &= 0 \\ (x+4)(x-2) &= 0 \\ \text{Hence} \quad x &= -4, \text{ or } 2 \end{aligned}$$

At $x = -4$, $\dfrac{d^2y}{dx^2} = 6 \times -4 + 6 = -18 < 0$. Substituting $x = -4$ into the original equation, gives $y = 80$. Therefore, $(-4, 80)$ is a **maximum** point.

At $x = 2$, $\dfrac{d^2y}{dx^2} = 6 \times 2 + 6 = 18 > 0$. Also, at $x = 2$, $y = -28$. Hence $x = 2$, $y = -28$ is a **minimum** point.

If $x = 0$, $y = 0$, and so we have three points on the curve, which can now be sketched, see Figure 5.8.

We need to look carefully at the situation where $\dfrac{d^2y}{dx^2} = 0$.

Consider the graph of $y = x^3 + 1$:

$$\frac{dy}{dx} = 3x^2, \quad \frac{d^2y}{dx^2} = 6x$$

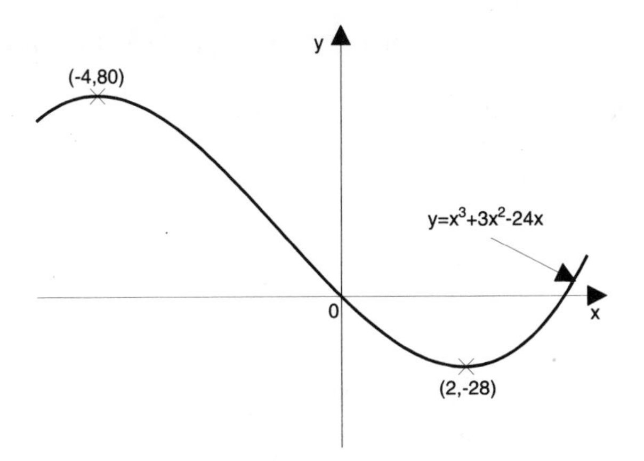

Figure 5.8

At turning points $\frac{dy}{dx} = 0$, so $3x^2 = 0$.
Hence $x = 0$.

$$\text{Now:} \quad \frac{d^2y}{dx^2} = 6 \times 0 = 0$$

Also if $x=0$, $y=1$. What can we say about the point $(0, 1)$?

This is neither a maximum nor a minimum, and is referred to as a **point of inflexion** (see Figure 5.9).

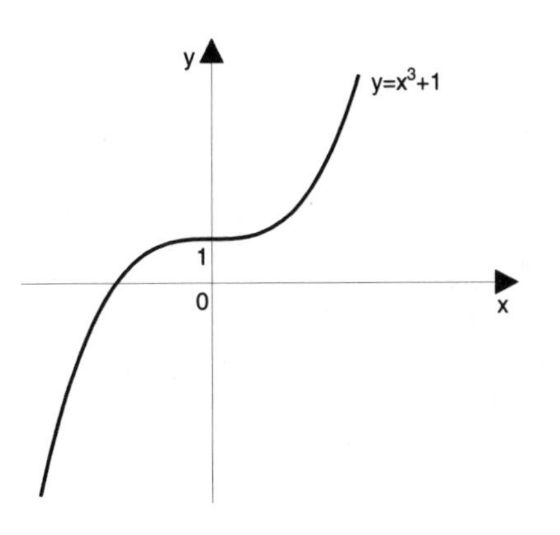

Figure 5.9

However, if you look at $y = x^4$:

$$\frac{dy}{dx} = 4x^3, \quad \frac{d^2y}{dx^2} = 12x^2$$

When $x = 0$, $\frac{dy}{dx} = 0$ and $\frac{d^2y}{dx^2} = 0$, but clearly the graph in Figure 5.10 shows the curve to have a minimum point, not a point of inflexion.

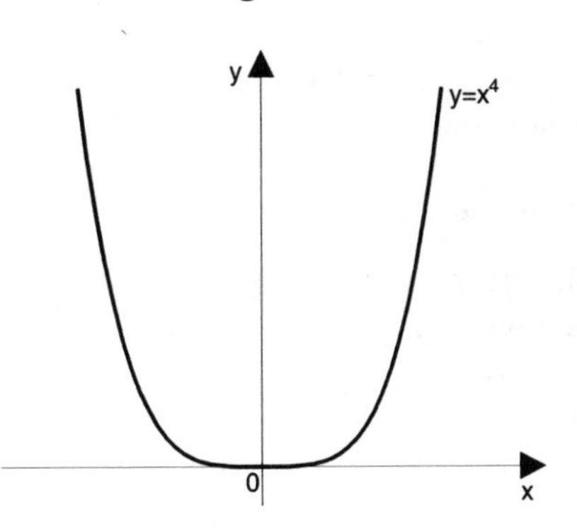

Figure 5.10

You need to investigate the **sign** of $\frac{dy}{dx}$ on either side of the stationary point to see what type it is.

$$\text{In this case, if } x = -0.1, \quad \frac{dy}{dx} = 4 \times (-0.1)^3 = -0.004$$

$$\text{if } x = 0.1, \quad \frac{dy}{dx} = 4 \times (0.1)^3 = 0.004$$

Hence the gradient changes from negative to positive. In other words, it is increasing, hence it is a minimum.

If you go back and look at $y = x^3 + 1$ again,

$$\text{at } x = -0.1, \quad \frac{dy}{dx} = 3 \times (-0.1)^2 = 0.03$$

$$\text{at } x = 0.1, \quad \frac{dy}{dx} = 3 \times (0.1)^2 = 0.03$$

The gradient remains positive, and so it is a point of inflexion.

MEMORY JOGGER

At a point of inflexion:

(i) $\frac{dy}{dx}$ need not be zero;

(ii) $\frac{d^2y}{dx^2}$ must be zero.

You cannot be sure it is a point of inflexion without checking the gradient.

Although the maximum and minimum technique is useful in curve sketching, its main use comes when applied in problem solving. At this stage, you will be able to find the maximum or minimum value of any quantity that can be expressed as a function of **one variable** only. Look carefully at the following examples.

Example 5.12

A farmer wishes to fence a rectangular area with 100 metres of fencing. One side of the area is a fixed wall. What will be the dimensions of the rectangle that gives a maximum area.

Solution

If the measurements are x by y, as shown in Figure 5.11, then:

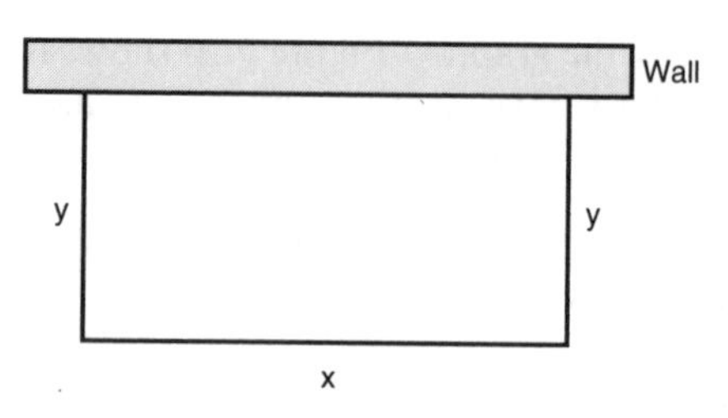

Figure 5.11

$$\text{The perimeter} = x + y + y$$
$$= x + 2y$$

$$\therefore \quad 2y + x = 100$$

$$\text{So} \quad y = \frac{100 - x}{2}$$

$$\text{The area of the field } A = xy = \frac{x(100 - x)}{2}$$

$$\text{that is} \quad A = 50x - \tfrac{1}{2}x^2$$

MEMORY JOGGER

You must express the quantity in terms of only *one* variable in a maximum and minimum problem

So $\quad \dfrac{dA}{dx} = 50 - x = 0 \quad \text{if } x = 50$

Also $\quad \dfrac{d^2A}{dx^2} = -1 < 0 \quad \text{hence a maximum at } x = 50$

Substituting in (i), gives $y = 25$. The dimensions of the rectangle are 25 m by 50 m.

Example 5.13

The formula for calculating reading ages is:

$$r = \frac{2}{5}\left(\frac{N}{s} + \frac{100}{N}L\right)$$

where in a given passage, N is the number of words, s is the number of sentences, and L is the number of words containing three or more syllables (excluding 'ing' or 'ed').

Mary is tested on a passage containing 30 sentences, and her reading age was found to be 15. Find an expression for L in terms of N.

Hence find the maximum number of words containing three or more syllables. Comment on your answer.

Solution

$R = 15$ and $s = 30$.

$$\therefore \quad 15 = \frac{2}{5}\left(\frac{N}{30} + \frac{100L}{N}\right)$$

$$= \frac{N}{75} + \frac{40L}{N}$$

$$\therefore \quad \frac{40L}{N} = 15 - \frac{N}{75}$$

$$\text{and} \quad L = \frac{N}{40}\left(15 - \frac{N}{75}\right)$$

$$= \frac{3N}{8} - \frac{N^2}{3000}$$

$$\frac{dL}{dN} = \frac{3}{8} - \frac{N}{1500} = 0 \text{ for a maximum if } N = \frac{4500}{8} = 562.5.$$

Clearly N is a whole number, so put $N = 562$ into the formula for L, to give: $L = 105$.

This shows that the passage should contain 562 words if it is to contain the maximum of 105 three-syllable words. Since there are thirty sentences, the average sentence length would be $562 \div 30 = 18.7$. This would certainly be a difficult passage to read.

Example 5.14

A firework is made in the shape of a right cone, the sum of the height and the base radius is 10 cm. Find the measurements of the cone if it is to hold the greatest volume.

Solution

Let the base radius $= r$ cm.

Let the height $= h$ cm.

Then $h + r = 10$.

The volume of a cone V, is given by the formula:

$$V = \tfrac{1}{3}\pi r^2 h$$

$$\text{Now } h = 10 - r \qquad \text{(i)}$$

Hence V can be written in terms of r:

i.e. $V = \tfrac{1}{3}\pi r^2(10 - r)$

So: $V = \dfrac{10\pi}{3}r^2 - \dfrac{\pi}{3}r^3$

$$\frac{dV}{dr} = \frac{20\pi}{3}r - \pi r^2 \qquad \text{(ii)}$$

At maximum V, $\dfrac{dV}{dr} = 0$

So $\dfrac{20\pi}{3}r - \pi r^2 = 0$, that is, $\pi r\left(\dfrac{20}{3} - r\right) = 0$

$\therefore \quad r = 0$ or $\dfrac{20}{3} = r$

But r cannot be zero, and so:

$$r = \frac{20}{3}$$

We should check that this gives a maximum value for V.

Differentiating Equation (ii) $\dfrac{d^2V}{dr^2} = \dfrac{20\pi}{3} - 2\pi r$

if $r = \dfrac{20}{3}$ $\quad \dfrac{d^2V}{dr^2} = \dfrac{20\pi}{3} - 2\pi \times \dfrac{20}{3} = \dfrac{-20\pi}{3}$

Hence $\quad \dfrac{d^2V}{dr^2} < 0$ indicating that V is a maximum

Returning to equation (i)

$$h = 10 - \frac{20}{3} = \frac{10}{3}$$

Hence the dimensions of the cone are:

base radius $= 2\frac{2}{3}$ cm   height $= 3\frac{1}{3}$ cm

ACTIVITY 5

There are many problems concerned with the design of containers. Very often, the problem will be one of maximising or minimising a particular aspect of the container (for example, area or volume).

Figure 5.12 shows the net for constructing a scoop similar to that in the picture. It is cut from a piece of metal which measures 40 cm by 20 cm. Using the ideas of the calculus, try and find out the measurements of the net, if the scoop is to hold the maximum possible volume.

You could extend this investigation by looking at other shapes, or you could consider a container of a given volume, and investigate, say, the minimum net area needed.

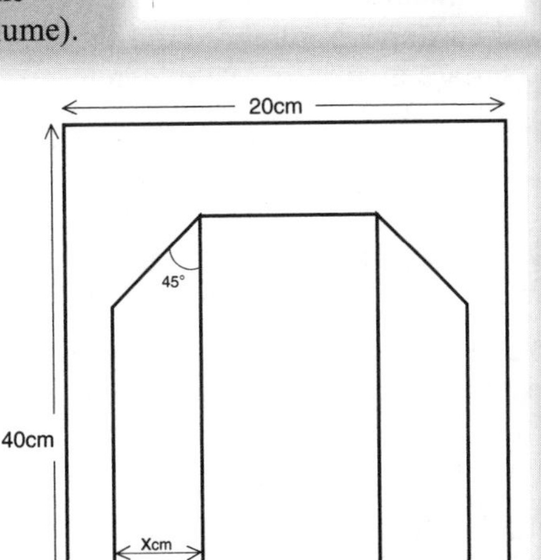

Figure 5.12

Exercise 5(d)

1 Find the turning points on the following curves, distinguishing between maxima, minima and points of inflexion.

(i) $y = 6x^2 - 5x + 2$

(ii) $y = 1 - 3x^3$

(iii) $y = x^4 - 4x^2$

(iv) $y = x + \dfrac{1}{x}$

(v) $y = \dfrac{x^2 + 1}{x - 2}$

(vi) $y = 3x + \dfrac{2}{2x + 1}$

2 The cost of running a machine is C pence/hour where $C = 84 + 17v(v - 3)$ and v is the speed of the machine in revolutions per second. Find the speed of running the machine to minimise the cost.

3 A cylindrical tank of radius r cm has a fixed volume of 200 cm^3. Find the value of r so that the total surface area of the cylinder is a minimum.

4 If $y = \dfrac{x}{x+1}$, find the value of $\dfrac{d^2y}{dx^2}$ when $x = 2$.

5. A cuboid is made so that the length is 20% greater than the width, and the volume is 10 cm^3. Find the measurements of the cuboid, if its total surface area is to be a minimum.

5.9 Rates of change

Very often a quantity varies with time, and you are required to differentiate with respect to time (t). This concept is usually referred to as a **rate of change.**

Example 5.15

The radius r cm of a circular ink spot, t seconds after the ink is spilt, is given by the formula:

$$r = \frac{2+5t}{3+t}$$

Calculate:

(i) The time it takes to reach a radius of 2 cm.
(ii) The rate of increase of the radius when $t = 2$.
(iii) The rate of increase of the area when $t = 2$.
(iv) The radius of the spot when the ink stops spreading.

Solution

(i) If $r = 2$,

$$2 = \frac{2+5t}{3+t}$$

$$\therefore \quad 2(3+t) = 2 + 5t$$

$$6 + 2t = 2 + 5t$$

$$4 = 3t, \quad t = \tfrac{4}{3}$$

$$\therefore \quad \text{time} = 1\tfrac{1}{3} \text{ seconds}$$

(ii) The rate of increase of r is calculated from $\frac{dr}{dt}$. Hence differentiating r using the quotient rule, we have:

$$\frac{dr}{dt} = \frac{(3+t) \times 5 - (2+5t) \times 1}{(3+t)^2}$$

$$= \frac{15 + 5t - 2 - 5t}{(3+t)^2} = \frac{13}{(3+t)^2}$$

when $t = 2$ $\quad \frac{dr}{dt} = \frac{13}{5^2} = 0.52$

The rate of increase of the radius is therefore: 0.52 cm/s

(iii) Since the shape is a circle, its area A is given by $A = \pi r^2$

$\therefore$ differentiate this with respect to time:

$$\frac{dA}{dt} = \frac{dA}{dr} \times \frac{dr}{dt} \quad \text{(using the chain rule)}$$

$$= 2\pi r \frac{dr}{dt}$$

Now when $t = 2$, $r = \frac{2 + 5 \times 2}{3 + 2} = 2.4$

Hence: $\frac{dA}{dt} = 2\pi \times 2.4 \times 0.52$

$$= 7.84$$

Hence the rate of increase in area is: 7.84 cm^2/sec

(iv) as $t \to \infty$

$$5t + 2 \approx 5t$$

and $\quad t + 3 \approx t$

$$r \to \frac{5t}{t} = 5$$

Hence the final radius of the blot is 5 cm.

5.10 Distance, speed and acceleration

DO YOU KNOW?

(i) Displacement means distance in a given direction.

(ii) Average speed $= \dfrac{\text{distance travelled}}{\text{time taken}}$

(iii) Velocity is speed in a given direction.

(iv) Acceleration $= \dfrac{\text{change in velocity}}{\text{time taken}}$.

An important application of rates of change is in dealing with speed and acceleration problems.

If displacement is not changing at a constant rate, then you can only find the velocity (v) at any particular instant of time. It will be the slope of the displacement (s)–time (t) graph (see Figure 5.13):

that is, $v = \dfrac{ds}{dt}$

Similarly, the acceleration (a) of a body whose velocity is not changing at a constant rate will be given by the slope of the velocity–time graph.

Hence $a = \dfrac{dv}{dt}$, or $\dfrac{d^2s}{dt^2}$

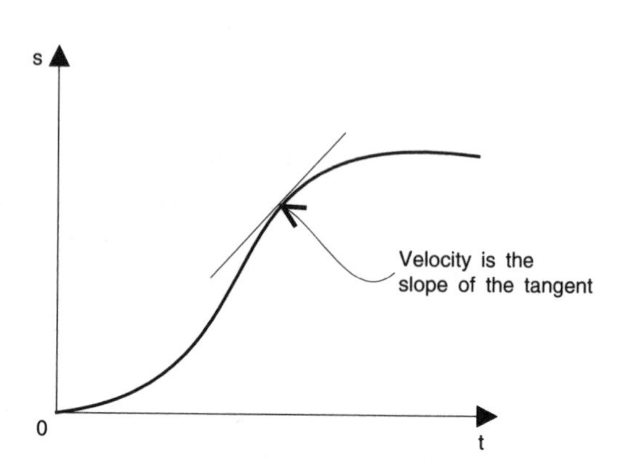

Figure 5.13

Example 5.16

A small model car travels in a straight line so that its displacement (s) metres from a fixed point O, t seconds after passing O is given by the formula:

$s = 6t - t^2$

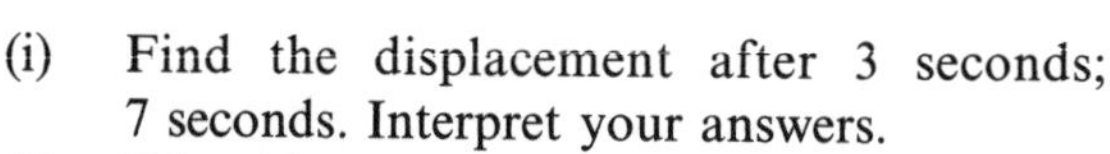

(i) Find the displacement after 3 seconds; 7 seconds. Interpret your answers.
(ii) When does the car stop?
(iii) Find the distance travelled between 2 and 4 seconds.
(iv) Find the acceleration of the car after 1, 3 and 5 seconds. Comment on your results.
(v) Draw (a) displacement–time; (b) velocity–time; and (c) acceleration–time graphs for the motion of the car between 0 and 8 seconds.

Solution

(i) If $t = 3$, $s = 18 - 9 = 9$

If $t = 7$, $s = 42 - 49 = -7$

Therefore, the car has moved 9 metres from O after 3 seconds, but has returned to O and moved 7 metres in the opposite direction to the original direction after 7 seconds.

(ii) It stops when $v = \frac{ds}{dt} = 0$

$$\frac{ds}{dt} = 6 - 2t = 0 \quad \text{if } t = 3$$

Hence it has stopped after 3 seconds.

(iii) You need to be aware that the car has *changed direction* between 2 and 4 seconds (see Figure 5.14a).

If $t = 2$: $s = 12 - 4 = 8$ m
If $t = 3$: $s = 9$ m
If $t = 4$: $s = 24 - 16 = 8$ m

The total distance travelled $= (9 - 8) + (9 - 8) = 2$ metres

(iv) Acceleration $a = \frac{d^2s}{dt^2} = -2$.

At $t = 1$, the acceleration $= -2 \text{ ms}^{-2}$

Since this is negative, it means the car is slowing down.

At $t = 3$, $v = 0$ but $a = -2$

This means the car is about to move backwards.

At $t = 5$, $v = 6 - 10 = -4$ and $a = -2$

This time the car is moving in the opposite direction, and so it is accelerating at 2 ms^{-2}

(v)

The graphs are shown as (a), (b) and (c) respectively in Figure 5.14.

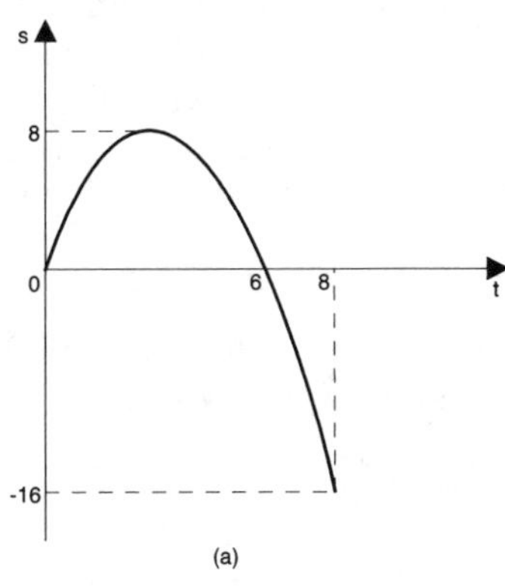

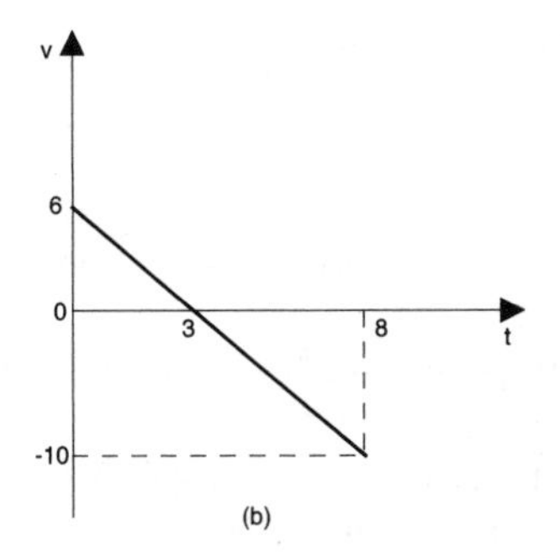

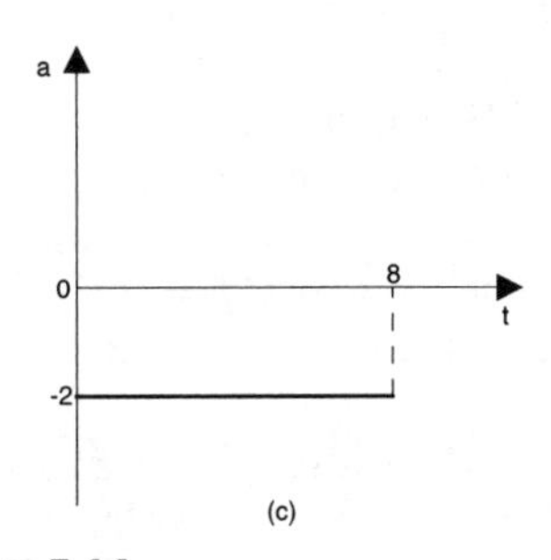

Figure 5.14

Exercise 5(e)

1 The radius of a circular ink blot is increasing at a rate of 0.2 cm/s. Find the rate of increase of the area of the blot when $r = 3$ cm.

2 A balloon is inflated at a constant rate of 100 cm^3/s. Find the rate at which the surface area is increasing when $r = 10$ cm.

3 If $s = 30t + 5t^2$, find: (a) the velocity; (b) the acceleration, when $t = 4$.

4 A metal block is in the shape of a cuboid with sides of length x cm, $3x$ cm and h cm respectively. The total surface area of the block is 98 cm^2. Show that:

$$h = \frac{49 - 3x^2}{4x}$$

The block has to have the maximum volume possible subject to these conditions. Find the measurements of the block needed.

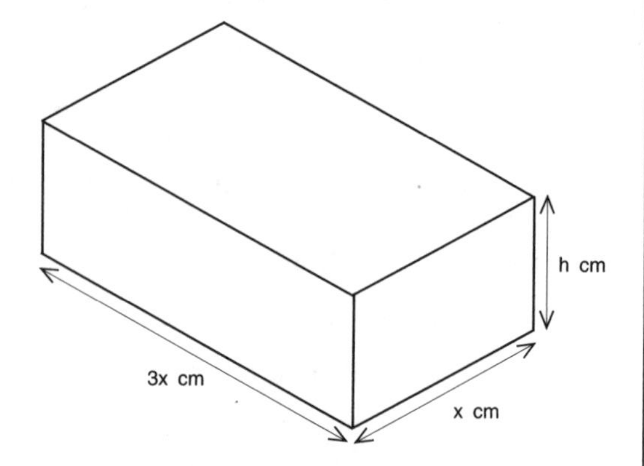

Figure 5.15

5 The distance s metres from a point O in t seconds is given by the equation $s = t^3 - 7t^2 + 12$. Find:
(i) the distance travelled in the fourth second;
(ii) the time when the body comes to rest;
(iii) the maximum speed reached by the body.

Miscellaneous Examples 5

1 Differentiate the following functions with respect to x and simplify your answers where possible.

(i) $\dfrac{1}{4x^3}$ (ii) $\dfrac{3}{(x+1)^2}$ (iii) $\sqrt{1 - 2x^2}$

(iv) $\dfrac{x}{\sqrt{1+x}}$ (v) $\dfrac{2x-1}{1-x^3}$ (vi) $\left(1 + \dfrac{1}{\sqrt{x}}\right)\left(1 - \dfrac{1}{x^2}\right)$

2 Find the equation of the tangent and normal to the curve $y = (2 - 3\sqrt{x})^3$ at the point where $x = 1$.

3 Find the equation of the tangent to the curve $y = 6 - 3x^2$ which is parallel to the line $2x + 3y - 5 = 0$.

4 Locate the turning points on the graph $y = 4x(8x^2 - 6)$ and hence sketch the curve.

5 The surface area of an ice cube is increasing at a rate of 20 cm^2/minute. Find the rate at which its volume is increasing when the volume of the cube is 800 cm^3.

6 The surface area of a solid cylinder is 200 cm^2. Find the radius of the cylinder if the volume is a maximum.

7 The distance s metres moved by a vehicle which travels in a straight line between two points A and B is given by $s = 48t - 2t^2$, where t is the time in seconds after passing A. Find the distance AB.

Revision Problems 5

1 Figure 5.16 shows a brick in the shape of a cuboid with base x cm by $2x$ cm and height h cm. The total surface area of the brick is 300 cm^2.

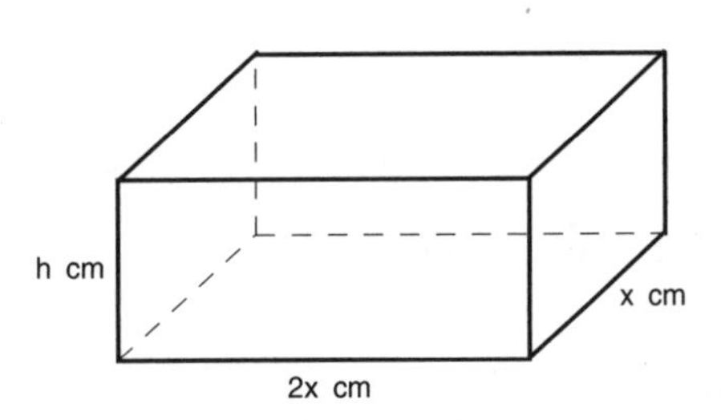

Figure 5.16

(a) Show that $h = \frac{50}{x} - \frac{2x}{3}$

(b) The volume of the brick is V cm^3. Express V in terms of x only.

(c) Given that x can vary, find the maximum value of V.

(d) Explain why the value of V you have found is the maximum.

(LEAG)

2 The function f is defined by

$f(x) = 4x^2 - 3 - \frac{1}{x}$ for $x \neq 0$.

A sketch (not to scale) of the graph of f(x) is shown in Figure 5.17.

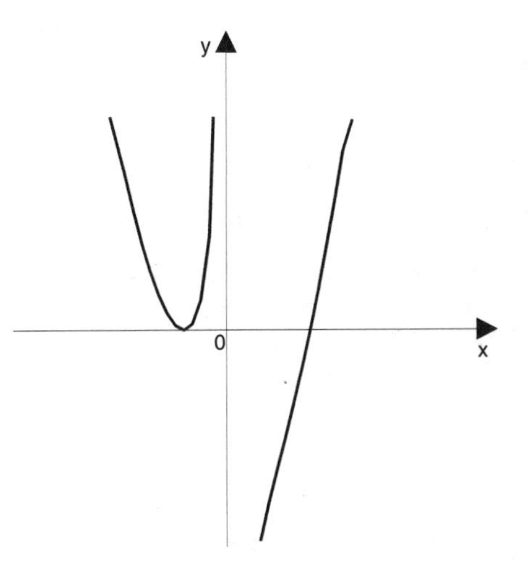

Figure 5.17

(i) Find $f'(x)$ and $f''(x)$.

(ii) Find the value of a so that $f'(a) = 0$, and use your value of a to calculate $f''(a)$. What information does this give you about the graph of $y = f(x)$?

(iii) Find the values of x where the graph of $y = f(x)$ meets the x-axis.

(iv) Explain how the shape of the graph is related to the terms in the expression for $f(x)$.

(a) when x is near zero.

(b) when x is large (positive or negative).

(Oxford & Cambridge)

6 Curves

In Chapter 4, we looked at the linear equation $y = mx + c$, and in Chapter 2 the quadratic equation $y = ax^2 + bx + c$. By looking in more detail at the cubic equation, you should then be in a position to sketch any polynomial equation.

6.1 Cubic function $y = ax^3 + bx^2 + cx + d$

We shall illustrate the techniques for sketching a polynomial with a specific example.

Suppose $y = x^3 + 4x^2 - 3x + 1$

It would be nice to be able to find the points where the graph crosses the axes, but this is not always possible.

If $x = 0,\ y = 1$

However, if $y = 0$, we get the equation:

$$x^3 + 4x^2 - 3x + 1 = 0$$

It is not easy to find a linear factor, and so a different technique must be used. Look at the gradient function:

$$\frac{dy}{dx} = 3x^2 + 8x - 3$$

$$\frac{d^2y}{dx^2} = 6x + 8$$

If $\frac{dy}{dx} = 0,\ 3x^2 + 8x - 3 = 0$, that is $(3x - 1)(x + 3) = 0$

$$\therefore \quad x = -3 \quad \text{or} \quad x = \tfrac{1}{3}$$

Now at $x = -3,\quad \frac{d^2y}{dx^2} = 6 \times -3 + 8 = -10 < 0$

$$y = (-3)^3 + 4(-3)^2 - 3(-3) + 1 = 19$$

Since $\frac{d^2y}{dx^2} < 0$, then $(-3, 19)$ will be a maximum turning point.

$$\text{At } x = \tfrac{1}{3}, \quad \frac{d^2y}{dx^2} = 6 \times \tfrac{1}{3} + 8 = 10 > 0$$

$$y = \left(\tfrac{1}{3}\right)^3 + 4\left(\tfrac{1}{3}\right)^2 - 3\left(\tfrac{1}{3}\right) + 1$$

$$= \tfrac{13}{27}$$

Since $\frac{d^2y}{dx^2} > 0$, then $\left(\frac{1}{3}, \frac{13}{27}\right)$ will be a minimum turning point.

The graph can now be sketched (see Figure 6.1). **Note:** We still do not know where the graph crosses the x-axis. However, the graph shows there is only one point and gives a rough indication where it is. The values could be found by numerical methods.

It is also worth noting that, if x is very large, the dominant term in the equation is x^3

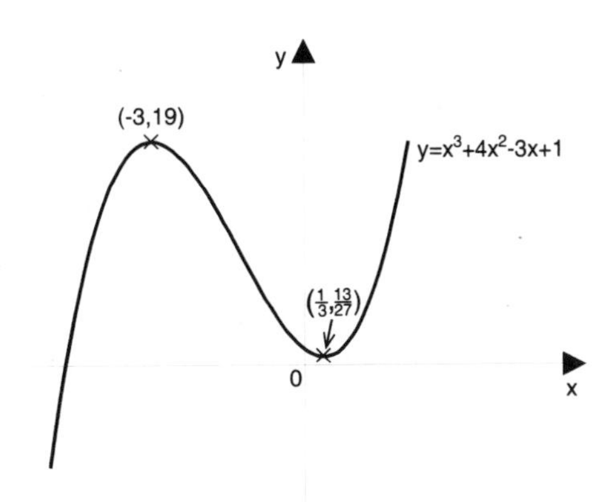

Figure 6.1

$$\text{Hence as} \quad x \to +\infty, \quad y \to +\infty$$
$$\text{and} \quad x \to -\infty, \quad y \to -\infty$$

Now look at one more worked example and you should be able to try Exercise 6(a) on page 123.

Example 6.1

Sketch the graph of $y = 4x^2 - x^4$, giving as much information as you can on the graph.

Solution

$$y = 4x^2 - x^4 = x^2(4 - x^2)$$
$$= x^2(2 - x)(2 + x)$$

MEMORY JOGGER

Always factorise the equation of a curve if possible. This helps to find the points where the graph crosses the x-axis at $y = 0$

If $x = 0$, $y = 0$
If $y = 0$, we need to solve: $0 = x^2(2 - x)(2 + x)$
hence, $x = 0$ (twice), $x = 2$, $x = -2$

This means that the curve *touches* the x-axis at $x = 0$.

$$\frac{dy}{dx} = 8x - 4x^3$$

$$\frac{d^2y}{dx^2} = 8 - 12x^2$$

$$\text{If } \frac{dy}{dx} = 0, \qquad 8x - 4x^3 = 0$$

$$\therefore \qquad 4x(2 - x^2) = 0$$

$$4x\left(\sqrt{2} - x\right)\left(\sqrt{2} + x\right) = 0$$

$$\therefore \quad x = 0, \sqrt{2}, -\sqrt{2}$$

At $x = 0$, $\dfrac{d^2y}{dx^2} = 8 > 0$

$\therefore$ (0, 0) is a minimum turning point

At $x = \sqrt{2}$, $\dfrac{d^2y}{dx^2} = -16 < 0$ $\quad \therefore$ maximum, $y = 4$

At $x = -\sqrt{2}$, $\dfrac{d^2y}{dx^2} = -16 < 0$ $\quad \therefore$ maximum, $y = 4$

There are five points that can be put on to the diagram, and so the sketch is now relatively straightforward.

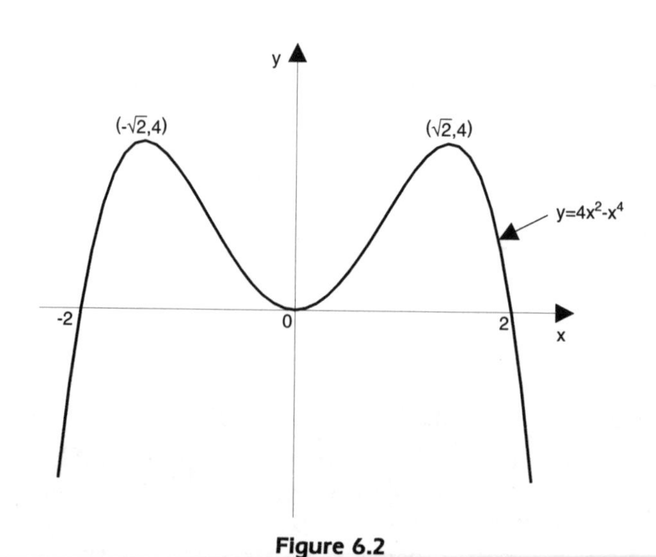

Figure 6.2

6.2 Rational functions

An expression of the form $y = \frac{f(x)}{g(x)}$ where $f(x)$ and $g(x)$ are polynomials, is called a **rational function**. The techniques of sketching this type of curve are quite straightforward.

Example 6.2

Sketch the graph of $y = \frac{x+1}{x-1}$.

Solution

If $x = 1$, then $y = \frac{2}{0}$ which is not defined. As x gets nearer and nearer to 1, the curve gets nearer and nearer to the line $x = 1$, without touching it.

We say that $x = 1$ is an **asymptote** for the graph.

If x is just less than 1, say 0.99, then

$$y = \frac{1.99}{-0.01} = -199$$

If x is just greater than 1, say 1.01, then

$$y = \frac{2.01}{0.01} = 201$$

You can see that the curve approaches the asymptote at $y = +\infty$ on the right, and at $y = -\infty$ on the left (Figure 6.3).

The curve **cannot cross a vertical asymptote** and will actually be **discontinuous** at $x = 1$.

If $x = 0$, $\quad y = -1$

If $y = 0$, $\quad 0 = \frac{x+1}{x-1}$,

so $\quad 0 = x + 1$

and $\quad x = -1$

To find whether the graph has a horizontal asymptote (x very large), observe that if x is very large, $x + 1 \approx x$ and also $x - 1 \approx x$.

Hence $\quad y \approx \frac{x}{x} = 1$

If you add all this information to the preceding graph, the final graph can be drawn (see Figure 6.4).

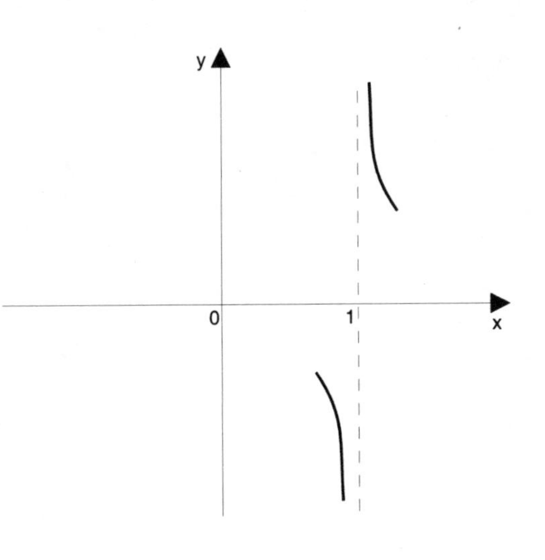

Figure 6.3

Although this curve does not cut the horizontal asymptote, there are situations when this can occur, see **Examples 6.3** and **6.4**.

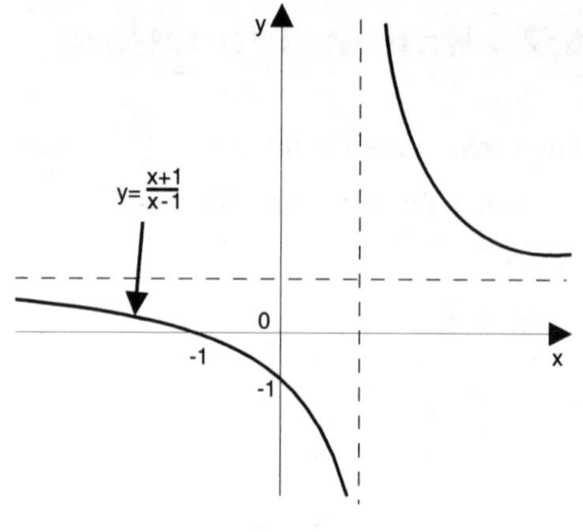

Figure 6.4

Example 6.3

Sketch the graph of $y = \dfrac{x-1}{x^2+1}$.

Solution

There is no vertical asymptote here, because if you try to make the bottom line zero, i.e. $x^2 + 1 = 0$, there is no real solution.

If $x = 0$, $\quad y = -1$

If $y = 0$, $\quad 0 = \dfrac{x-1}{x^2+1} \quad \therefore \quad x - 1 = 0$, that is, $\quad x = 1$

If $x \to \infty$, $\quad y \approx \dfrac{x}{x^2} = \dfrac{1}{x} \to 0$

Hence the horizontal asymptote is $y = 0$.

The information gained so far is shown in Figure 6.5. This is not really enough to complete the sketch.

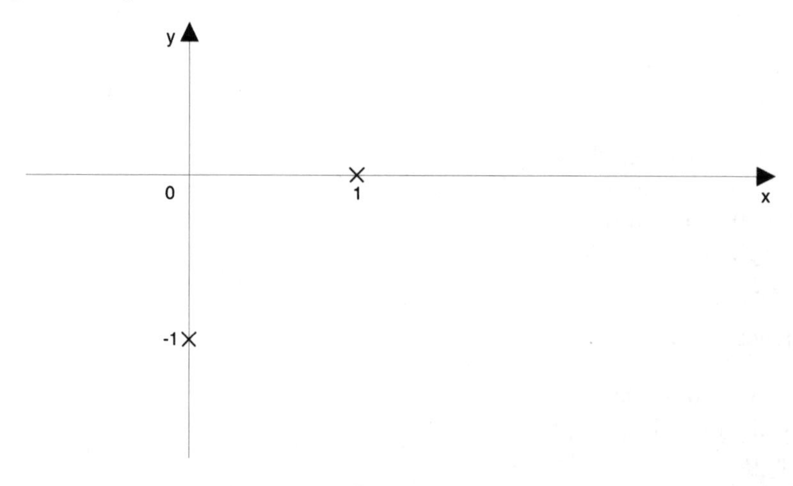

Figure 6.5

We need to look for turning points.

$$\frac{dy}{dx} = \frac{(x^2+1) \times 1 - (x-1) \times 2x}{(x^2+1)^2} = \frac{-x^2+2x+1}{(x^2+1)^2}$$

$$= 0 \quad \text{if} \quad -x^2+2x+1 = 0$$

giving $x = 2.4$ and -0.4 (using the quadratic formula.)

for which $y = 0.2$ and -1.2

With these turning points added to the graph, it can now be completed (see Figure 6.6).

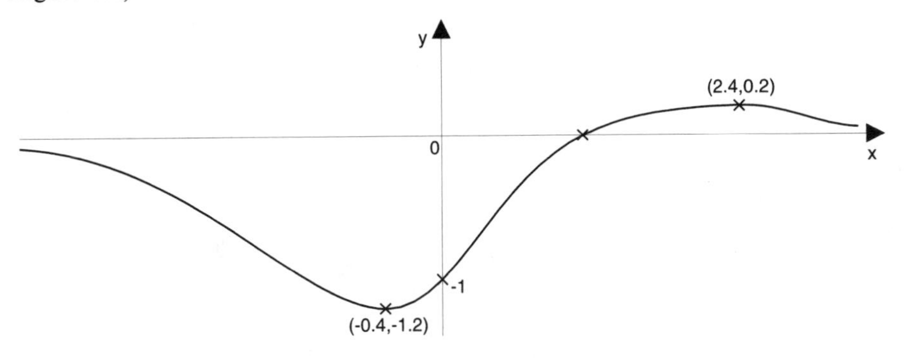

Figure 6.6

MEMORY JOGGER

Example 6.3 shows that a graph can cross a horizontal asymptote.

Example 6.4

Sketch the graph of $y = \dfrac{x^2+1}{(x-1)(x-2)}$

Solution

This is slightly more involved, but can be solved by following similar steps.

If $x = 1$, $x = 2$ then $y \to \infty$ giving vertical asymptotes

If $x = 0$, $y = 0.5$

If $y = 0$, $x^2 + 1 = 0$, giving no solutions

Hence the graph does *not* cut the x-axis.

$$\text{If } x \to \infty, \quad y \approx \frac{x^2}{x \times x} = 1$$

Hence $y = 1$ is a horizontal asymptote.

It is worth substituting $y = 1$ into the original equation, to see if the curve cuts this horizontal asymptote.

$$1 = \frac{x^2 + 1}{x^2 - 3x + 2}$$
$$\therefore \quad x^2 - 3x + 2 = x^2 + 1$$
$$\therefore \quad 1 = 3x, \ x = \tfrac{1}{3}$$

This is a useful point to add to the diagram. The information so far is shown in Figure 6.7. Clearly, there is not enough information to complete the graph.

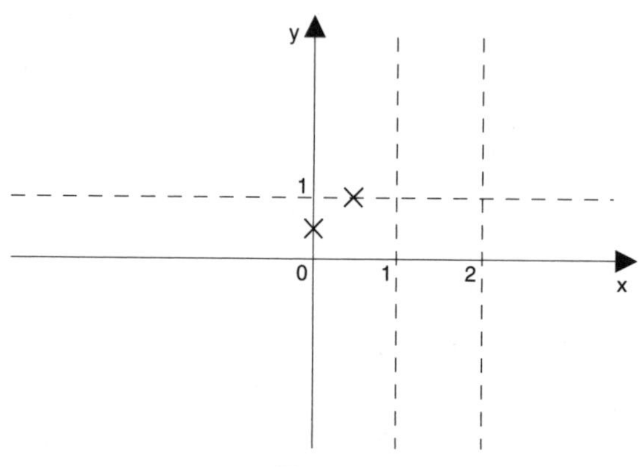

Figure 6.7

Again, we need to look for turning points:

$$y = \frac{x^2 + 1}{x^2 - 3x + 2}$$
$$\frac{dy}{dx} = \frac{(x^2 - 3x + 2)2x - (x^2 + 1)(2x - 3)}{(x^2 - 3x + 2)^2}$$
$$= \frac{2x^3 - 6x^2 + 4x - 2x^3 + 3x^2 - 2x + 3}{(x^2 - 3x + 2)^2}$$
$$= \frac{-3x^2 + 2x + 3}{(x^2 - 3x + 2)^2}$$
$$= 0 \quad \text{if} \quad -3x^2 + 2x + 3 = 0$$

giving $x = -0.7$ and 1.4 (by the quadratic formula)

hence $y = 0.3$ and -12

The curve can now be completed. If you are not convinced, try and argue against any other answer by realising that you cannot cross a vertical asymptote. There is only one point where $y = 1$, and there are no points when $y = 0$.

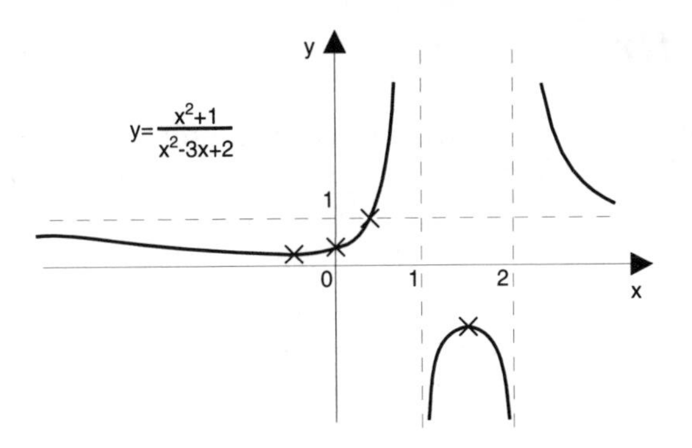

Figure 6.8

Exercise 6(a)

Draw a sketch to show that you understand the shape of the graph represented by the following equations. Always label the coordinates of any points where the graph crosses the coordinate axes where you can.

1 $y = 4x - 3$

2 $3x + 4y = 24$

3 $y = 4x^2 - 6x$

4 $y = 8 - x^2$

5 $y = x^2 - 6x + 8$

6 $y = x^2 + 3$

7 $y = x^3 + x$

8 $y = 2x^3 - 3x^2$

9 $y = 1 + x^4$

10 $y = 3x^3 - 2x + 7$

11 $y = \dfrac{x}{x-2}$

12 $y = \dfrac{x^2}{x^2+1}$

6.3 e^x and $\ln x$

The graphs of the functions e^x, e^{-x} and $\ln x$ are shown in Figure 6.9.

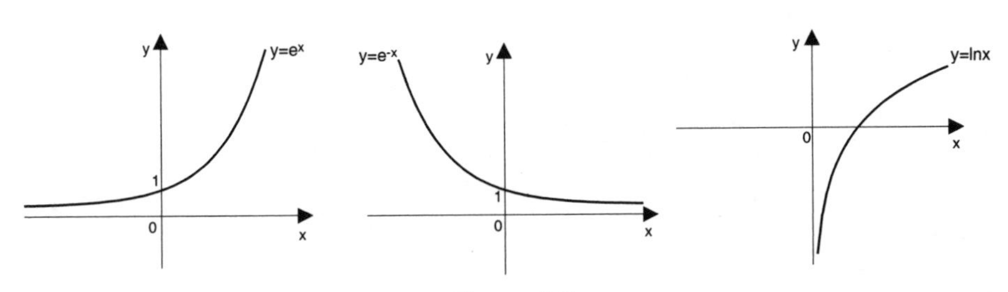

Figure 6.9

Note: $\ln x$ does not exist if $x < 0$, as you cannot have the logarithm of a negative number, or zero.

6.4 $y^2 = f(x)$

Relationships involving $y^2 = f(x)$ lead to certain problems not associated with a simple expression of the type $y = f(x)$. The technique will be illustrated with a particular example.

Example 6.5

Sketch the graph of the relationship $y^2 = x(x+1)(x-2)$.

Solution

(i) $y = \pm\sqrt{x(x+1)(x-2)}$

Note: $\pm$ sign. This means that, for each value of x, there are two values of y, giving a graph which is symmetrical about the x-axis.

(ii) The function $x(x+1)(x-2)$ should be sketched first (see Figure 6.10).

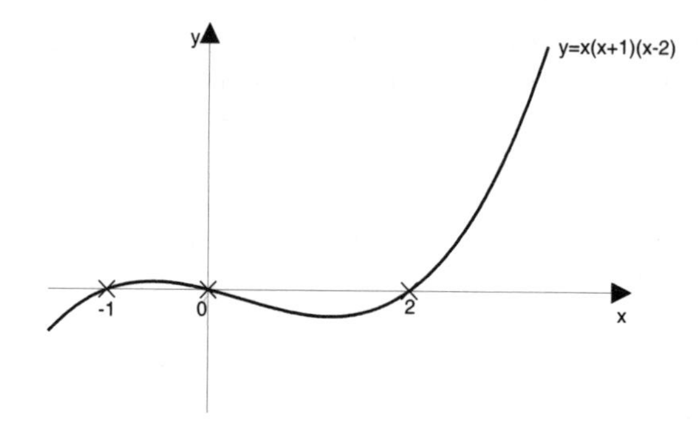

Figure 6.10

The part of this curve for which $y \geq 0$ is now used as a guide line (shown dotted in Figure 6.11).

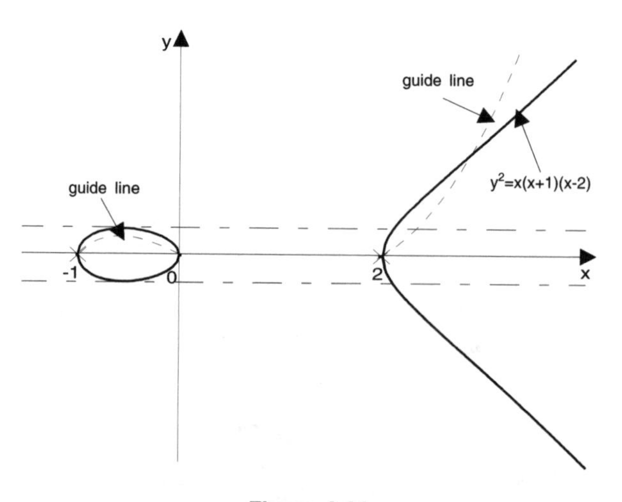

Figure 6.11

(iii) Now add the lines $y = \pm 1$ to the diagram (in roughly the correct positions).

Since you are plotting $\sqrt{x(x+1)(x-2)}$, this is the square root of the y values on our guide line. Since the square root of a number less than one increases, the curve will be above the guide line for $y < 1$, and the curve will be below the guide line for $y > 1$. Remember to make the graph symmetrical about the x-axis. The graph is shown as a continuous line. *These three steps must be applied to all graphs of this type.*

In order to plot this type of relationship on a graphics calculator, you will probably need to graph $y = +\sqrt{f(x)}$, leave this on the screen, and then graph $y = -\sqrt{f(x)}$. Try it out on your calculator for the equation given in the above worked example.

Example 6.6

Sketch the graph given by the equation $x^2 + y^2 = 9$.

Solution

(i) This can be written $y^2 = 9 - x^2$

Hence $\qquad y = \pm\sqrt{9 - x^2}$

(ii) The function $9 - x^2$ is sketched in Figure 6.12, and the part of this curve where $y \geq 0$ can be used as a guide line.

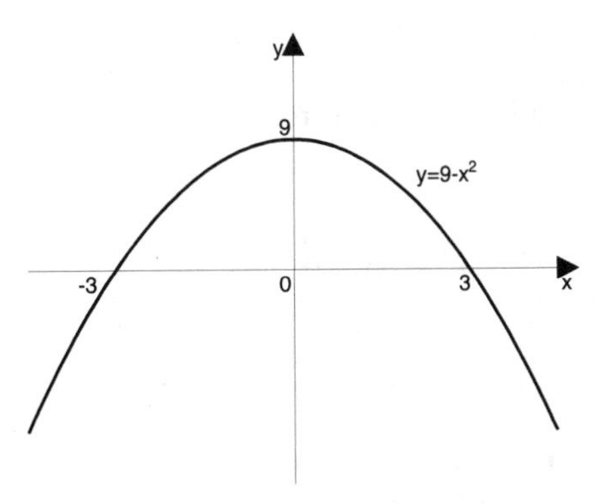

Figure 6.12

(iii) This guide line is shown in Figure 6.13 together with $y = \pm 1$.

The square root is plotted, and it can be seen that the curve is a circle of radius 3.

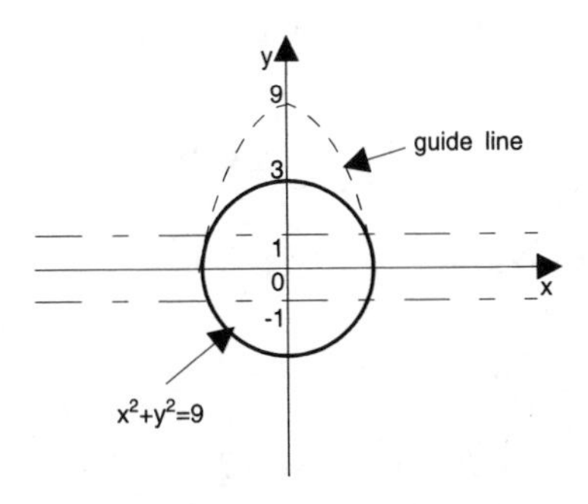

Figure 6.13

This leads us to the result:

> **MEMORY JOGGER**
>
> The equation of a circle of radius r, centre the origin 0 (0, 0), is given by
> $x^2 + y^2 = r^2$ (C8)

Example 6.7

The straight line $y = x - 2$ cuts the circle $x^2 + y^2 = 16$, at the points P and Q. Show that the distance $PQ = 2\sqrt{14}$.

Solution

This question involves simultaneous equations.

$$y = x - 2 \quad \text{(i)}$$
$$x^2 + y^2 = 16 \quad \text{(ii)}$$

These are shown in Figure 6.14

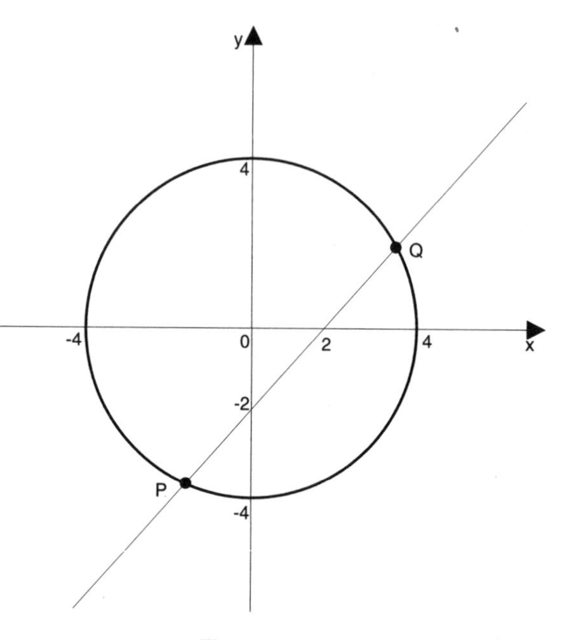

Figure 6.14

Substitute in (ii) for y:

$$x^2 + (x-2)^2 = 16$$
$$x^2 + x^2 - 4x + 4 = 16$$
$$2x^2 - 4x - 12 = 0$$
$$x^2 - 2x - 6 = 0$$

Using the formula for a quadratic equation:

$$x = \frac{2 \pm \sqrt{(-2)^2 - 4x - 6}}{2}$$

$$x = \frac{2 \pm \sqrt{28}}{2} = \frac{2 \pm 2\sqrt{7}}{2}$$

that is, $x = 1 \pm \sqrt{7}$

Since $y = x - 2$:

At P: $x = 1 - \sqrt{7}$, $y = -1 - \sqrt{7}$

At Q: $x = 1 + \sqrt{7}$, $y = -1 + \sqrt{7}$

Using the formula $\sqrt{(x_2 - x_1)^2 + (y_2 - y_1)^2}$ for the distance between two points:

$$\begin{aligned} PQ &= \sqrt{(2\sqrt{7})^2 + (2\sqrt{7})^2} \\ &= \sqrt{28 + 28} = \sqrt{56} = 2\sqrt{14} \end{aligned}$$

Note that you have to work in surds throughout this question, otherwise you would not get the exact answer of $2\sqrt{14}$, but a decimal which you could not say with certainty was $2\sqrt{14}$.

If the centre of the circle is not at the origin, its equation alters in form slightly. The following section looks at the equation of any circle.

6.5 General equation of a circle

The equation $(x - h)^2 + (y - k)^2 = r^2$ (C9) represents a circle, centre (h, k) and radius r.

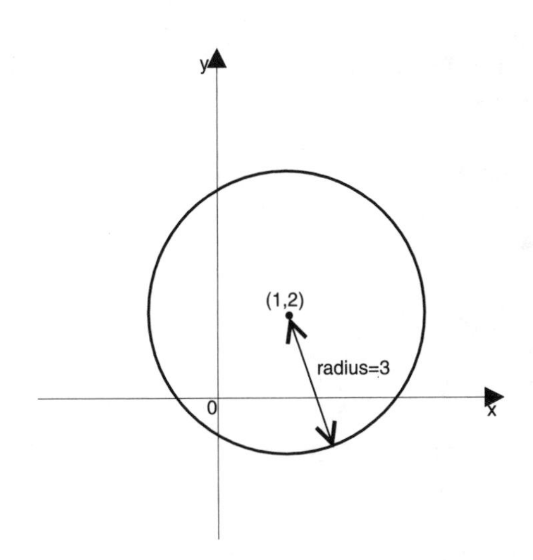

Figure 6.15

In Figure 6.15, the circle has radius 3, and centre $(1, 2)$. Its equation is:

$$\begin{aligned} & (x - 1)^2 + (y - 2)^2 &&= 3^2 \\ \text{or} \quad & x^2 - 2x + 1 + y^2 - 4y + 4 &&= 9 \\ \text{that is,} \quad & x^2 + y^2 - 2x - 4y - 4 &&= 0 \end{aligned}$$

Example 6.8

The equation of a circle is known to be $x^2 + y^2 - 6x + 8y - 11 = 0$. Find the coordinates of the centre of the circle, and its radius.

Solution

$x^2 + y^2 - 6x + 8y - 11 = 0$ can be rewritten:

$$x^2 - 6x + y^2 + 8y - 11 = 0$$

or $\quad (x-3)^2 - 9 + (y+4)^2 - 16 - 11 = 0$, using completing the square,

that is, $\quad (x-3)^2 + (y+4)^2 = 36$

The centre is $(3, -4)$, the radius $= \sqrt{36} = 6$.

6.6 The ellipse

Another equation worth learning, similar to that of the circle, is the equation of an ellipse.

If you sketch the curve:

$$\frac{x^2}{9} + \frac{y^2}{16} = 1$$

it is the ellipse shown in Figure 6.16a.

In general the equation is:

$$\frac{x^2}{a^2} + \frac{y^2}{b^2} = 1 \quad \text{(Figure 6.16b)}$$

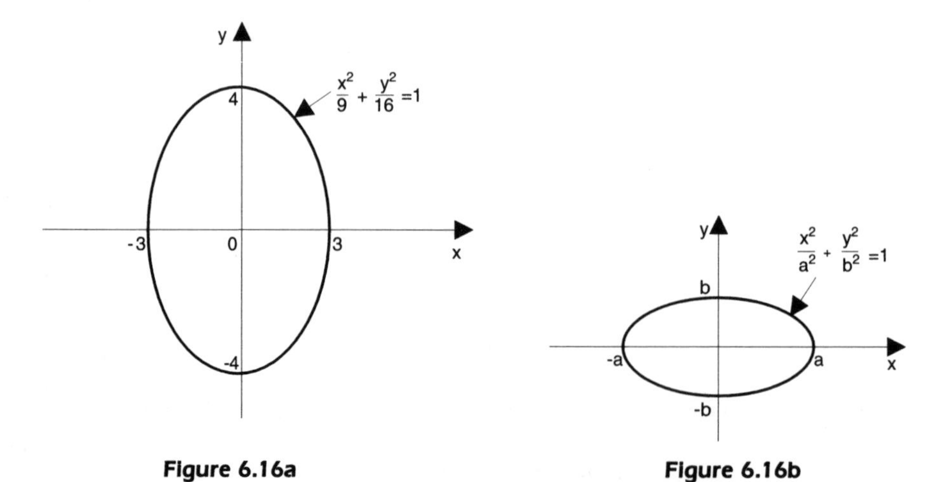

Figure 6.16a

Figure 6.16b

It is interesting to note that the area of the ellipse $= \pi ab$. The equation of an ellipse which does not have the origin as the centre is beyond the scope of this book.

Exercise 6(b)

1 Sketch graphs of the functions:

(i) $y = e^{2x}$ (ii) $y = e^{-x} + 2$

(iii) $y = e^{-x^2}$ (iv) $y = \ln(x+1)$

(v) $y = x + \ln x \quad (x > 0)$ (vi) $y = 1 - e^x$

2 Sketch graphs of the relations given by:

(i) $y^2 = x^2 + 1$ (ii) $y^2 = x(x-1)$

(iii) $y^2 = e^x$ (iv) $y^2 = x(x-1)(x+2)$

(v) $y^2 = 24 - x^2$ (vi) $y^2 = 2x + 1$

3 Find the equation of the circle with the following centres and radii:

(i) $(0,0)$; 4 (ii) $(0,1)$; 3 (iii) $(1,-2)$; 6

(iv) $(-2,-1)$; $\frac{1}{4}$ (v) $(-3,6)$; 8

4 Find the coordinates of the centre of the following circles, and the radius of the circle:

(i) $x^2 + y^2 = 36$ (ii) $x^2 + y^2 - 8 = 0$

(iii) $2x^2 + 2y^2 = 25$ (iv) $(x+1)^2 + (y-2)^2 = 25$

(v) $x^2 + y^2 - 4x - 6y - 3 = 0$

5 Draw a sketch of the following ellipses and find the area of each one.

(i) $\dfrac{x^2}{4} + \dfrac{y^2}{25} = 1$ (ii) $9x^2 + 4y^2 = 36$

(iii) $2x^2 + 3y^2 = 12$ (iv) $x^2 + 4y^2 = 2$

6.7 Parametric coordinates

There are many occasions when the Cartesian equation of a curve is difficult to use. To help overcome this problem, an extra variable (called a parameter) is often introduced, which enables x and y to be written in a simpler fashion.

Consider the following example:

$$x = 2t^2 - t + 1$$
$$y = 1 + t^3$$

where t can take any value. We will set up a table of values for t from -3 to 3, by way of example.

t	-3	-2	-1	0	1	2	3
x	22	11	4	1	2	7	16
y	-26	-7	0	1	2	9	28

Having completed the table, ignore the values of the parameter t, and plot the x and y values as a point for each value of t.

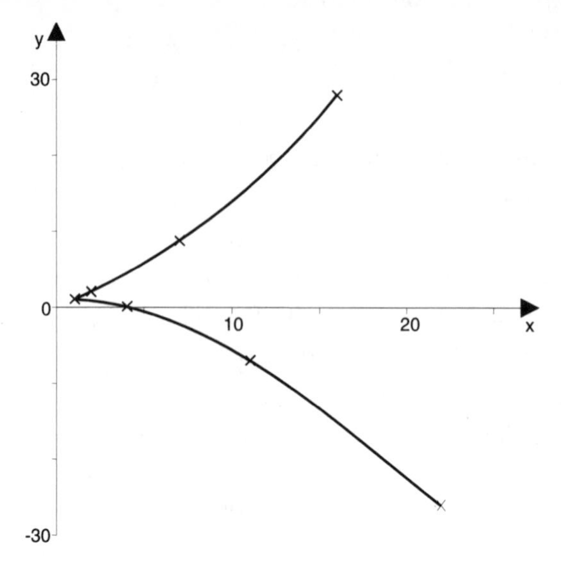

Figure 6.17

The resultant curve is shown in Figure 6.17. The curve comes to a point at $(1, 1)$ called a **cusp**.

It is possible to get the Cartesian equation of the curve in the following way:

since $y = 1 + t^3$, $t^3 = y - 1$

$$\therefore \quad t = \sqrt[3]{y-1}$$

Substitute into $x = 2t^2 - t + 1$ to get

$$x = 2(y-1)^{\frac{2}{3}} - (y-1)^{\frac{1}{3}} + 1$$

This is not an explicit function for y in terms of x, and it would be extremely difficult to plot a graph from it.

Example 6.9

Find the Cartesian equations of the curves given parametrically by:

(i) $x = 2t^2$, $y = t + 1$ (ii) $x = \dfrac{1+t}{1-t}$, $y = \dfrac{2t}{1-t}$ (iii) $x = 4t$, $y = \dfrac{6}{t}$

Solution

(i) If $y = t + 1$, then $t = y - 1$.

Hence: $x = 2t^2$

$= 2(y-1)^2$

that is, $x = 2y^2 - 4y + 2$

(ii) It is tempting to rearrange $x = \dfrac{1+t}{1-t}$ to make t the subject:

$$
\begin{aligned}
x(1-t) &= 1+t \\
x - xt &= 1+t \\
\therefore \quad x - 1 &= t(1+x) \\
t &= \frac{x-1}{x+1}
\end{aligned}
$$

$$\therefore \quad y = \frac{2t}{1-t} = \frac{2\left(\dfrac{x-1}{x+1}\right)}{1-\left(\dfrac{x-1}{x+1}\right)} = \frac{2(x-1)}{(x+1)-(x-1)}$$

$$y = \frac{2(x-1)}{2} = x - 1$$

This result is surprisingly simple. However, it could have been achieved straight away if you had observed that:

$$x - y = \frac{1+t-2t}{1-t} = \frac{1-t}{1-t} = 1$$

that is, $y = x - 1$.

(iii) Here, you can see that:

$$x \times y = 4t \times \frac{6}{t} = 24$$

$\therefore$ the equation is $xy = 24$.

When plotting a curve using parameters, the parameter can be used to indicate a direction of movement along the graph. Study the next example carefully.

Example 6.10

Draw a reasonable sketch to show what the curves given by the following parametric coordinates look like. On each curve, indicate with an arrow the direction indicated by increasing the parameter.

(i) $x = 2t^2 + 1,\ y = t - 2 \qquad -2 \leq t \leq 2$
(ii) $x = 3 \sin t,\ y = 2 \cos 2t \qquad 0° \leq t \leq 360°$

Solution

(i) You need to set out the results in a table. Substitute values of t into the expressions for x and y. You can then plot the point for each value of t chosen. The graph is shown in Figure 6.18.

t	-2	-1	0	1	2
x	9	3	1	3	9
y	-4	-3	-2	-1	0

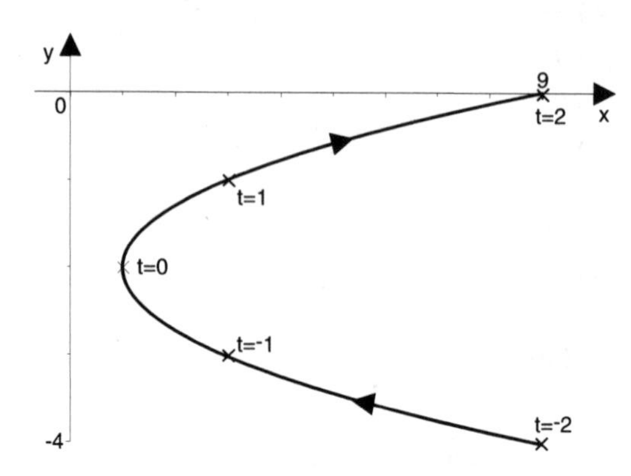

Figure 6.18

t	0	45	90	135	180	225	270	315	360
x	0	2.1	3	2.1	0	-2.1	$-3.$	-2.1	0
y	2	0	-2	0	2	0	-2	0	2

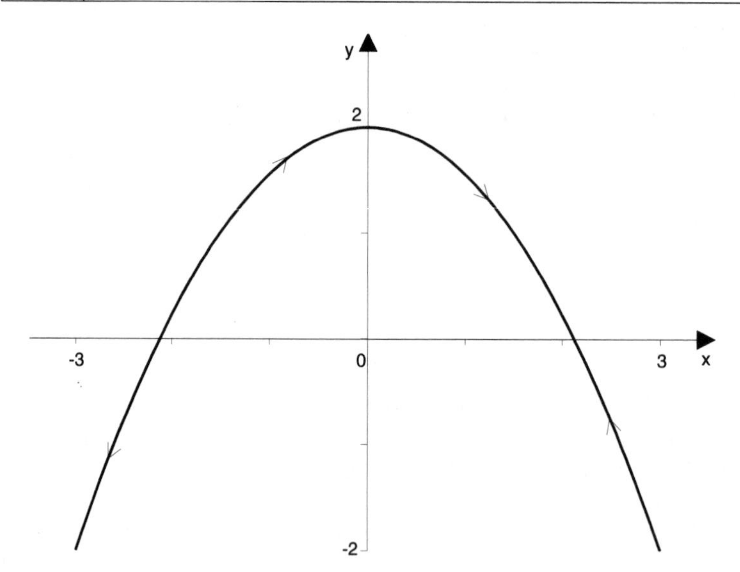

Figure 6.19

(ii) The final result is quite surprising. It is once again a parabola, and the curve is travelled along from $(0, 2)$ to $(3, -2)$, then back round to $(-3, -2)$ before returning to $(0, 2)$ (see Figure 6.19).

One *disadvantage* of parametric coordinates is that it is not always clear what range of values to choose for t to get an idea of what the complete graph looks like. At the end of the day, only experience will help you decide.

The following example illustrates how parametric equations are particularly useful in graphs which contain loops.

Example 6.11

(i) Plot the graph of $x = 2\sin 2\theta$, $y = \cos\theta$ for values of θ between $0°$ and $360°$.

(ii) $x = 2\cos 2\theta$, $y = 3\sin 3\theta$.

Solution

(i) Here, θ is the parameter. **Note** that here θ is in degrees, but you need to be careful if you are going to differentiate this (see Chapter 14). The table of values is shown below:

θ	0	30	60	90	120	150	180	210	240	270	300	330	360
x	0	1.7	1.7	0	−1.7	−1.7	0	1.7	1.7	0	−1.7	−1.7	0
y	1	0.87	0.5	0	−0.5	−0.87	−1	−0.87	−0.5	0	0.5	0.87	1

The graph can now be plotted as shown in Figure 6.20. Make sure you join up the points in the order of the parameter. The Cartesian equation of this sort of curve is complicated.

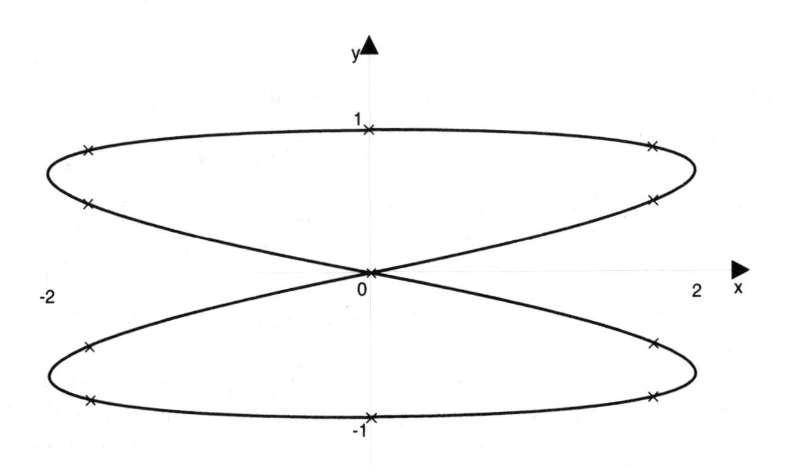

Figure 6.20

(ii) Here we have not been given any idea as to the range of values of θ to be used. As a general rule, it is worth trying θ from $0°$ to $360°$ in this case.

θ	0	30	60	90	120	150	180	210	240	270	300	330	360
x	2	1	−1	−2	−1	1	2	1	−1	−2	−1	1	2
y	0	3	0	−3	0	3	0	−3	0	3	0	−3	0

The points can now be plotted as shown in Figure 6.21. You would find it almost impossible to draw the graph from this diagram.

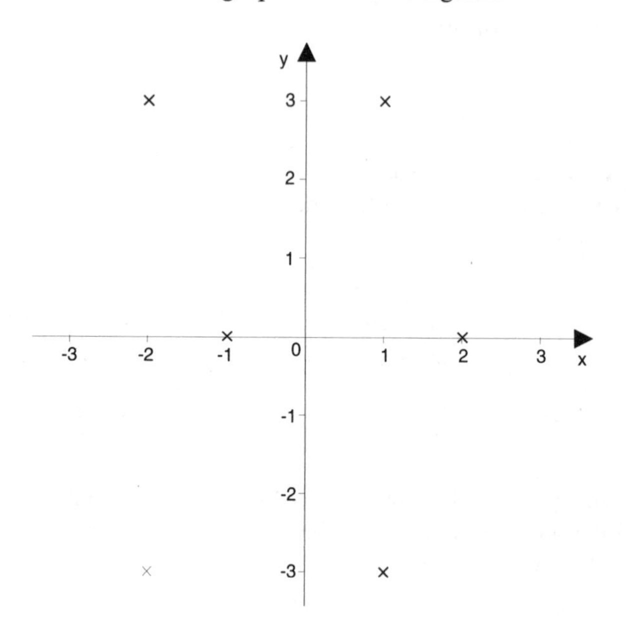

Figure 6.21

We shall work out x and y for more values of the parameter.

θ	15	45	75	105	135	165	195	225	255	285	315	345
x	1.7	0	−1.7	−1.7	0	1.7	1.7	0	−1.7	−1.7	0	1.7
y	2.1	2.1	−2.1	−2.1	2.1	2.1	−2.1	−2.1	2.1	2.1	−2.1	−2.1

These points can now be plotted. When you join up the points, make sure you are following θ in sequence.

You will be surprised to see the curve goes over itself twice as you are drawing it. The final graph is shown in Figure 6.22. This is, of course, quite a difficult example (one hopes you would not get such a difficult example in an examination), and you may be able to use a graphics calculator to help you. However, it should be said that the more interesting curves are bound to have more difficult equations.

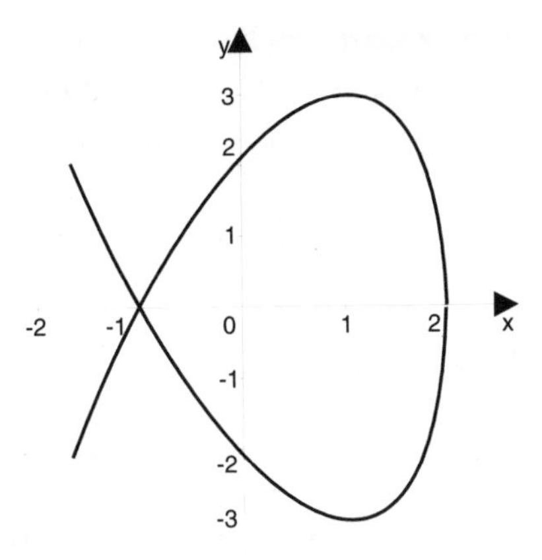

Figure 6.22

Exercise 6(c)

Plot the following parametrically defined curves.

1	$x = 2t + 1,$	$y = t^2 - 2$	$-3 \leq t \leq 3$
2	$x = \dfrac{4}{t+1},$	$y = t + 2$	$0 \leq t \leq 6$
3	$x = 2\sin\theta,$	$y = 5\cos\theta$	$0° \leq \theta \leq 360°$
4	$x = \dfrac{1-t}{1+t},$	$y = 2t$	$0 \leq t \leq 6$
5	$y = 4\sin 2\theta,$	$x = 3\cos\theta$	$0° \leq \theta \leq 360°$
6	$x = q^3 - 1,$	$y = q^2 + 1$	$-4 \leq q \leq 4$

In each case, indicate a direction on the curve as the parameter increases. Also find the cartesian equation of the curve.

Miscellaneous Examples 6

1 Sketch the graphs:

(i) $y = 4 - 2x^3$ (ii) $y = x^4 + 4x^2$ (iii) $y = \dfrac{x}{x+1},\ x \neq -1$

(iv) $y = x + \dfrac{1}{x},\ x \neq 0$ (v) $y = xe^x$ (vi) $y = \dfrac{x+1}{x^2},\ x \neq 0$

2 Sketch the graphs of the following relations:

(i) $y^2 = 2x + 5$ (ii) $y^2 = x^3$ (iii) $y^2 = 4x - x^2$

(iv) $y^2 = \dfrac{x}{1+x},\ x \neq -1$ (v) $y^2 = e^{-x^2}$

3 Plot the graphs given parametrically by:

(i) $y = 4t^2$, $\quad x = 3t^3 \quad -2 \le t \le 2$

(ii) $y = \cos 2\theta$, $\quad x = 1 + \cos\theta \quad 0° \le \theta \le 360°$

(iii) $y = \dfrac{t}{1+t}$, $\quad x = \dfrac{t}{1+t^2} \quad 0 \le t \le 6$

Revision Problems 6

1 The equation of a curve is $y = \dfrac{3x}{(x+2)^2}$. State the equations of the asymptotes of the curve, and use calculus to find the coordinates of any turning points.

Sketch the graphs of (i) $y = \dfrac{3x}{(x+2)^2}$; (ii) $y^2 = \dfrac{3x}{(x+2)^2}$

2 Sketch on a separate diagram, the curves:

(i) $y = \dfrac{x-2}{2x}$ $\quad$ (ii) $y = \dfrac{x^2-4}{4x^2}$

Hence sketch the graphs of (iii) $y = \left|\dfrac{x-2}{2x}\right|$ and (iv) $y^2 = \dfrac{x^2-4}{4x^2}$

3 Find the equation of the straight line that passes through the points of intersection of the two circles:

$x^2 + y^2 - 4x - 6 = 0$ and $x^2 + y^2 + 2x + 4y - 8 = 0$.

4 Plot the graph of the curve given parametrically by $x = 3\sin 2\theta$, $y = 4\sin\theta$. Find the Cartesian equation in the form which does not contain a square root.

7 Functions

The words 'function', 'mapping' and 'relationship' are often confused and used incorrectly. The differences are, in fact, not always easy to understand, and although it is desirable to use the correct words when describing something, at the end of the day, the *ideas* developed in this chapter are more important than how they are described.

7.1 Basic definitions

A statement such as $y = 4x + 1$ gives a **relationship** between x and y. This relationship can be shown by using a **mapping** diagram as shown in Figure 7.1. The values of x used form the **domain** of this relationship, and can be chosen in any way you wish. Each value of y obtained is called the **image** of that value of x. Hence $y = -7$ is the image of $x = -2$.

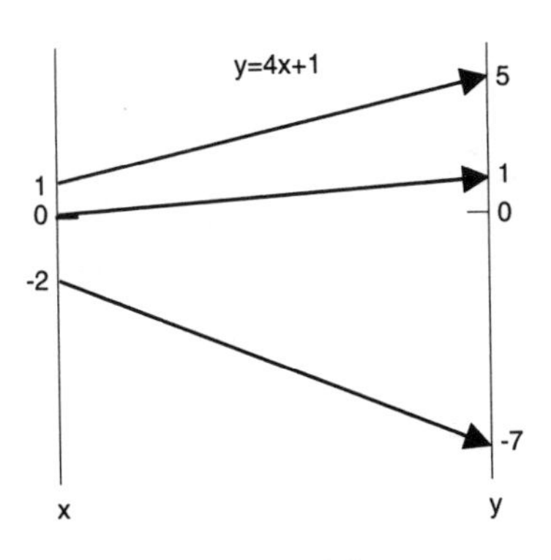

Figure 7.1

The complete set of y values obtained is called the **range** of the mapping. The relationship $y = 4x + 1$ is also a **function**, because for each value of x in the domain, there is only one value of y. If, however, you consider the relationship $y^2 = x + 4$ for the domain $x = \{-2, -1, 0, 1\}$, then Figure 7.2 shows that, for example, 0 maps on to both 2 and -2. This is because, if you substitute $x = 0$ into the relationship, you get:

$$y^2 = 4 \quad \text{so } y = \pm 2$$

This type of mapping or relationship is *not* a function.

> **MEMORY JOGGER**
> A relationship (or mapping) is a function if it assigns only one value to each number in the domain.

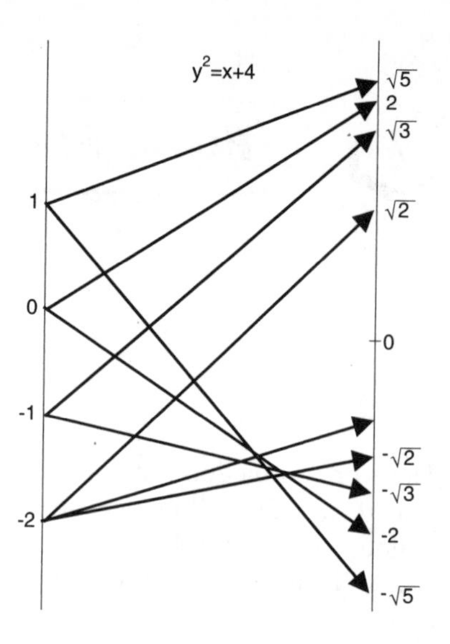

Figure 7.2

Example 7.1

Draw a sketch of the function $y = x^2 - x$ for $-1 \le x \le 3$, and hence state the range of the function.

Solution

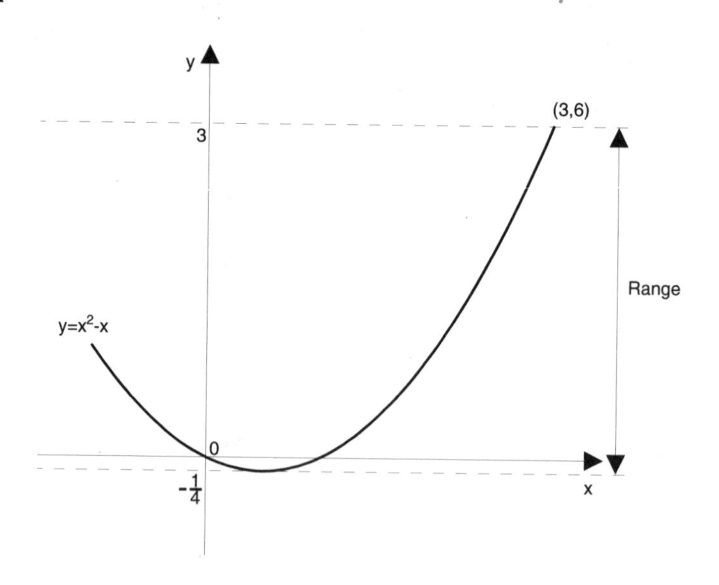

Figure 7.3

Figure 7.3 shows the graph (part of a parabola). The range of the function cannot be stated until the minimum point is found:

$y = x^2 - x = \left(x - \tfrac{1}{2}\right)^2 - \tfrac{1}{4}$

Hence if $x = \tfrac{1}{2}$, $y = -\tfrac{1}{4}$

and $x = 3$, $y = 6$

The range of the function is therefore:

$-\tfrac{1}{4} \leq y \leq 6$

Don't forget the technique of completing the square when dealing with the quadratic function

MEMORY JOGGER

Looking at the diagram will often help you to find the range.

7.2 Alternative notation

The function we used in Example 7.1 can be written in two other ways:

(i) $f(x) = x^2 - x \quad -1 \leq x \leq 3$

(ii) $f : x \mapsto x^2 - x \quad -1 \leq x \leq 3$

For convenience, we shall refer to $f(x)$ as the function notation, and (ii) as the mapping function notation. Type (ii) is often used in computer software design. (Any letter other than f can be used).

Example 7.2

Draw a sketch to show the function:

$$g : x \mapsto x + \frac{1}{x} \qquad \tfrac{1}{2} \leq x \leq 3$$

(i) Find the range of the function

(ii) Solve the equation $g(x) = 4$.

Solution

(i) The graph is shown in Figure 7.4. It is tempting to say that the minimum point occurs when $x = 1$. You must not assume this, however, and need to use the calculus, to find it exactly.

If $\quad y = x + \dfrac{1}{x}$

$$\frac{dy}{dx} = 1 - \frac{1}{x^2}$$

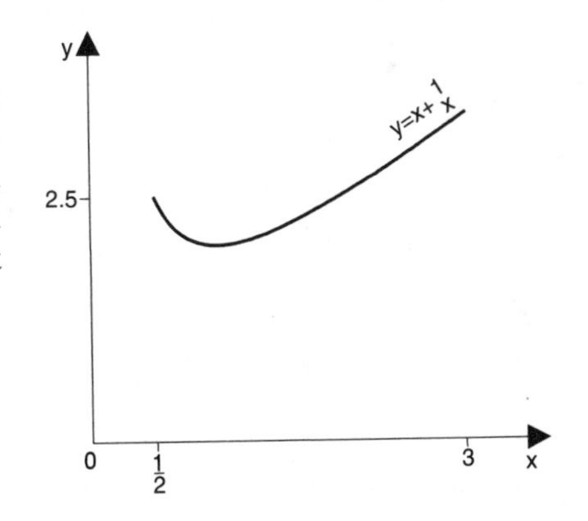

Figure 7.4

At the turning point, $\dfrac{dy}{dx} = 0$

$$\therefore \quad 0 = 1 - \frac{1}{x^2}, \quad x^2 = 1$$

hence $\quad x = \pm 1$

Now $x = -1$ is **not** in the domain, and so the minimum point is when $x = 1$, and hence $y = 1 + \dfrac{1}{1} = 2$

When $x = 3$, $y = 3 + \frac{1}{3} = 3\frac{1}{3}$

The range of g is $2 \leq y \leq 3\frac{1}{3}$ (see Figure 7.4).

(ii) If $g(x) = 4$

$$x + \frac{1}{x} = 4$$

$$\therefore \quad x^2 + 1 = 4x$$

$$x^2 - 4x + 1 = 0$$

$$x = \frac{4 \pm \sqrt{(-4)^2 - 4}}{2}$$

$$= 3.73,\ 0.27$$

You need to realise that both of these values do not lie in the domain of g, and so, in fact, the equation $g(x) = 4$ does not have a solution

7.3 Odd and even functions

A function is said to be **odd** if $f(-x) = -f(x)$ for all values of x. This implies that the graph has rotational symmetry of order 2 about the origin 0. Some examples of odd functions are shown in Figure 7.5.

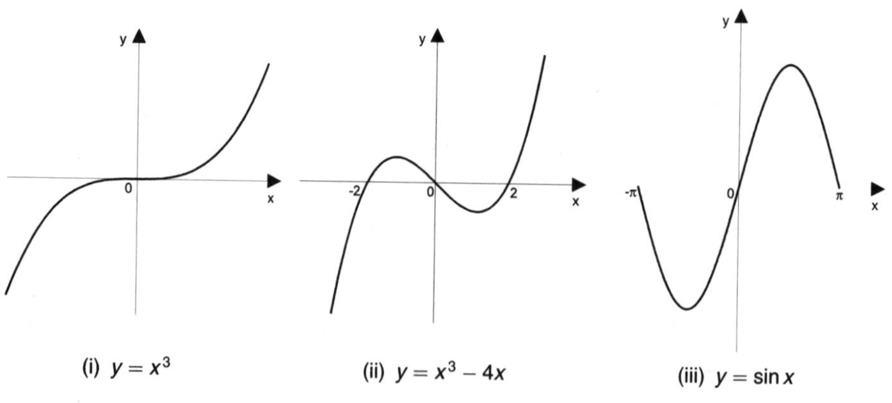

Figure 7.5

A function is said to be **even** if $f(-x) = f(x)$ for all values of x. This implies that the graph is symmetrical about the y-axis. Examples of even functions are shown in Figure 7.6.

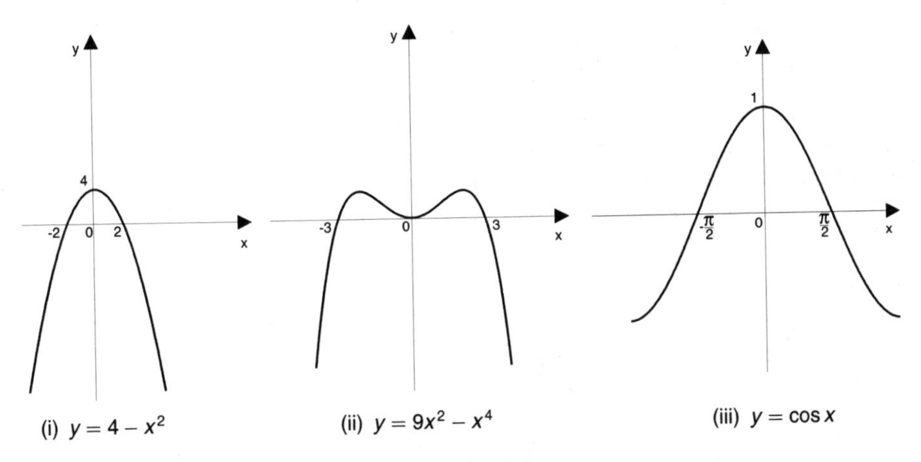

(i) $y = 4 - x^2$ (ii) $y = 9x^2 - x^4$ (iii) $y = \cos x$

Figure 7.6

Exercise 7(a)

Sketch the graph of the following functions and where possible state the range of the function.

1 $f : x \mapsto 2x + 1$ $-1 \le x \le 3$

2 $g : x \mapsto x^2 - 2x + 5$ $-3 \le x \le 3$

3 $h : x \mapsto 4 + x^3$ $-2 \le x \le 2$

4 $i : x \mapsto \dfrac{1}{x}$ $1 \le x \le 4$

5 $k : x \mapsto 4 - 3x^2$ $-4 \le x \le 4$

6 $m : x \mapsto 1 - 3x - 2x^2$ $0 \le x \le 1$

7 $n : x \mapsto 2^x - 1$ $0 \le x \le 4$

8 $p : x \mapsto \dfrac{1 + x}{1 - x}$ $2 \le x \le 4$

State whether the following functions are odd, even, or neither.

9 $f(x) = x^3 + 4$

10 $g(x) = (x - 2)^2 + 4$

11 $h(x) = \dfrac{1 - x^2}{1 + x^2}$

12 $k(x) = 2^x + 4$

13 $m(x) = x(2 - 4x^2)$

14 $n(x) = x - \dfrac{1}{x}$

15 $p(x) = x^4 + 6x^2 + 1$

7.4 Inverse functions

Consider the graphs of the following two functions:

(i) $f : x \mapsto x^2 + 1 \qquad -2 \le x \le 2$

(ii) $g : x \mapsto x^2 + 1 \qquad 0 \le x \le 2$

You will notice that the *relationships* are the same, but because each relation has a different domain, they are *different* functions. Hence we will need a different letter to distinguish them. The graphs are shown in Figure 7.7.

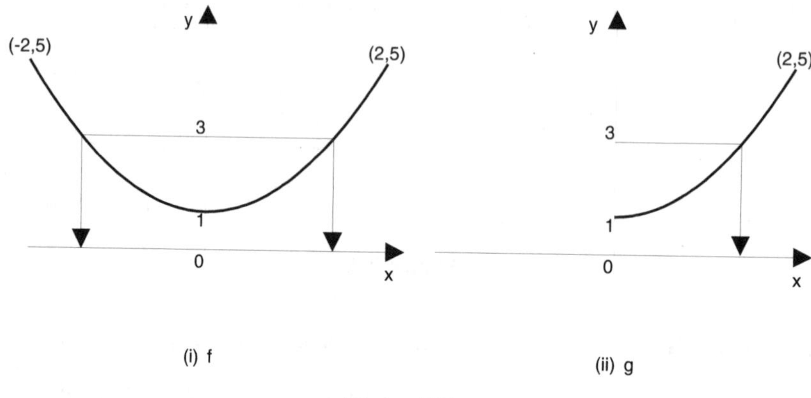

Figure 7.7

Look at the equations $f(x) = 3$, $g(x) = 3$.

If $x^2 + 1 = 3$ then $x^2 = 2$, so $x = \pm\sqrt{2}$

Hence, for $f(x)$, there are two values of x which give $y = 3$. This would be true for any value of y in the range.

For $g(x)$, there is always only one value of x for every value of y in the range. When this happens, the function is said to be a **1:1 function**.

Now writing

$$y = x^2 + 1$$
$$y - 1 = x^2$$
$$\therefore \quad x = \pm\sqrt{y - 1}$$

For $g(x)$, x can only be positive, and hence:

$$x = +\sqrt{y - 1}$$

This is called the **inverse relation**, written g^{-1}, or $g^{-1}(y)$.

The y values in the range now become the domain of this inverse function. It is perhaps somewhat confusing that this inverse relation is then written $g^{-1}(x) = \sqrt{x - 1}$ or $g^{-1} : x \mapsto \sqrt{x - 1}$, by convention. Now since the range of g is $1 \le y \le 5$, this becomes the domain of g^{-1}. Hence the function g^{-1} in full is:

$$g^{-1} : x \mapsto \sqrt{x - 1} \qquad 1 \le x \le 5$$

Example 7.3

The function $f : x \mapsto 3x + x^2$, $x \in R$, $x \geq h$, is known to have an inverse f^{-1}. Indicate the implications of this in a diagram, and find out what you can about h ($x \in R$ means x **belongs** to the set of all real numbers, that is, x can be any value between $-\infty$ and ∞).

Solution

The given function has a graph which is part of a parabola. In order for f to have an inverse function, the graph must be a 1:1 relation. Hence it must be only on one side of the line of symmetry.

If $y = 3x + x^2$

$= x(3 + x)$

then $y = 0$ when $x = 0$ or -3. And so the line of symmetry is $x = -\frac{3}{2}$.

Hence $h \geq -\frac{3}{2}$

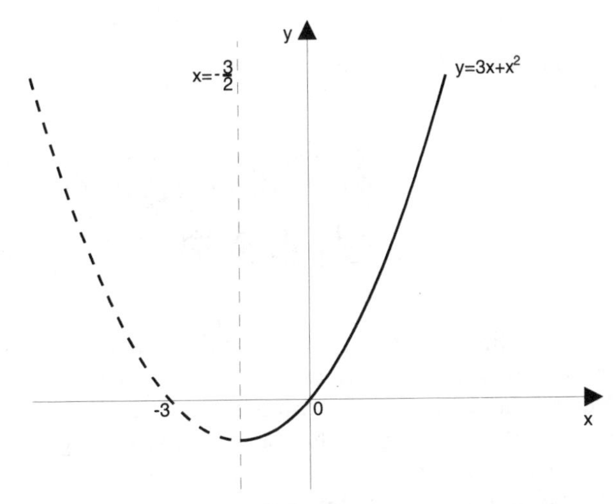

Figure 7.8

Exercise 7(b)

For the following functions, determine whether or not the inverse function exists. Where it does, give an algebraic form; where it does not, give a reason.

1 $f : x \mapsto x^2$ $\quad -2 \leq x \leq 2$

2 $g : x \mapsto 3x + 4$ $\quad -1 \leq x \leq 4$

3 $h : x \mapsto x^3 + 1$ $\quad -4 \leq x \leq 4$

4 $k : x \mapsto \dfrac{1}{x+1}$ $\quad 2 \leq x \leq 4$

5 $l : x \mapsto 4 - 2x^2$ $\quad 0 \leq x$

6 $m : x \mapsto 4 + 2x^2$ $\quad -1 \leq x$

7 $n : x \mapsto 1 - 2x^2$ $\quad x \in R.$

8 $p : x \mapsto x - x^3$ $\quad x \geq 0$

9 $q : x \mapsto e^x$ $\quad x \in R$

10 $r : x \mapsto \ln(x + 1)$ $\quad x > -1$

7.5 Composite functions

It is possible to combine two or more functions into a single function.

So if $f(x) = x^2 + 1$, and $g(x) = 2x - 1$, then $f(g(x))$ means replace x in the function f by $2x - 1$. In other words, g **operates** on x first, and then f operates on the answer.

$$\text{Hence}\quad f(g(x)) = (2x-1)^2 + 1 = 4x^2 - 4x + 2$$
$$\text{Now}\quad g(f(x)) = 2(x^2+1) - 1 = 2x^2 + 1$$

> **MEMORY JOGGER**
> In most cases, $f(g(x)) \neq g(f(x))$

Sometimes the notation is abbreviated, and $f(g(x))$ might just be written fg. If the mapping function notation is used, a symbol such as $\circ$ or $*$ might be used. Hence, the composite functions $f \circ g$ or $g * h$ might occur. When evaluating a numerical example, note that fg(2) means apply g first to $x = 2$, and then apply f to the answer g(2).

So, in the above example:

$$g(2) = 2 \times 2 - 1 = 3, \text{ f now operates on 3}$$
$$\text{and}\quad f(3) = 3^2 + 1 = 10,$$
$$\text{so}\quad fg(2) = 10$$

Example 7.4

A function f is defined by:

$$f : x \mapsto 2 - 3x^{-1},\ x \in \mathrm{R},\ x \neq 0,\ x \neq 1.5$$

Evaluate: (i) $ff(x)$ (ii) $f^{-1}(x)$

Solution

(i) $ff(x)$ means $f(f(x)) = f(2 - 3x^{-1})$

$$= 2 - \frac{3}{(2 - 3x^{-1})} = \frac{2(2 - 3x^{-1}) - 3}{(2 - 3x^{-1})}$$

$$= \frac{4 - 6x^{-1} - 3}{2 - 3x^{-1}} = \frac{1 - 6x^{-1}}{2 - 3x^{-1}} = \frac{x - 6}{2x - 3}$$

(ii) To find $f^{-1}(x)$, let $y = f(x)$ and rearrange the formula to make x the subject.

$$\text{So:}\quad y = 2 - \frac{3}{x}$$

$$\therefore\quad \frac{3}{x} = 2 - y$$

$$\therefore\quad \frac{x}{3} = \frac{1}{2-y}$$

$$\therefore\quad x = \frac{3}{2-y}$$

Conventionally, the domain of a function is always denoted by x, and so although the inverse relation operates on a y value, conventionally, x and y are now interchanged when stating the inverse.

$$\text{Hence}\quad f^{-1}(x) = \frac{3}{2-x}$$

Example 7.5

(a) Functions g and h are defined by:

$$g : x \mapsto \ln x, \qquad x \in R, x > 0$$
$$h : x \mapsto 1 + x, \qquad x \in R$$

The function f is defined by: $f : x \mapsto gh(x), \quad x \in R, x > -1$

(i) Sketch the graph of $y = f(x)$.

(ii) Write down the expressions for $g^{-1}(x)$ and $h^{-1}(x)$.

(iii) Write down an expression for $g^{-1}h^{-1}(x)$.

(b) The function q is defined by: $q : x \mapsto x^2 - 4x, \quad x \in R, |x| \leq 1$
Show by means of a graphical argument or otherwise, that q is 1:1, and find an expression for $q^{-1}(x)$.

(UCLES)

Solution

(a) (i)
$$gh(x) = g(1 + x)$$
$$= \ln(1 + x)$$

The graph of $y = \ln(1 + x)$ is shown in Figure 7.9.

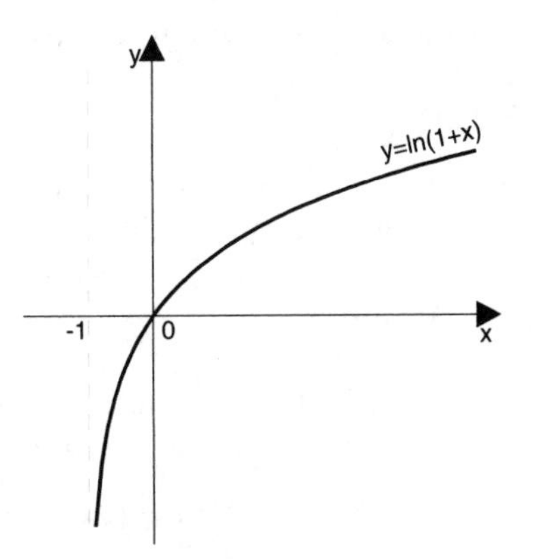

Figure 7.9

(ii) If $y = \ln x$, then $x = e^y$

If $y = 1 + x$, then $x = y - 1$

$$\therefore\quad g^{-1}(x) = e^x$$

and $h^{-1}(x) = x - 1$

(iii) $g^{-1}h^{-1}(x) = g^{-1}(x - 1) = e^{x-1}$

(b) The turning point on $y = x^2 - 4x = x(x-4)$ is at the point where $x = 2$, the line of symmetry. It is $(2, -4)$. You can see on the graph in Figure 7.10 that the function is 1:1 because the turning point has not been reached.

Completing the square:

$$y = (x-2)^2 - 4$$
$$\therefore \quad y + 4 = (x-2)^2$$
$$x - 2 = \pm\sqrt{y+4}$$
$$\therefore \quad x = 2 \pm \sqrt{y+4}$$

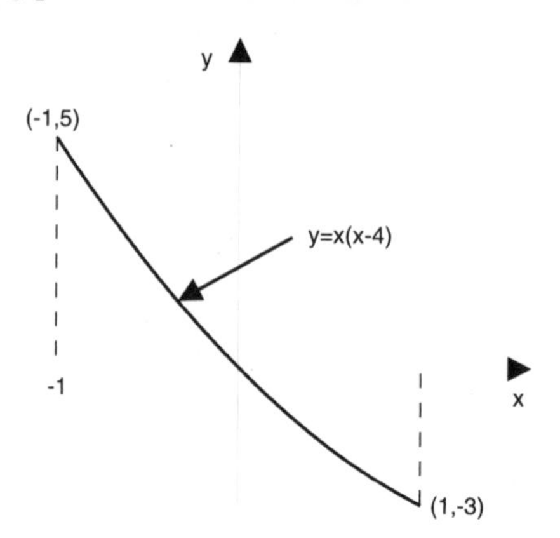

Figure 7.10

On the part of the curve we are concerned with, when $y = 0$, $x = 0$

$$\therefore \quad 0 = 2 \pm \sqrt{0+4}$$

Hence the $-$ sign is required to make this true.

Hence: $\quad x = 2 - \sqrt{y+4}$

that is, $\quad q^{-1}(x) = 2 - \sqrt{x+4}$

It is important to realise that $fg(x)$ might not always exist. The following rule tells you when it does:

> **MEMORY JOGGER**
>
> $fg(x)$ only exists, if the range of g is a part (subset) of the domain of f

Example 7.6

If $f(x) = 2x + 1$, $0 \le x \le \frac{1}{2}$, and $g(x) = x^2 + 6$, $0 \le x \le 2$, investigate whether the functions (i) $fg(x)$ (ii) $gf(x)$ exist.

Solution

(i) $fg(x)$ means 'apply g first to x'. Now the range of g is $6 \le y \le 10$. All these values are outside the domain of f. For example, if you try to find $fg(1)$ then $g(1) = 7$, and $f(7)$ does not exist. Hence the composite function $fg(x)$ cannot be formed.

(ii) $gf(x)$ means 'apply f first to x'. Now the range of f is $1 \le y \le 2$. All these values lie inside (a subset of) the domain of g. For example, $gf\left(\frac{1}{4}\right) = g\left(1\frac{1}{2}\right) = 8\frac{1}{4}$. Hence the composite function $gf(x)$ can be found.

In fact, $gf(x) = (2x+1)^2 + 6 = 4x^2 + 4x + 7 \qquad 0 \le x \le \frac{1}{2}$

The range of gf is $7 \le y \le 10$.

Exercise 7(c)

In the following examples, investigate whether the composite function fg exists. In the cases where it does not, give a reason. Where it does, give an explicit expression for the function.

1 $f : x \mapsto x^2$
$g : x \mapsto 2x + 1$

2 $f : x \mapsto x + 1 \quad 0 \leq x \leq 4$
$g : x \mapsto \sqrt{2x - 1} \quad 1 \leq x \leq 9$

3 $f : x \mapsto \dfrac{1}{x+1} \quad x < -1$
$g : x \mapsto x + 2$

4 $f : x \mapsto 2x + 1 \quad x \leq 3$
$g : x \mapsto 2 - x^2$

5 $f : x \mapsto e^x$
$g : x \mapsto \ln(x + 1)$

6 $f : x \mapsto e^{-x}$
$g : x \mapsto \ln(1 - x)$

7.6 Modulus functions

In section 3.4, we briefly introduced the idea of the modulus sign $|x|$. We can now look at some examples that involve this function.

You need to be aware that:

(i) if $f(x) > 0 \quad |f(x)| = f(x)$
(ii) if $f(x) < 0 \quad |f(x)| = -f(x)$

When considering functions that involve $|f(x)|$ you must consider separately the situations $f(x) > 0$, and $f(x) < 0$. These can then be combined to give the complete graph.

For example, if a function contains the expression $|2x + 1|$, then you must look at $2x + 1 > 0$, that is $x > -\frac{1}{2}$ and $2x + 1 < 0$, that is $x < -\frac{1}{2}$ separately. The following example illustrates how this is done.

Example 7.7

Sketch graphs of the following functions:

(i) $f : x \mapsto |x + 1| + x$

(ii) $f : x \mapsto \dfrac{1 + |x|}{1 - |x|}$

Solution

(i) We shall adopt the technique of combining the graphs of $|x + 1|$ and x.

For $x < -1$, $\quad |x + 1| = -(x + 1)$
$\therefore \quad |x + 1| + x = -(x + 1) + x = -1$
For $x > -1$, $\quad |x + 1| = x + 1$
$\therefore \quad |x + 1| + x = (x + 1) + x = 2x + 1$

The finished graph is shown in Figure 7.11.

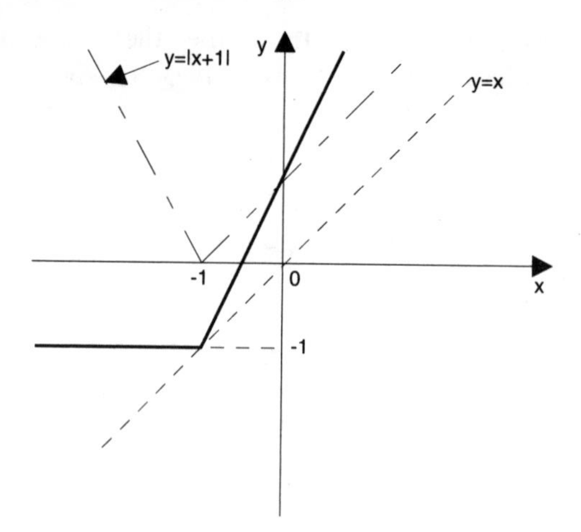

Figure 7.11

(ii) At first sight, this might look quite difficult. But try to see how to get rid of the modulus sign.

If $x \geq 0 \quad |x| = x$

Hence: $\quad \text{f} : x \mapsto \dfrac{1+x}{1-x} \qquad x \geq 0, x \neq 1$

If $x < 0 \quad |x| = -x$

$\therefore \quad \text{f} : x \mapsto \dfrac{1-x}{1+x} \qquad x < 0, x \neq -1$

The graph can now be sketched. See Figure 7.12.

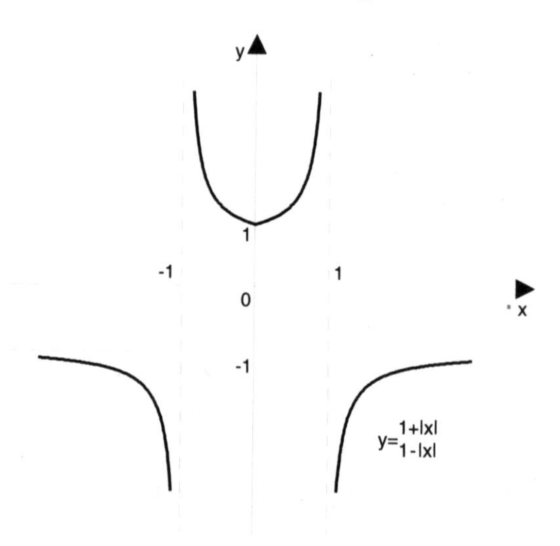

Figure 7.12

7.7 Transformation of functions

When looking at the graph of a given function, very often its shape can be determined by considering a basic curve, and using a transformation of the variable. For example, if we know the shape of $f(x) = x^2$, we can work out the shape of $g(x) = (x + 2)^2$.

Look at the following curves, all of which are related to $f(x) = x^2$.

(i) $f(x) = x^2$ is shown in Figure 7.13.

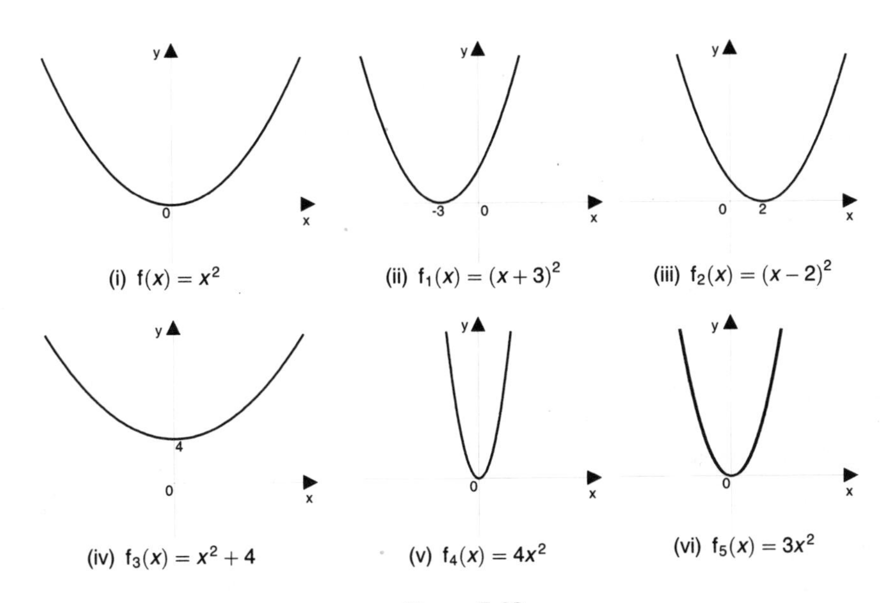

Figure 7.13

(ii) Here $f_1(x) = f(x + 3)$

The graph of $f(x)$ has been translated -3 parallel to the x-axis.

This could be written using the vector $\begin{pmatrix} -3 \\ 0 \end{pmatrix}$

(iii) $f_2(x) = f(x - 2)$

The graph of $f(x)$ has been translated $+2$ parallel to the x-axis.

This could be written using the vector $\begin{pmatrix} 2 \\ 0 \end{pmatrix}$.

(iv) $f_3(x) = 4 + f(x)$.

The graph of $f(x)$ has been translated $+4$ parallel to the y-axis.

This could be written using the vector $\begin{pmatrix} 0 \\ 4 \end{pmatrix}$.

(v) $f_4(x) = f(2x)$: **Note** $4x^2 = (2x)^2$

The graph of $f(x)$ has been stretched by a factor $\frac{1}{2}$ parallel to the x-axis.

(vi) $f_5(x) = 3f(x)$

The graph of $f(x)$ has been stretched by a factor 3 parallel to the y-axis.

These results can in general be summarised as follows:

(i) $f(x+a)$ translates $f(x)$ by $\begin{pmatrix} -a \\ 0 \end{pmatrix}$

(ii) $f(x)+a$ translates $f(x)$ by $\begin{pmatrix} 0 \\ a \end{pmatrix}$

(iii) $af(x)$ is a sketch parallel to the y-axis factor a

(iv) $f(ax)$ is a sketch parallel to the x-axis factor $\frac{1}{a}$

Example 7.8

The function $f(x)$ is represented by the graph shown in Figure 7.14. Elsewhere, $f(x) = 0$.

Sketch the graph of:

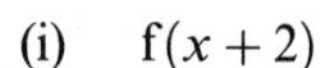

(i) $f(x+2)$

(ii) $f(\frac{1}{2}x)$

(iii) $f(x)+2$

In each case, state the coordinates of A.

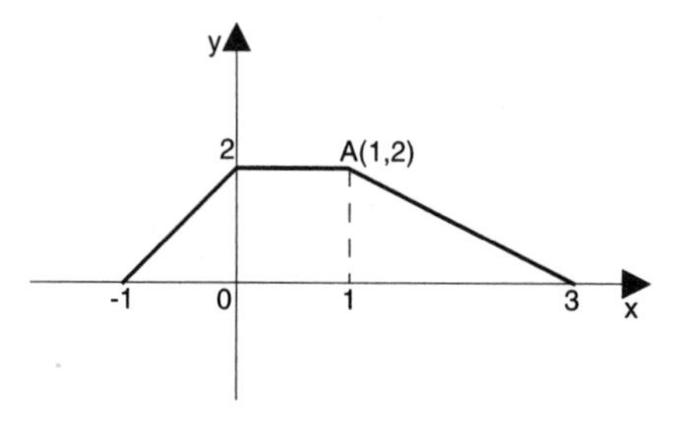

Figure 7.14

Solution

The graphs are shown in Figure 7.15.

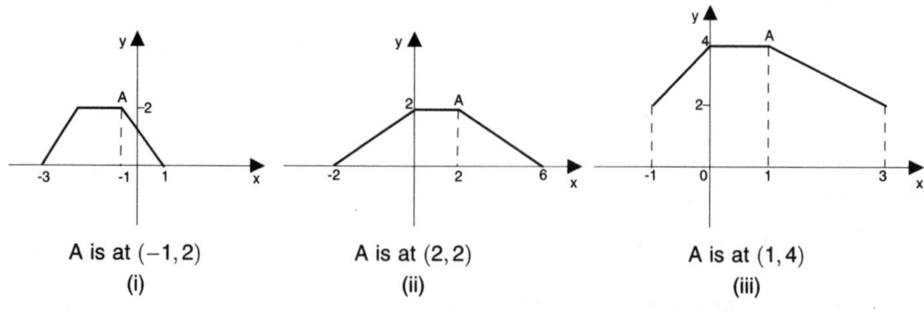

A is at $(-1, 2)$
(i)

A is at $(2, 2)$
(ii)

A is at $(1, 4)$
(iii)

Figure 7.15

Exercise 7(d)

1 For the following functions, sketch the graphs of $f(2x)$, $f(x-2)$, $2f(x)$:

(i) $f(x) = x^3$ (ii) $f(x) = x + 1$

(iii) $f(x) = x^2$ (iv) $f(x) = \frac{1}{x}$

(v) $f(x) = e^x$ (vi) $f(x) = \ln(x+2)$

(vii) $f(x) = |x+1|$ (viii) $f(x) = x^2 - x$

2 The coordinates of three points on a curve $y = f(x)$ are $A\,(1, 1)$, $B\,(2, -1)$ and $C\,(3, 0)$. Find the coordinates of these points after transformations.

(i) $f(x+2)$ (ii) $f(x-1)$ (iii) $3f(x)$ (iv) $f(x) - 2$

7.8 Inequalities

(i) Linear types

The statement $x \leq 4$ tells us that x can be any number less than or equal to 4 (see Figure 7.16). This can be represented on the number line, and clearly, there are an infinite number of possible values of x (solutions).

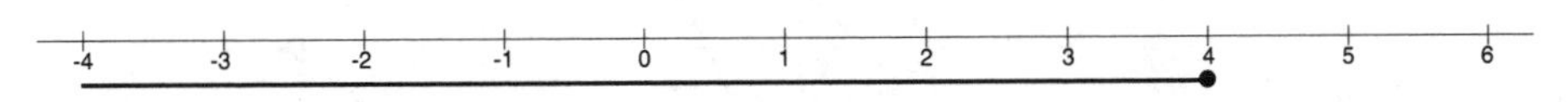

Figure 7.16

The statement $2x + 1 > 4x - 3$ is not quite so obvious. To solve it, proceed in a similar fashion to solving an equation, but with some care:

$$\begin{aligned} \therefore \quad 1 + 3 &> 4x - 2x \\ 4 &> 2x \\ 2 &> x \end{aligned}$$

Hence, x can be any value less than, but not including, two.

Why do we need to take care?

Suppose you had put the x terms on the left:

$$\begin{aligned} 2x - 4x &> -1 - 3 \\ \therefore \quad -2x &> -4 \end{aligned}$$

If you now divide this by -2, you get $x > 2$, which is clearly wrong.

This leads to the rule stated in the following memory jogger box.

> **MEMORY JOGGER**
>
> If you divide or multiply an inequality by a negative number, you must reverse the direction of the inequality sign.

It is worth looking again at the graphical implications of the inequality we have just solved:

$$2x + 1 > 4x - 3$$

On the same diagram, draw the lines $y = 2x + 1$, and $y = 4x - 3$. They intersect at $x = 2$, and it is clear that the line $y = 2x + 1$ is **above** the line $y = 4x - 3$ if $x < 2$ and so $2x + 1 > 4x - 3$ if $x < 2$.

A graphical approach to inequalities is often the best way of understanding what is required (see Figure 7.17).

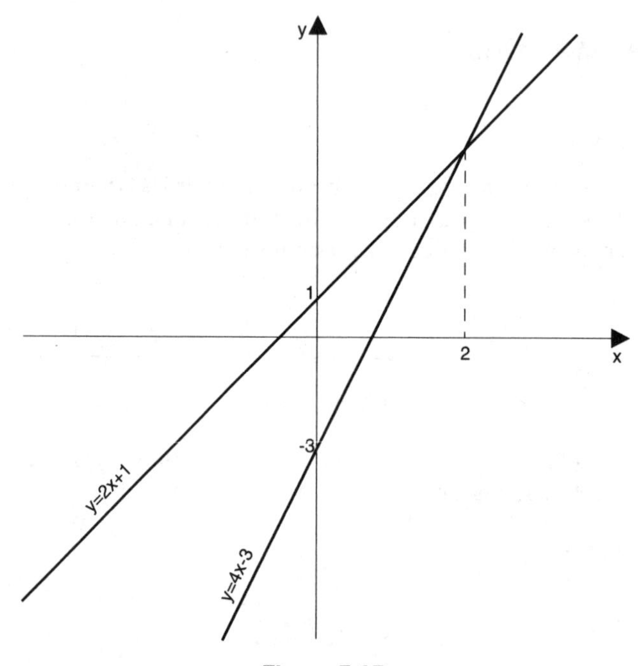

Figure 7.17

(ii) Quadratic types

Consider the inequality $x^2 + x - 6 > 0$. Factorising gives $(x + 3)(x - 2) > 0$. The **critical values** in this expression are where $x = 2$ and $x = -3$ (obtained by solving $(x + 3)(x - 2) = 0$).

Look at the number line in Figure 7.18. It is divided into **three** regions by these critical points.

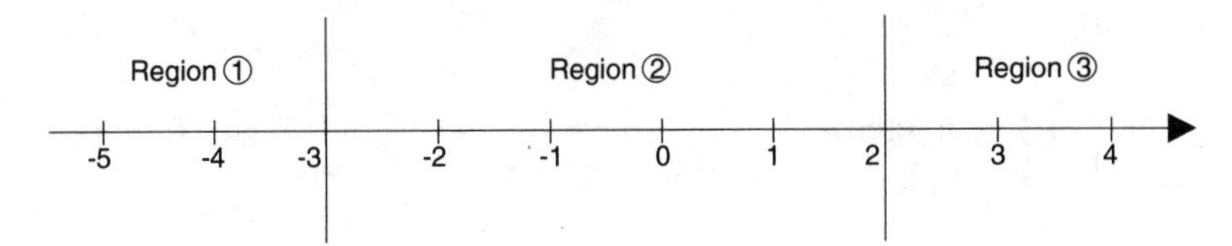

Figure 7.18

Take any value in Region ①, say $x = -5$, and substitute it into the inequality:

$(-5 + 3)(-5 - 2)$ is positive

Take any value in Region ②, say $x = 0$, and substitute this into the inequality:

$(0 + 3)(0 - 2)$ is negative

Choose any value in Region ③, say $x = 4$ and substitute it into the inequality:

$(4 + 3)(4 - 2)$ is positive

We want the inequality to be positive: this occurs in Region ① and Region ③. The solution is $x < -3$ or $x > 2$.

A graphical solution to this problem is much easier to follow. If you draw the graph $y = x^2 + x - 6$ (see Figure 7.19), you can see that $y > 0$ (above the x-axis), if $x > 2$ or $x < -3$ straight away.

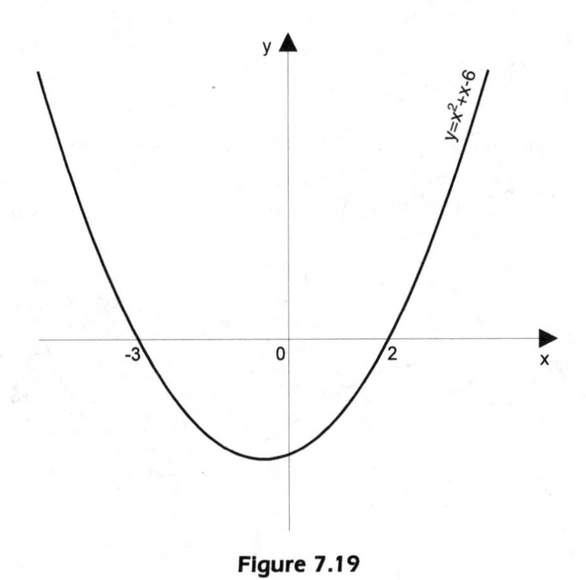

Figure 7.19

(iii) Polynomial types

Any polynomial type inequality can be solved in the same sort of way. Look at the following example.

Example 7.9

Show that $x+4$ is a factor of $2x^3 - x^2 - 41x - 20$ and hence find the set of values of x for which $2x^3 - x^2 - 41x - 20 < 0$.

Solution

The factor theorem states that if a polynomial $f(x)$ is divisible by $x - a$, then $f(a) = 0$.

In this case: $f(-4) = 2 \times (-4)^3 - (-4)^2 - 41 \times -4 - 20 = 0$

Hence: $x - -4$, that is $x + 4$, is a factor

By division:

$$\begin{array}{r l}
 & 2x^2 - 9x - 5 \\
x+4 \,) & 2x^3 - x^2 - 41x - 20 \\
 & \underline{2x^3 + 8x^2} \\
 & -9x^2 - 41x - 20 \\
 & \underline{-9x^2 - 36x} \\
 & -5x - 20 \\
 & \underline{-5x - 20} \\
 & 0
\end{array}$$

Hence $f(x) = (x+4)(2x^2 - 9x - 5)$

$= (x+4)(2x+1)(x-5)$ after factorisation

A sketch of $y = f(x)$ is shown in Figure 7.20.

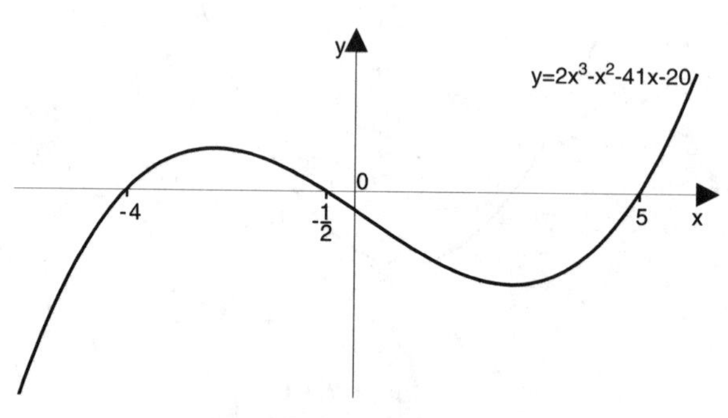

Figure 7.20

We want $f(x) < 0$, where the graph is below the x-axis:

$\therefore \quad x < -4$ or $-\frac{1}{2} < x < 5$

(iv) Rational function types

If an inequality contains the variable x in the denominator, you must make sure you multiply through by a quantity that is always positive (for example, $(x+2)^2$ rather than $x+2$) otherwise the inequality sign may have to be reversed.

Example 7.10

Solve the inequality $x + 1 > \frac{2}{x}$, illustrating your answer with a diagram.

Solution

You cannot multiply both sides by x, because you do not know whether x is positive or negative and hence whether the inequality sign should be reversed. However, x^2 is always positive, so multiply both sides by x^2.

Hence: $\quad x^2(x+1) > 2x$

$\therefore \quad x^3 + x^2 - 2x > 0$

$x(x^2 + x - 2) > 0$

if $\quad E = x(x+2)(x-1) > 0$

Solving $x(x+2)(x-1) = 0$ gives us the critical points $x = -2, 0, 1$.

These divide the number line into four regions (see Figure 7.21).

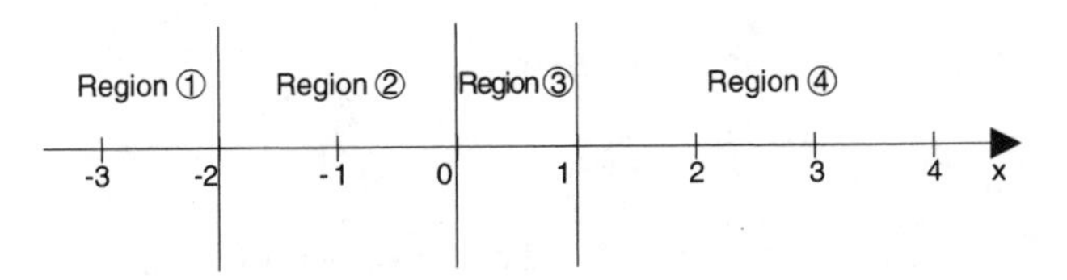

Figure 7.21

Take a value from each region, and substitute into the expression E.

Region ①, $\quad x = -3; \quad E = -12 \quad < 0$

Region ②, $\quad x = -1; \quad E = 2 \quad > 0$

Region ③, $\quad x = \frac{1}{2}; \quad E = -\frac{5}{8} \quad < 0$

Region ④, $\quad x = 2; \quad E = 6 \quad > 0$

We require $E > 0$

$\therefore \quad -2 < x < 0$ or $x > 1$

In order to show this graphically, draw the graphs of $y = \frac{2}{x}$ and $y = x + 1$. The shaded region (see Figure 7.22) shows where the straight line is above the curve, that is, where $x + 1 > \frac{2}{x}$.

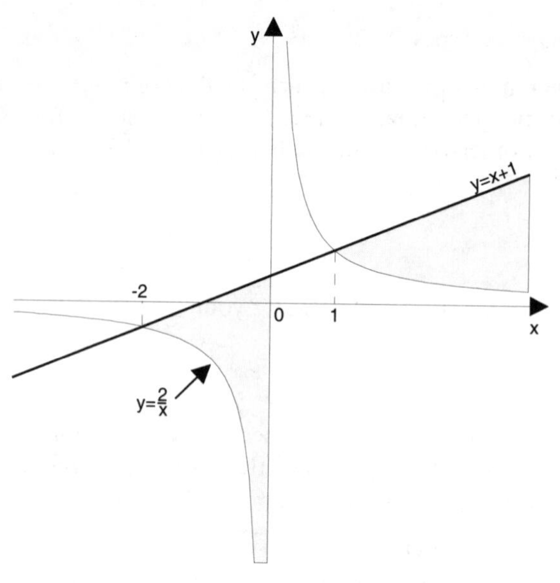

Figure 7.22

Example 7.11

Find the set of real numbers, for which:

$$|3x+1| < |4-2x|$$

Illustrate your answer with a suitable diagram.

Solution

The graphs of $y = |3x+1|$ and $y = |4-2x|$ are shown in Figure 7.23.

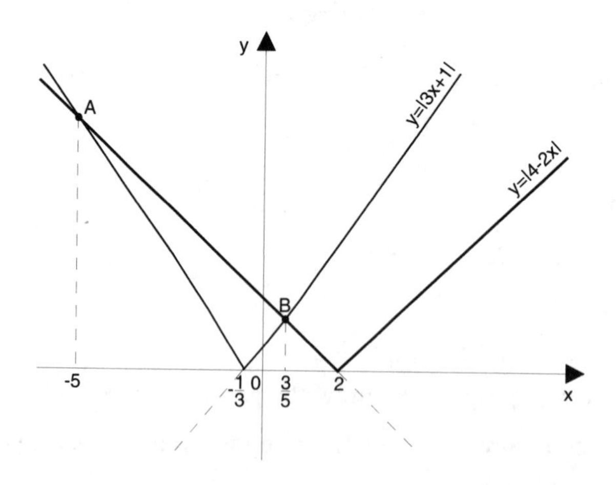

Figure 7.23

In order to find the coordinates of B, solve simultaneously $y = 3x + 1$ with $y = 4 - 2x$.

Eliminating y, $3x + 1 = 4 - 2x$

$\therefore \quad 5x = 3, \; x = \frac{3}{5}$

To find the coordinates of A, solve:

$y = -(3x + 1)$ with $y = 4 - 2x$

↑

The reflection of $y = (3x + 1)$ in the x-axis

$\therefore \quad 4 - 2x = -(3x + 1) = -3x - 1$

$\therefore \quad x = -5$

$|3x + 1| < |4 - 2x|$ when the graph of $y = |3x + 1|$ is below $y = |4 - 2x|$. This happens between A and B. Hence the solution to the problem is $-5 < x < \frac{3}{5}$.

Alternative Method

Because $|x|$ is a positive number, then we can square both sides of the inequality, to get:

$$(3x + 1)^2 < (4 - 2x)^2$$

$$9x^2 + 6x + 1 < 16 + 4x^2 - 16x$$

Hence: $5x^2 + 22x - 15 < 0$

$$(5x - 3)(x + 5) < 0$$

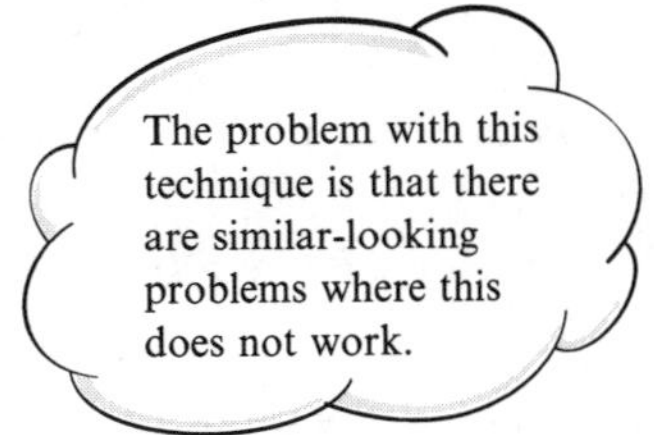

As before, this is solved to give $-5 < x < \frac{3}{5}$

Example 7.12

Solve the inequality:

$1 - x < |x + 2|$

Solution

The inequality is illustrated in Figure 7.24.

To find A, solve $y = 1 - x$ with $y = x + 2$

$\therefore \quad 1 - x = x + 2$

$-1 = 2x$

$x = -\frac{1}{2}$

$y = |x + 2|$ is above $y = 1 - x$, if $x > -\frac{1}{2}$.

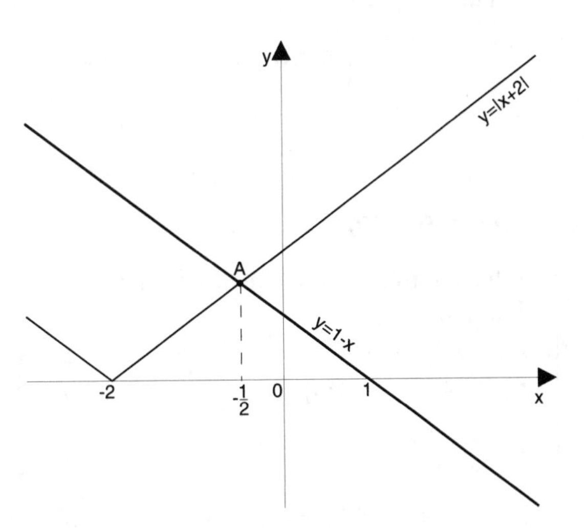

Figure 7.24

It is worth a reminder here that completing the square can be a useful method in the solution of inequalities. Look at the following example.

Example 7.13

Write the expression $3x^2 + 3x + h$ in the form $a(x+b)^2 + c$. Hence find the set of values of h if $3x^2 + 3x + h > 0$ for all real values of x.

Solution

Using the completing the squares technique:

$$\begin{aligned} 3x^2 + 3x + h &= 3\left[x^2 + x + \frac{h}{3}\right] \\ &= 3\left[\left(x + \tfrac{1}{2}\right)^2 + \frac{h}{3} - \tfrac{1}{4}\right] \\ &= 3\left(x + \tfrac{1}{2}\right)^2 + h - \tfrac{3}{4} \end{aligned}$$

We are now required to consider:

$$3\left(x + \tfrac{1}{2}\right)^2 + h - \tfrac{3}{4} > 0$$

Since $3\left(x + \tfrac{1}{2}\right)^2 \geq 0$ for all values of x, then $h - \tfrac{3}{4}$ must also be greater than zero,

that is, $h - \tfrac{3}{4} > 0$

$\therefore \quad h > \tfrac{3}{4}$ is the required condition

Exercise 7(e)

Solve the following inequalities, in each case illustrating your answer with a diagram.

1 $3x + 1 < x - 2$

2 $x^2 + 1 < 5(x - 1)$

3 $2x^2 - x - 6 \leq 0$

4 $|x + 1| < 1 - \tfrac{1}{2}x$

5 $\dfrac{2}{x} < 1$

6 $\dfrac{1}{x+1} > \dfrac{2}{x-3}$

7 $(x - 1)(x + 2)(x + 1) \geq 0$

8 $x^2(x + 1) > 0$

Miscellaneous Examples 7

1 Sketch the function $f : x \mapsto x^2 - 4$. State the range of the function and solve the equation $f(x) = 10$.

State the range of the following functions:

(i)	$f : x \mapsto \ln x$	$x > 2$
(ii)	$g : x \mapsto e^{2x}$	$x \in R$
(iii)	$h : x \mapsto x^3 - x$	$-3 \leq x \leq 3$
(iv)	$k : x \mapsto xe^{-x}$	$x > 0$

3 Find the inverse where it exists of the following functions:

(i) $f : x \mapsto 2x - 3$ (ii) $g : x \mapsto e^{-x}$

(iii) $h : x \mapsto x^2 - 1$ (iv) $k : x \mapsto 3x^3$

4 If $f : x \mapsto x^2 - 1$, and $g : x \mapsto 2x + 1$, find expressions for fg and gf. Is there any solution to the equation $fg(x) = gf(x)$.

5 Solve the inequality $|x + 1| < |2x - 3|$. Illustrate your solution with a diagram.

6 Solve the inequality $\dfrac{1}{x+1} > \dfrac{2}{3-x}$. Illustrate your answer with a diagram.

Revision Problems 7

1 The functions f and g are defined by:

$$f : x \mapsto x^2 - 10, \quad x \in R$$

$$g : x \mapsto |x - 2|, \quad x \in R$$

(a) Show that $f \circ f : x \mapsto x^4 - 20x^2 + 90$, $x \in R$. Find all the values of x for which $f \circ f(x) = 26$.

(b) Show that $g \circ f(x) = |x^2 - 12|$. Sketch the graph of $g \circ f$. Hence, or otherwise, solve the equation: $g \circ f(x) = x$.

(AEB)

2 Figure 7.25 shows a sketch of a graph of $y = f(x)$ for $0 \leq x \leq 7$. Elsewhere, $f(x) = 0$.

On separate axes, sketch the graphs of the functions:

(i) $f(x + 2)$

(ii) $f(\frac{1}{2}x)$

(iii) $1 - f(x)$

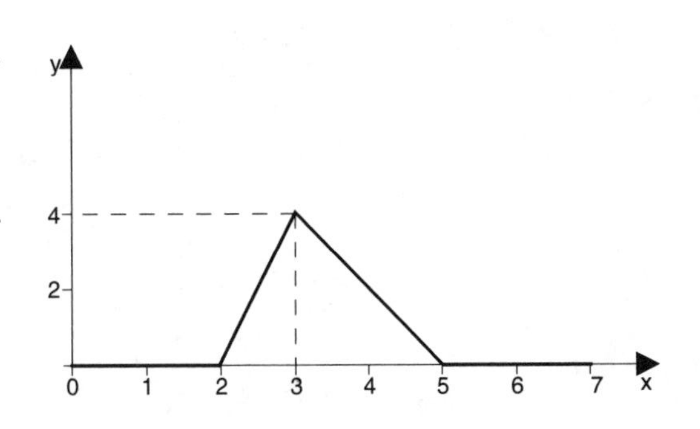

Figure 7.25

3 The functions a, b and c are defined by:

$a : x \mapsto \ln 2x \qquad x > 0$

$b : x \mapsto \dfrac{2}{x} \qquad x > 0$

$c : x \mapsto 3x^2$

(i) Evaluate expressions for ab, and a^{-1}.

(ii) What is the relationship between the graphs of a and a^{-1}?

(iii) The function q is defined by:

$$q : x \mapsto ca(x)$$

Sketch the graph of q^{-1}, and give an explicit expression for $q^{-1}(x)$.

8 The integral calculus

In Chapter 5, you were introduced to the ideas of differentiation. The reverse of this process is known as **integration**. It plays a very important part in mathematics and there are many uses of it in science.

8.1 The basic ideas

We have shown that if you are given a function, such as

$$y = 6x^3 + 2x^2 + 3x$$

then:

$$\frac{dy}{dx} = 18x^2 + 4x + 3$$

We know that this process is called differentiation, and that $18x^2 + 4x + 3$ is the derivative of $6x^3 + 2x^2 + 3x$ with respect to x.

If we reverse this process, we could ask the following question. If the derivative of a function $f(x)$ is $18x^2 + 4x + 3$, what was the original function? The process of finding that function is called **integration**. The sign for integration is $\int$, and dx is used to tell us that the original differentiation process had been with respect to x.

We write this question in mathematics as:

$$\int 18x^2 + 4x + 3\,dx$$

This means integrate (or find the integral of) the function $18x^2 + 4x + 3$ with respect to x. It would be tempting to write that:

$$\int 18x^2 + 4x + 3\,dx = 6x^3 + 2x^2 + 3x$$

Unfortunately, this is not quite correct, as any constant when differentiated gives zero, an unknown constant (called the **constant of integration**) must be added on to the answer.

Hence $\int 18x^2 + 4x + 3\,dx = 6x^3 + 2x^2 + 3x + c$ (c is the constant of integration)

Because we do not quite know exactly what the answer is, this is called an **indefinite integral**. Let us now make the process a little more systematic.

If $y = \frac{1}{4}x^4$ then $\frac{dy}{dx} = \frac{1}{4} \times 4x^3 = x^4$

If $y = \frac{1}{10}x^{10}$ then $\frac{dy}{dx} = \frac{1}{10} \times 10x^9 = x^9$

Hence in order to get an answer of x^n, that is, if $\frac{dy}{dx} = x^n$, you can see that:

$$y = \frac{1}{n+1}x^{n+1} + c$$

that is, $$\int x^n\,dx = \frac{1}{n+1}x^{n+1} + c \qquad \text{(I1)}$$

Note: This rule does not work if $n = -1$.
If you multiply x^n by a constant k, it only multiplies the answer by k.

So: $$\int kx^n\,dx = \frac{k}{n+1}x^{n+1} + c \qquad \text{(I1a)}$$

Example 8.1

Integrate the following functions with respect to x:

(i) $4x^7$ (ii) $\sqrt[3]{x}$ (iii) $\frac{1}{4x^3}$ (iv) $\frac{4}{\sqrt{x}}$

MEMORY JOGGER

Always try to write a simple function in the form kx^n before attempting to integrate.

Solution

(i) $\int 4x^7\,dx = 4 \times \frac{x^8}{8} + c = \frac{1}{2}x^8 + c$

(ii) $\int \sqrt[3]{x}\,dx = \int x^{\frac{1}{3}}\,dx = \frac{x^{\frac{1}{3}+1}}{\frac{4}{3}} + c = \frac{3}{4}x^{\frac{4}{3}} + c$

(iii) $\int \frac{1}{4x^3}\,dx = \int \frac{1}{4}x^{-3} = \frac{1}{4} \times \frac{x^{-3+1}}{-2} + c = \frac{x^{-2}}{-8} + c$

or: $-\frac{1}{8x^2} + c$

Students make a lot of mistakes with this type of function. It is only $\frac{1}{x^3}$ which becomes x^{-3}, because $\frac{1}{4x^3} = \left(\frac{1}{4}\right) \times \left(\frac{1}{x^3}\right) = \frac{x^{-3}}{4}$

(iv) $\int \frac{4}{\sqrt{x}}\,dx = \int 4x^{-\frac{1}{2}}\,dx = 4 \times \frac{x^{\frac{1}{2}}}{\frac{1}{2}} + c = 8x^{\frac{1}{2}} + c$

or: $8\sqrt{x} + c$

If a function consists of a sum or difference of functions, then the integral consists of the sum or difference of the integrals of each function.

Example 8.2

Integrate the following functions with respect to x:

(i) $6x^4 + 7x^2 + 1$ (ii) $\frac{1}{\sqrt{x}} + \frac{2}{x^2}$

Solution

(i) $\int (6x^4 + 7x^2 + 1)\,dx = 6 \times \frac{x^5}{5} + 7 \times \frac{x^3}{3} + 1 \times x + c = \frac{6}{5}x^5 + \frac{7}{3}x^3 + x + c$

(ii) $\int \left(\frac{1}{\sqrt{x}} + \frac{2}{x^2}\right) dx = \int \left(x^{-\frac{1}{2}} + 2x^{-2}\right) dx$

$$= \frac{x^{\frac{1}{2}}}{\frac{1}{2}} + 2 \times \frac{x^{-1}}{-1} + c = 2\sqrt{x} - \frac{2}{x} + c$$

There are no simple rules for integrating products or quotients. You should try and write them in similar ways to the previous questions.

So: $\int (2x^2 + 1)^2$ can be evaluated if you square out the bracket first,

that is, $\int (4x^4 + 4x^2 + 1)\,dx = \frac{4x^5}{5} + \frac{4x^3}{3} + x + c$

$\int\left(\frac{x^4+1}{x^2}\right) dx$ is the same as:

MEMORY JOGGER

$$\frac{a+c}{b}=\frac{a}{b}+\frac{c}{b}$$

$$\int\left(\frac{x^4}{x^2}+\frac{1}{x^2}\right) dx = \int(x^2+x^{-2})\,dx$$

$$=\frac{x^3}{3}+\frac{x^{-1}}{-1}+c=\tfrac{1}{3}x^3-\frac{1}{x}+c$$

As with differentiating, the variable is unimportant. It is the underlying process that is important. Look at the following integral.

Example 8.3

Integrate the function $\frac{(2t+1)^2}{t^4}$ with respect to t. If the integral equals 2 for $t = 1$, evaluate the integral when $t = -1$.

Example

Since the integral is with respect to t, write:

$$\int\frac{(2t+1)^2}{t^4}\,dt$$

Expand the top line first, and then divide by t^4.

So: $$\int\left(\frac{4t^2+4t+1}{t^4}\right) dt = \int\left(\frac{4t^2}{t^4}+\frac{4t}{t^4}+\frac{1}{t^4}\right) dt$$

$$=\int(4t^{-2}+4t^{-3}+t^{-4})\,dt$$

$$=4\times\frac{t^{-1}}{-1}+4\times\frac{t^{-2}}{-2}+\frac{t^{-3}}{-3}+c$$

$$=\frac{-4}{t}-\frac{2}{t^2}-\frac{1}{3t^3}+c$$

We shall refer to this answer as $f(t)$.

So: $$f(t)=-\frac{4}{t}-\frac{2}{t^2}-\frac{1}{3t^3}+c$$

Now we know that $f(1) = 2$

So: $2=-4-2-\tfrac{1}{3}+c$, hence: $c=8\tfrac{1}{3}$

Hence: $$f(t)=-\frac{4}{t}-\frac{2}{t^2}-\frac{1}{3t^3}+8\tfrac{1}{3}$$

We can now find $f(-1)=4-2+\tfrac{1}{3}+8\tfrac{1}{3}=10\tfrac{2}{3}$.

Exercise 8(a)

Integrate the following functions with respect to x, simplifying your answers where possible.

1 x^4	**2** $4x^5$	**3** $\dfrac{3}{x^2}$
4 $2\sqrt{x}$	**5** $x+\dfrac{1}{x^2}$	**6** $\dfrac{x+1}{x^3}$
7 $(x-1)^4$	**8** $\left(x+\dfrac{1}{x}\right)^2$	**9** $\dfrac{4}{\sqrt[3]{x}}$
10 $\sqrt{2x}$	**11** $\dfrac{(x+1)^2}{x^4}$	**12** $\dfrac{3}{\sqrt{x^3}}$
13 $\dfrac{2}{\sqrt[4]{x}}$	**14** $7x^3+6x^2-1$	**15** $x^3+\dfrac{2}{x^2}$
16 $7\sqrt{x^7}$	**17** $\sqrt{x}+\dfrac{1}{\sqrt{x}}$	**18** $\dfrac{(\sqrt{x}+1)^2}{\sqrt{x}}$
19 $(2x-1)^3$	**20** $(ax+b)^2$ where a and b are constants.	

8.2 Function derivative rule

There is one rule in integration that is extremely useful and is worth making a special effort to understand. It looks complicated to start with, but if you follow the worked examples carefully, you should not have too many problems.

If: $$\int \mathrm{f}(x)\,\mathrm{d}x = \mathrm{g}(x)$$

then: $$\int \mathrm{k}\mathrm{h}'(x)\mathrm{f}(\mathrm{h}(x))\,\mathrm{d}x = \mathrm{k}\mathrm{g}(\mathrm{h}(x)) + c \qquad \text{(I2)}$$

where $$\mathrm{h}'(x) = \frac{\mathrm{dh}}{\mathrm{d}x}$$

In the particular case where $\mathrm{h}(x) = ax + b$ (a linear function), then the rule becomes:

$$\int \mathrm{f}(ax+b)\,\mathrm{d}x = \frac{1}{a}\mathrm{g}(ax+b) \qquad \text{(I2a)}$$

(i) Consider $\int 3x^2(x^3+1)^9\,dx$

since $\dfrac{d}{dx}(x^3+1) = 3x^2$, then $h(x) = x^3+1$

and $f(x) = x^9$, giving $f(h(x)) = (x^3+1)^9$

So: $$\int 3x^2(x^3+1)^9\,dx = \frac{(x^3+1)^{10}}{10} + c$$

This is simply the reverse of the chain rule (see Section 5.2).

> **MEMORY JOGGER**
> The function derivative integration rule is the reverse of the chain rule.

(ii) Consider $\int 4x\sqrt{1+x^2}\,dx$

Now $\dfrac{d}{dx}(1+x^2) = 2x$, so $4x = 2 \times 2x$

hence $k = 2$ in Equation I2

So: $$\int 4x\sqrt{1+x^2}\,dx = 2 \times \frac{(1+x^2)^{\frac{3}{2}}}{\frac{3}{2}} = \tfrac{4}{3}(1+x^2)^{\frac{3}{2}}$$

(iii) $$\int \frac{5}{(1+2x)^5}\,dx$$

$$= \int 5(1+2x)^{-5}\,dx = 5 \times \frac{(1+2x)^{-4}}{-4} \div 2$$

$$\uparrow \qquad\qquad\qquad\qquad \uparrow$$

$$a = 2 \qquad\qquad\qquad\qquad \div a$$

$$= -\tfrac{5}{8}(1+2x)^{-4} \quad \text{or} \quad \frac{-5}{8(1+2x)^{-4}}$$

This rule is really a special case of the method of substitution, which is shown in full in Chapter 15.

Example 8.4

Evaluate the integrals of the following functions with respect to x, using the function derivative rule.

(i) $5x^2(x^3+4)^8$ (ii) $\dfrac{5x}{\sqrt{1+x^2}}$ (iii) $\dfrac{4x+2}{(x^2+x+1)^3}$

Solution

(i) $\dfrac{d}{dx}(x^3+4)=3x^2$

Since $5x^2=\frac{5}{3}\times 3x^2$

then: $\displaystyle\int 5x^2(x^3+4)^8\,dx=\frac{5}{3}\times\frac{(x^3+4)^9}{9}+c$

$$=\frac{5(x^3+4)^9}{27}+c$$

(ii) $\dfrac{d}{dx}(1+x^2)=2x$

Here: $5x=\frac{5}{2}\times 2x$

So: $\displaystyle\int\frac{5x}{\sqrt{1+x^2}}\,dx=\int 5x(1+x^2)^{-\frac{1}{2}}\,dx$

$$=\frac{5}{2}\times\frac{(1+x^2)^{\frac{1}{2}}}{\frac{1}{2}}=5(1+x^2)^{\frac{1}{2}}+c$$

(iii) $\dfrac{d}{dx}(x^2+x+1)=2x+1$

Now $4x+2=2\times(2x+1)$

So: $\displaystyle\int\frac{4x+2}{(x^2+x+1)^3}\,dx=\int 2\times(2x+1)(x^2+x+1)^{-3}\,dx$

$$=\frac{2(x^2+x+1)^{-2}}{-2}=\frac{-1}{(x^2+x+1)^2}+c$$

Exercise 8(b)

Evaluate the following integrals using the function derivative rule:

1 $3x^2(1+x^3)^5$ **2** $(4x+1)^9$ **3** $\dfrac{3x^2}{(x^3-1)^2}$

4 $\dfrac{2x+1}{\sqrt{x^2+x-5}}$ **5** $\dfrac{1}{(3x-1)^{\frac{2}{3}}}$ **6** $x\sqrt{1+2x^2}$

7 $\dfrac{3x^2}{(1+x^3)^4}$ **8** $\dfrac{3x-6}{(x^2-4x+1)^3}$ **9** $5x(1-3x^2)^4$

10 $\sqrt{x}\left(1+x^{\frac{3}{2}}\right)^3$

8.3 Definite integration

If we know $\int g(x)\,dx = f(x)$, then $\int_a^b g(x)\,dx$ is defined as $f(b) - f(a)$ and is called the **definite integral** between a and b of the function $g(x)$. The answer does not contain x.

For example, we know $\int 3x^2 + 4x\,dx$

Hence: to find $\int_1^2 3x^2 + 4x\,dx = [x^3 + 2x^2 + c]_{\text{at } x=2}$

$$- [x^3 + 2x^2 + c]_{\text{at } x=1}$$

This is written $[x^3 + 2x^2 + c]_1^2$ to avoid writing the same function twice.

$$= [8 + 8 + c] - [1 + 2 + c] = 13$$

You will notice that $+c$ cancels out, and so in future, in definite integration, the $+c$ will be omitted. The numbers at the top and bottom of the integral sign are called the **limits**: the number at the bottom is called the **lower limit**, and the number at the top the **upper limit**. **Note:** there is no reason why the upper limit should be numerically greater than the lower limit (although in practice it usually is).

Example 8.5

Evaluate:

(i) $\int_1^4 \frac{1}{4\sqrt{x}}\,dx$ (ii) $\int_{-2}^{-1} \left(\frac{t^4 + 1}{t^2}\right) dt$ (iii) $\int_4^1 (t^2 - 3t)\,dt$

Solution

(i) $\int_1^4 \frac{1}{4\sqrt{x}}\,dx = \int_1^4 \frac{1}{4}x^{-\frac{1}{2}}\,dx = \left[\frac{1}{2}x^{\frac{1}{2}}\right]_1^4$

$$= \left[\tfrac{1}{2} \times 2\right] - \left[\tfrac{1}{2} \times 1\right] = \tfrac{1}{2}$$

(ii) The integral is with respect to t, and so the limits refer to values of t.

$$\int_{-2}^{-1} \left(\frac{t^4 + 1}{t^2}\right) dt = \int_{-2}^{-1} (t^2 + t^{-2})\,dt$$

$$= \left[\frac{t^3}{3} - t^{-1}\right]_{-2}^{-1} = \left[-\tfrac{1}{3} + 1\right] - \left[-\tfrac{8}{3} + \tfrac{1}{2}\right] = \tfrac{2}{3} - \tfrac{-13}{6} = \tfrac{17}{6} = 2\tfrac{5}{6}$$

(iii) Here, you will see that the upper limit is numerically less than the lower limit. This does not make any difference, and the integral is carried out in the conventional way:

$$\therefore \quad \int_4^1 (t^2 - 3t)\,dt = \left[\frac{t^3}{3} - \frac{3t^2}{2}\right]_4^1$$

$$= \left[\tfrac{1}{3} - \tfrac{3}{2}\right] - \left[\tfrac{64}{3} - 24\right] = 1\tfrac{1}{2}$$

MEMORY JOGGER

Note: $\int_a^b f(x)\,dx = -\int_b^a f(x)\,dx$

Exercise 8(c)

Evaluate the following definite integrals:

1 $\int_1^2 x^2\,dx$ **2** $\int_1^4 \sqrt{x}\,dx$ **3** $\int_{-1}^{-2} x + 2\,dx$

4 $\int_4^9 \frac{1}{\sqrt{x}}\,dx$ **5** $\int_0^2 t^2 - 1\,dt$ **6** $\int_0^1 (2x+1)^3\,dx$

7 $\int_{\frac{1}{2}}^1 \left(x + \frac{1}{x}\right)^2 dx$ **8** $\int_1^2 \frac{x^2+1}{x^2}\,dx$ **9** $\int_1^2 \frac{(x+1)^2}{x^4}\,dx$

10 $\int_1^2 1\,d\theta$ **11** $\int_0^1 x\sqrt{1+x^2}\,dx$ **12** $\int_0^1 x^2(1+x^3)^4\,dx$

13 $\int_2^3 \frac{x}{(1-x^2)^2}\,dx$ **14** $\int_2^1 \frac{4}{3\sqrt{x+1}}\,dx$ **15** $\int_0^1 \frac{A^4}{\sqrt{1+A^5}}\,dA$

8.4 Understanding the definite integral (area and volume)

(i) Area

Figure 8.1 shows part of the graph of $y = f(x)$ between $x = a$ and $x = b$.

Suppose we want to find the area A between the curve, the x-axis and the lines $x = a$ and $x = b$. The area A is divided up into strips of width δx and height y (these are *approximately* rectangles).

The area of each small strip $\delta A = y\delta x$ approximately,

$$\therefore \quad \frac{\delta A}{dx} \approx y$$

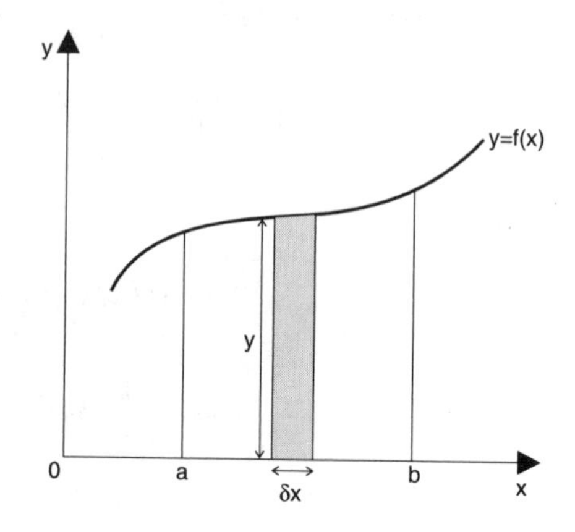

Figure 8.1

The closer δx gets to zero, the nearer each strip becomes to a rectangle and the better the approximation gets.

$$\text{Hence:} \quad \frac{dA}{dx} = y$$

Now $y = f(x)$, and so:

$$\frac{dA}{dx} = f(x)$$

$$\therefore \quad A = \int f(x)\,dx$$

Let the integral be

$$A = g(x), \text{ say}$$

Now if $x = a$, $\quad A = 0 \qquad \therefore \quad g(a) = 0$

$\quad$ if $x = b$, $\quad A =$ the required area

Hence: $\qquad A = g(b)$

But because $g(a) = 0$, we can write

$$A = g(b) - g(a)$$

This is simply $\displaystyle\int_a^b f(x)\,dx$, usually written $\displaystyle\int_a^b y\,dx$ (I3)

In other words, the definite integral gives us the area under the graph. It must be used with care.

MEMORY JOGGER

Always check that a graph does not cross the axis between the two limits in a definite integral

Example 8.5

Find the area enclosed between the curve $y = 4 - x^2$, and the x-axis.

Solution

A sketch is required (see Figure 8.2), otherwise you cannot see that the limits are -2 and 2. Hence:

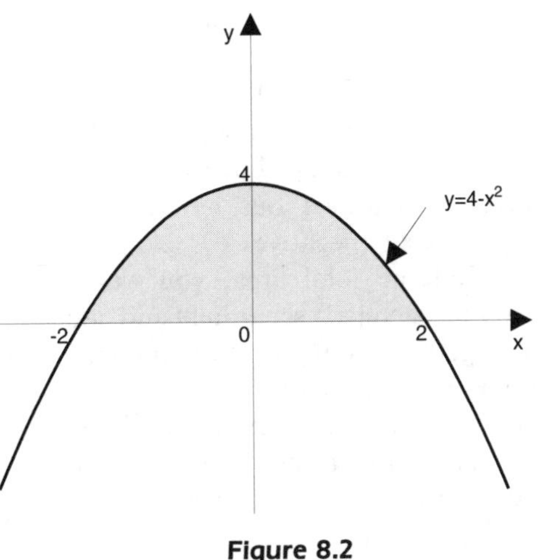

Figure 8.2

$$\text{area} = \int_{-2}^{2} (4 - x^2)\,dx$$

$$= \left[4x - \frac{x^3}{3}\right]_{-2}^{2}$$

$$= \left[8 - \tfrac{8}{3}\right] - \left[-8 + \tfrac{8}{3}\right]$$

$$= 10\tfrac{2}{3}\ \text{unit}^2 \quad (1 \text{ square measures } 1 \text{ unit} \times 1 \text{ unit})$$

Example 8.6

Find the area enclosed by the curve $y = x^3$, the x-axis and the lines $x = -3$ and $x = -1$.

Solution

See Figure 8.3.

$$\text{Area} = \int_{-3}^{-1} x^3\,dx$$

$$= \left[\frac{x^4}{4}\right]_{-3}^{-1} = \tfrac{1}{4} - \tfrac{81}{4} = -20$$

The area is negative because it is below the x-axis. We should say the area $= 20$ unit^2.

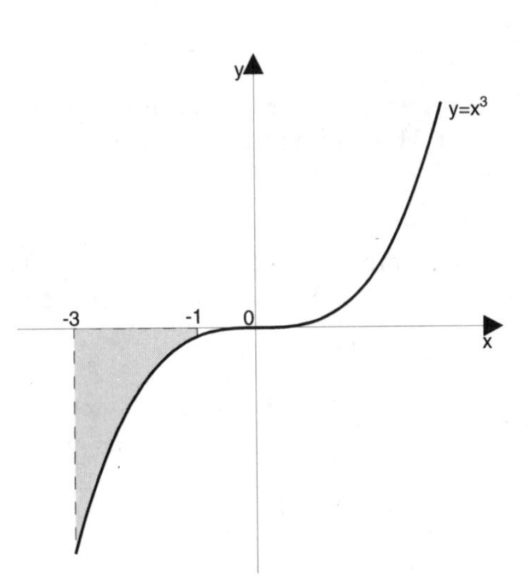

Figure 8.3

Example 8.7

Evaluate $\int_{-1}^{1} x - x^3\,dx$ and explain your answers geometrically.

Solution

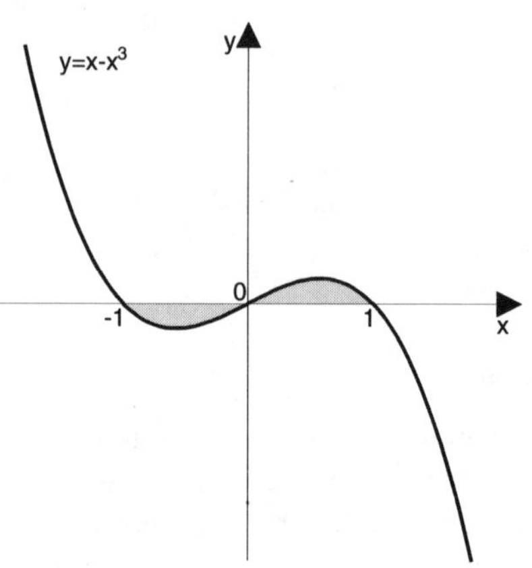

Figure 8.4

$$\int_{-1}^{1} x - x^3\,dx = \left[\frac{x^2}{2} - \frac{x^4}{4}\right]_{-1}^{1}$$

$$= \left[\tfrac{1}{2} - \tfrac{1}{4}\right] - \left[\tfrac{1}{2} - \tfrac{1}{4}\right] = 0$$

The area shown in Figure 8.4 is clearly not zero, but the area above the x-axis (positive) is cancelled out by the area below the x-axis (negative).

To find the total area, you would have to find each part separately and add them together. You would find the answer $= 2 \times \frac{1}{4} = \frac{1}{2}$ unit^2.

If you want to find the area enclosed between two functions $y_1 = f(x)$ and $y_2 = g(x)$, it is easier to do this as one integral:

$$\textbf{Area between curves} = \int_a^b (y_2 - y_1)\,dx \qquad \text{(I4)}$$

Here it is assumed that y_2 is always above y_1. If you get y_2 and y_1 the wrong way round, the answer will become negative.

Example 8.8

Find the area enclosed between the graphs $y = 1 + 3x - 2x^2$ and $y = x + 1$.

Solution

Clearly, both lines pass through $A\,(0, 1)$ (see Figure 8.5). To find the other point B:

$$1 + 3x - 2x^2 = 1 + x$$

$$\therefore\quad 2x = 2x^2, \text{ that is, } 2x(1 - x) = 0$$

$$\therefore\quad x = 0 \text{ at } A$$

$$x = 1 \text{ at } B$$

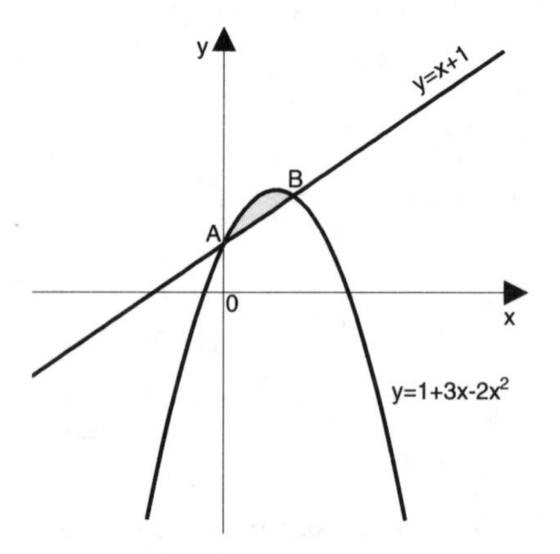

Figure 8.5

$$\text{Hence area} = \int_0^1 (1 + 3x - 2x^2) - (1 + x)\,dx$$

$$= \int_0^1 (2x - 2x^2)\,dx$$

$$= \left[x^2 - \tfrac{2}{3}x^3\right]_0^1 = \tfrac{1}{3}\text{ unit}^2$$

MEMORY JOGGER

It is worth simplifying the function to be integrated if possible, as it reduces the number of places where numerical mistakes can be made.

Example 8.9

Evaluate: $\int_1^8 \left(x^{\frac{2}{3}} - x^{-\frac{2}{3}}\right) dx$. Indicate in a diagram the area that this integral represents.

Solution

The graphs of $y = x^{\frac{2}{3}}$ and $y = x^{-\frac{2}{3}}$ are shown in Figure 8.6. They cross when $x = 1$. The integral represents the shaded area.

$$\int_1^8 \left(x^{\frac{2}{3}} - x^{-\frac{2}{3}}\right) dx = \left[\tfrac{3}{5}x^{\frac{5}{3}} - 3x^{\frac{1}{3}}\right]_1^8$$

$$= \left[\tfrac{3}{5} \times 8^{\frac{5}{3}} - 3 \times 8^{\frac{1}{3}}\right] - \left[\tfrac{3}{5} - 3\right]$$

$$= 13.2 - -2.4 = 15.6\text{ unit}^2$$

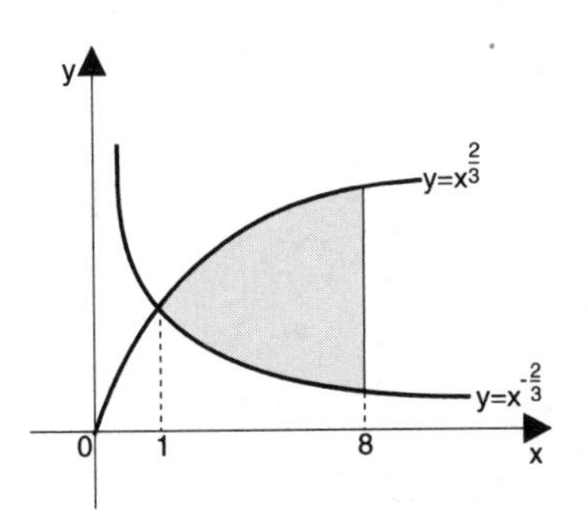

Figure 8.6

Example 8.10

Find the area enclosed between the graphs of $y = x^2 - 4$ and $y = 2x - 1$.

Solution

This is an excellent example to illustrate that you do not have to worry about where the curve is in relationship to the x-axis for this type of problem.

To find the coordinates of A and B, solve:

$$x^2 - 4 = 2x - 1$$

$$\therefore\ x^2 - 2x - 3 = 0$$

$$(x - 3)(x + 1) = 0$$

Hence $x = 3$ or -1 (see Figure 8.7).

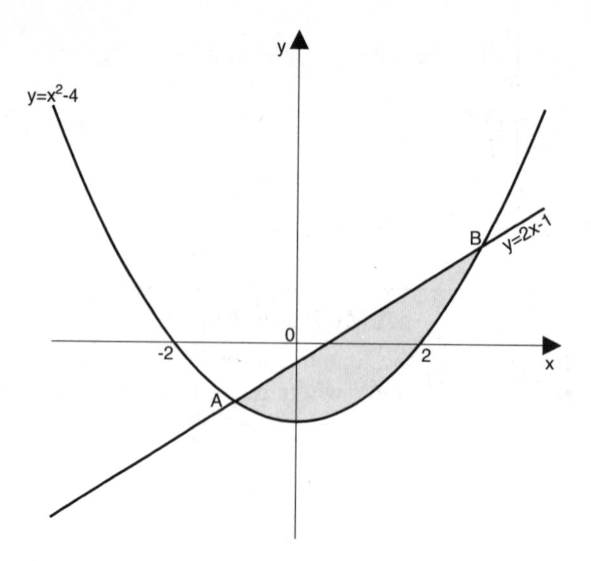

Figure 8.7

$$\text{The required area} = \int_{-1}^{3} (2x-1) - \left(x^2-4\right) \mathrm{d}x$$

$$= \int_{-1}^{3} \left(2x+3-x^2\right) \mathrm{d}x$$

$$= \left[x^2+3x-\frac{x^3}{3}\right]_{-1}^{3} = [9+9-9] - \left[1-3+\tfrac{1}{3}\right]$$

$$= 10\tfrac{2}{3} \text{ units}^2$$

If the area required is partly enclosed by the **y-axis**, then the following formula is required.

The area enclosed between the curve $y = \mathrm{f}(x)$, the lines $y = a$, $y = b$ and the y-axis, is given by:

$$\text{Area} = \int_{a}^{b} x \, \mathrm{d}y \qquad \text{(I5)}$$

This means rearranging the formula to make x the subject. If this produces a function you cannot easily integrate, you may need to find the area in a different way.

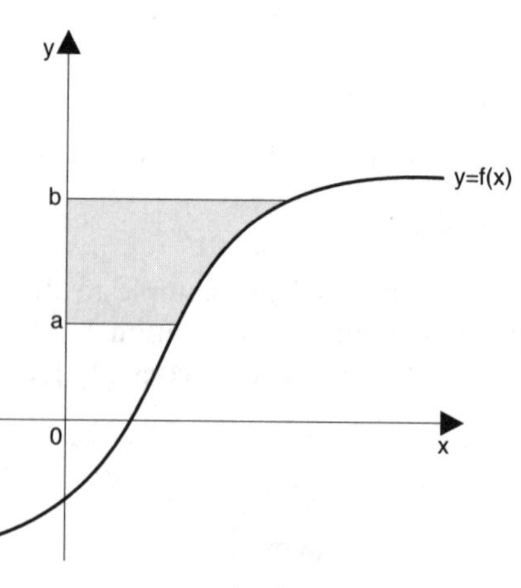

Figure 8.8

Example 8.11

Find the area of the region defined by the inequalities $x \geq 0$, $y \geq (x+1)^3$, $y \leq 8$.

Solution

The region is shown in Figure 8.9. First you have to make x the subject of the equation $y = (x+1)^3$.

So $y^{\frac{1}{3}} = x + 1$, $x = y^{\frac{1}{3}} - 1$

Note: The lower y limit is $y = 1$.

$$\text{The area} = \int_1^8 \left(y^{\frac{1}{3}} - 1\right) \mathrm{d}y$$

(using Equation I5)

$$= \left[\tfrac{3}{4} y^{\frac{4}{3}} - y\right]_1^8$$

$$= \left[\tfrac{3}{4} \times 8^{\frac{4}{3}} - 8\right] - \left[\tfrac{3}{4} \times 1^{\frac{4}{3}} - 1\right]$$

$$= 4 - -\tfrac{1}{4}$$

$$= 4\tfrac{1}{4} \text{ unit}^2$$

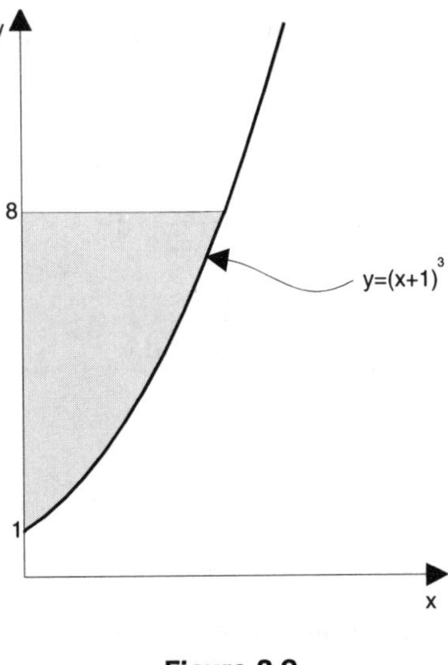

Figure 8.9

(ii) Volume

If the line $y = 2x$ $(0 \leq x \leq 3)$ is rotated by 360° about the x-axis, a **solid of revolution** is produced (in this case a cone) – see Figure 8.10.

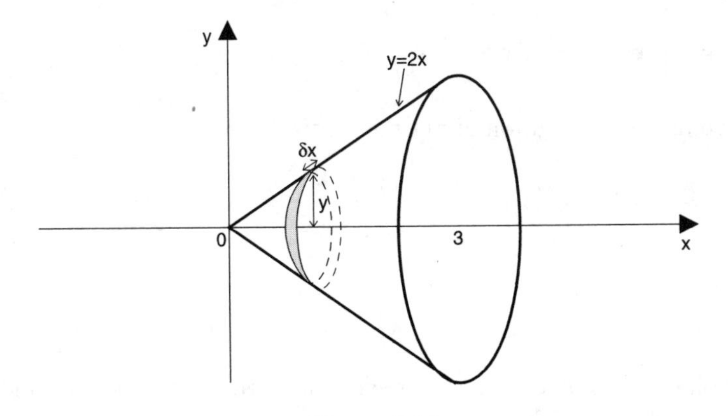

Figure 8.10

You can visualise that the cone is made up of a large number of discs of thickness δx, and radius y.

The volume of each disc $= \pi y^2 \delta x$ (using $\pi r^2 h$)

The total volume $= \sum \pi y^2 \delta x$

As $\delta x \to 0$, this becomes an integral

Hence, volume of revolution $= \int_a^b \pi y^2 \, dx$ (I6)

In the case of the cone,

$$\begin{aligned} \text{volume} &= \int_0^3 \pi(2x)^2 \, dx \\ &= \pi \int_0^3 4x^2 \, dx \\ &= \pi \left[\frac{4x^3}{3}\right]_0^3 \\ &= 36\pi \text{ unit}^3 \end{aligned}$$

If an area enclosed by a curve and the y-axis is rotated by 360° about the **y-axis**, the volume of revolution

$$= \int_a^b \pi x^2 \, dy \qquad \text{(I7)}$$

If an area **between two curves** is rotated, we have the formula:

(i) volume of revolution about the x-axis

$$= \pi \int_a^b (y_2{}^2 - y_1{}^2) \, dx \qquad \text{(I8)}$$

(ii) volume of revolution about the y-axis

$$= \pi \int_a^b (x_2{}^2 - x_1{}^2) \, dx \qquad \text{(I9)}$$

Example 8.12

Find the volume generated, when the region enclosed between $y = x^2$ and $y^2 = x$ is rotated by 360° about the y-axis.

Solution

The point of intersection at A is $(1, 1)$ (see Figure 8.11).

Now $x_2 = \sqrt{y}$ and $x_1 = y^2$,

$$\therefore \quad \text{the volume} = \pi \int_0^1 \left(\left(\sqrt{y}\right)^2 - \left(y^2\right)^2 \right) \mathrm{d}y$$

$$= \pi \int_0^1 \left(y - y^4\right) \mathrm{d}y$$

$$= \pi \left[\frac{y^2}{2} - \frac{y^5}{5} \right]_0^1$$

$$= \frac{3\pi}{10} \text{ unit}^3$$

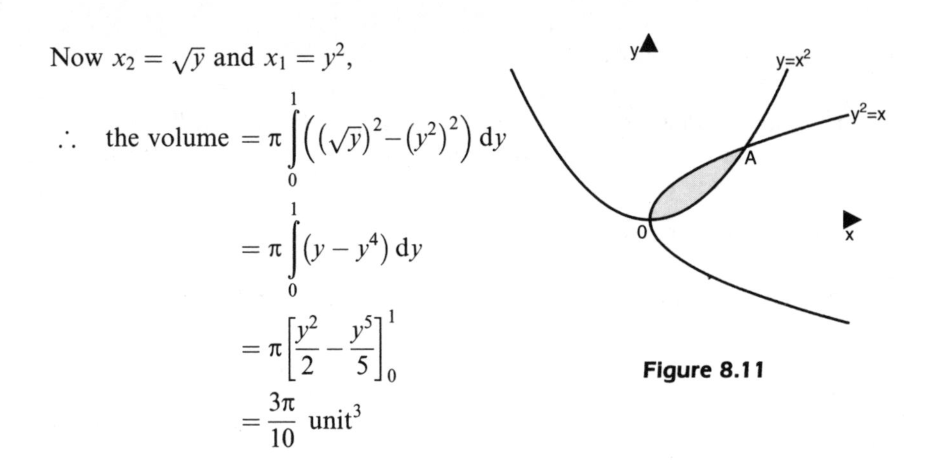

Figure 8.11

Exercise 8(d)

1 Find the area enclosed between the x-axis and:

(i) $y = x^2$; $x = 1$; $x = 2$
(ii) $y = 4 - x^2$
(iii) $y = x^3 + 4x - 1$; $x = 3$; $x = 4$
(iv) $y = \dfrac{1}{x^2}$; $x = -4$; $x = -2$

2 Find the area enclosed between the y-axis and:

(i) $y = x^2$; $y = 1$; $y = 2$
(ii) $y = x$; $y = 2 - x$
(iii) $y = x^3$; $y = 3$; $y = 4$
(iv) $y^2 = x + 3$

3 Find the area enclosed between the lines $y = 2x^2$ and $y = 6x$.
4 Find the area enclosed between the lines $y = x + 2$ and $y = x^2 - x - 6$.
5 Find the area enclosed between the lines $y = x^3$ and $y = 2x$.
6 Find the volume generated when the areas in question 1 are rotated by 360° about the x-axis.
7 Find the volume generated when the regions in question 2 are rotated by 360° about the y-axis.

8.5 Distance, speed, acceleration

In Section 5.10, we introduced the use of $\dfrac{\mathrm{d}s}{\mathrm{d}t}$ and $\dfrac{\mathrm{d}^2 s}{\mathrm{d}t^2}$. If the question involves integration, you must remember to include the constant of integration, which should be evaluated subject to the conditions of the problem.

Example 8.12

A particle moves in a straight line through O in the direction OR. At time $t = 0$, the particle is at rest at O. When it reaches R it stops, and the acceleration of the particle at time t is $12 - 4t$.

(a) Find t when the particle reaches R.
(b) Find the distance OR.
(c) Sketch the displacement–time and velocity–time graphs.

Solution

(a) If v is the velocity, then:

$$\text{acceleration } = \frac{dv}{dt} = 12 - 4t$$

$$\text{integrate } \quad v = 12t - 2t^2 + c$$

To find c, use the fact that when $t = 0$, $v = 0$.

$$\text{Hence:} \quad 0 = 0 - 0 + c \qquad \therefore \quad c = 0$$

$$\text{So:} \qquad v = 12t - 2t^2$$

at R, $v = 0$ (it has stopped),

$$\therefore \quad 0 = 12t - 2t^2 = 2t(6 - t) \qquad \therefore \quad t = 6 \text{ or } 0$$

But $t = 0$ at 0, hence $t = 6$.

(b) Now $v = \frac{ds}{dt}$

$$\text{Hence:} \quad \frac{ds}{dt} = 12t - 2t^2$$

$$\therefore \quad s = 6t^2 - \frac{2t^3}{3} + c_1 \quad \text{(a different constant is needed)}$$

Again, $s = 0$, when $t = 0$, hence $c_1 = 0$,

$$\therefore \quad s = 6t^2 - \frac{2t^3}{3}$$

To find OR, we know that $t = 6$.

$$\text{Hence:} \quad s = 6 \times 6^2 - \tfrac{2}{3} \times 6^3$$
$$= 72$$

Hence $OR = 72$ metres.

(c) The graphs are shown in Figure 8.12.

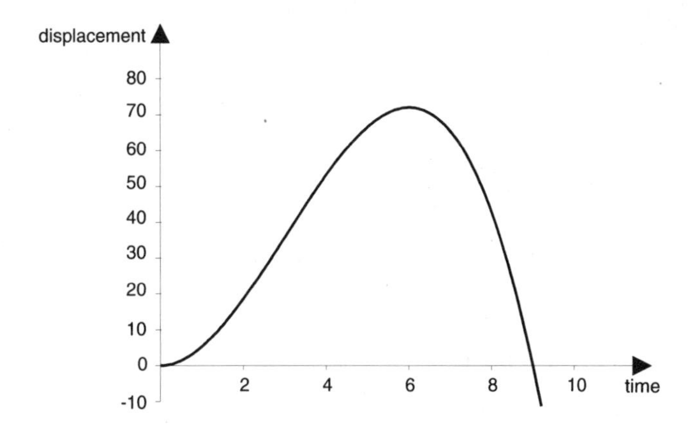

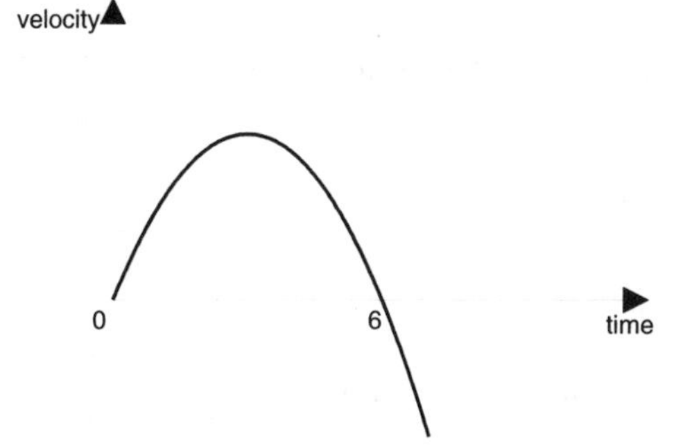

Figure 8.12

Example 8.13

A lift takes 30 seconds to travel from ground level to the top floor of a building. When the lift has been moving for t seconds, its speed v m/s is given by the equation:

$$v = 2t\left(1 - \frac{1}{900}t^2\right)$$

(a) Find the distance between ground floor level and top floor level.

(b) Plot an acceleration time graph for the motion.

Solution

(a) Since $$v = \frac{\mathrm{d}s}{\mathrm{d}t} = 2t\left(1 - \frac{1}{900}t^2\right)$$

$$= 2t - \frac{t^3}{450}$$

Integrating $$s = t^2 - \frac{t^4}{1800} + c$$

Now when $t = 0$, $s = 0$ $\quad\therefore c$ must be zero.

Hence $$s = t^2 - \frac{t^4}{1800}$$

when $t = 30$, $$s = 30^2 - \frac{30}{1800}$$

$$= 450$$

The distance between the floors is 450 metres.

$$\text{acceleration} = \frac{\mathrm{d}v}{\mathrm{d}t}$$

$$\therefore \quad a = 2 - \frac{3t^2}{450} = 2 - \frac{t^2}{150}$$

The acceleration time graph is shown in Figure 8.13.

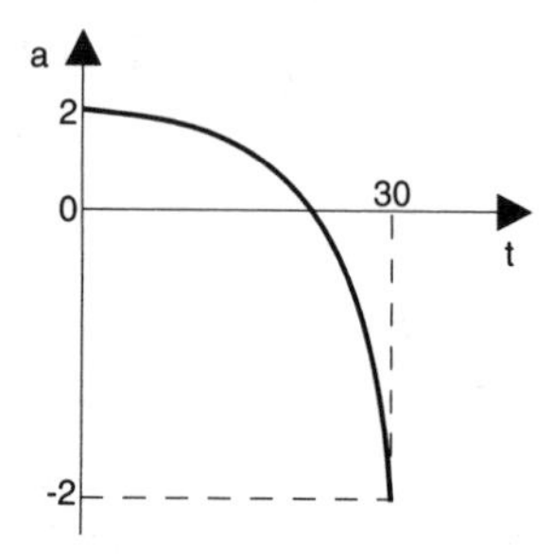

Figure 8.13

Exercise 8(e)

1 A body travels in a straight line passing O at time $t = 0$ seconds, and moving towards a point R. The acceleration of the body at time t is given by $\frac{\mathrm{d}^2s}{\mathrm{d}t^2} = 6 - 2t$ m/s^2. As the body passes O, its speed is measured at 4 m/s.

(i) When does the body come to rest?

(ii) Find how far the body has moved from O when it comes to rest.

2 The velocity v m/s of a small object is given by the formula: $v = t^4 + 1$.

(i) Find how far the object travels in the third second of its motion.

(ii) Find the acceleration of the body after 2 seconds.

8.6 Numerical integration

(i) The trapezium rule

When an integral cannot be evaluated exactly, a numerical technique can be used. To find the approximate area under $y = f(x)$, divide the graph into strips of width h, and join the points (a, y_0), $(a + h, y_1)$ etc. on the curve by straight lines, giving a number of trapezia (see Figure 8.14).

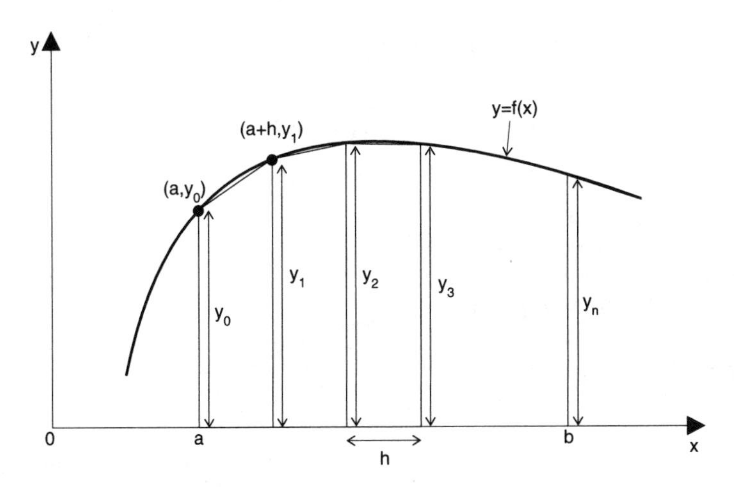

Figure 8.14

$$\text{The area of the trapezia} = \frac{h}{2}[y_0 + y_1] + \frac{h}{2}[y_1 + y_2] + \ldots + \frac{h}{2}[y_{n-1} + y_n]$$
$$= \frac{h}{2}[y_0 + 2y_1 + 2y_2 + \ldots + 2y_{n-1} + y_n]$$

Apart from a small space at the top, this must be approximately the area under the graph.

$$\text{Hence} \quad \int_a^b f(x)\,dx \approx \frac{h}{2}[y_0 + 2y_1 + 2y_2 + \ldots + 2y_{n-1} + y_n] \qquad \text{(I10)}$$

Since $(n + 1)$ ordinates produce n strips, it follows that $h = \dfrac{b - a}{n}$. This is known as the **trapezium rule**.

Example 8.14

Use the trapezium rule with 5 ordinates to evaluate: $\int_0^1 e^{-x^2}\,dx$. By using a clearly labelled diagram, indicate whether your answer is greater or less than the exact value.

Solution

If we are using 5 ordinates, then $n = 4$ giving 4 trapezia each of width $\frac{1}{4} = 0.25$.

Hence: $h = 0.25$

If $f(x) = e^{-x^2}$, the integral represents the area under the graph from $x = 0$ to $x = 1$.

$$\text{Hence} \quad \int_0^1 e^{-x^2}\,dx \approx \frac{0.25}{2}[f(0) + 2f(0.25) + 2f(0.5) + 2f(0.75) + f(1)]$$

$$= 0.125[1 + 1.8788 + 1.5576 + 1.1396 + 0.3679]$$

$$= 0.743 \quad (3 \text{ sig. figs}).$$

The graph of $y = e^{-x^2}$ for $x \geq 0$ is shown in Figure 8.15. You can see that the trapezia are all **below** the curve. Hence the answer of 0.743 will be less than the exact value.

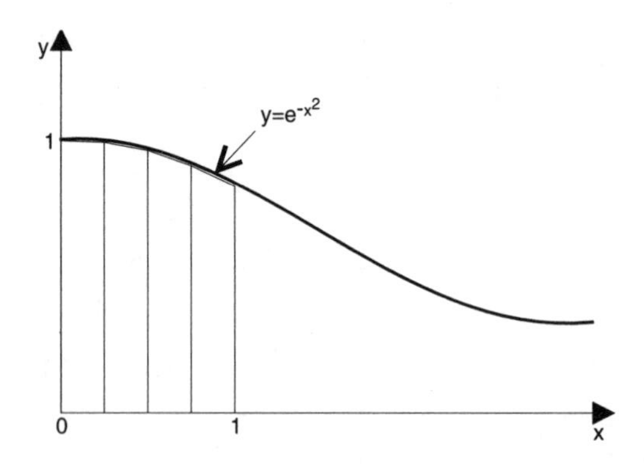

Figure 8.15

(ii) Simpson's rule

If, instead of joining the points on a curve with straight lines giving a series of trapezia (see Figure 8.14), the best parabola is drawn through sets of 3 consecutive points, it can be shown that another approximation to the area under the curve $y = f(x)$ between $x = a$ and $x = b$ is given by:

$$\int_a^b f(x) \approx \frac{h}{3}[y_0 + y_n + 4(y_1 + y_3 + \ldots) + 2(y_2 + y_4 + \ldots)]. \qquad \text{(I11)}$$

This is known as **Simpson's rule**; $n = 4$ gives a very accurate approximation.

Note that the only allowed values of n are $2, 4, 6, \ldots$. This is the same as saying the number of ordinates is odd, or the number of strips is even. Simpson's rule is more accurate than the trapezium rule, and is the formula used in a scientific calculator that has the facility to carry out numerical integration.

Example 8.15

Evaluate $\int_0^1 \sqrt[3]{1+e^x}$ correct to 4 decimal places, using Simpson's rule.

Solution

5 ordinates means 4 regions, hence $h = 1 \div 4 = 0.25$.

$$\text{The area} \quad = \frac{h}{3}[y_0 + 4(y_1 + y_3) + 2y_2 + y_4]$$

$$\text{Now} \quad f(x) = \sqrt[3]{1+e^x}$$

$$\text{So} \quad y_0 = f(0) = 1$$

$$y_1 = f(0.25) = 1.31694, \quad y_2 = f(0.5) = 1.38360$$

$$y_3 = f(0.75) = 1.46076, \quad y_4 = f(1) = 1.54922$$

$$\text{Hence: the integral} \approx \frac{0.25}{3}[1 + 4(1.31694 + 1.46076) + 2 \times 1.38360 + 1.54922]$$

$$= 1.3689 \text{ to 4 decimal places}$$

Exercise 8(f)

1 Evaluate the following integrals by means of the trapezium rule with 5 ordinates. Give your answers correct to 3 significant figures.

(i) $\int_0^1 \frac{x}{\sqrt{1+x^4}}$ (ii) $\int_1^2 x^3 \ln x \, dx$

(iii) $\int_1^2 \sqrt{1+e^x} \, dx$ (iv) $\int_0^1 \frac{x}{\sqrt{1+e^{-2x}}} dx$

2 Evaluate the integrals in Question 1 by using Simpson's rule with 5 ordinates.

Miscellaneous Examples 8

1 Evaluate

(i) $\int_0^{\frac{1}{2}} (x^2 + 1) \, dx$ (ii) $\int_1^2 \left(\frac{x+1}{x^4}\right) dx$ (iii) $\int_{\frac{1}{2}}^{1\frac{1}{2}} \sqrt{x} \, dx$

(iv) $\int_0^1 x\sqrt{1-x^2} \, dx$ (v) $\int_{0.1}^{0.2} \frac{x}{\sqrt{1-3x^2}} dx$

2 Find the area enclosed between the lines $y = x^4$ and $y = 1$.

3 Find the formula for the volume of a cone of base radius r and height h.

4 Use the trapezium rule with 7 ordinates, to evaluate: $\int_1^4 \sqrt{1 + \frac{1}{4}x^4}\,dx$. Give your answer to 3 significant figures.

5 Use Simpson's rule with 5 ordinates to evaluate: $\int_1^2 \frac{e^x}{1 + xe^x}\,dx$

Revision Problems 8

1 Figure 8.16 shows part of the graph of $y = x^2$, with rectangles approximating the area under the curve between $x = 0$ and $x = 1$. Prove that the total area of the four rectangles shown may be expressed as:

$$\frac{1}{5^3}\left(\sum_{r=1}^{4} r^2\right)$$

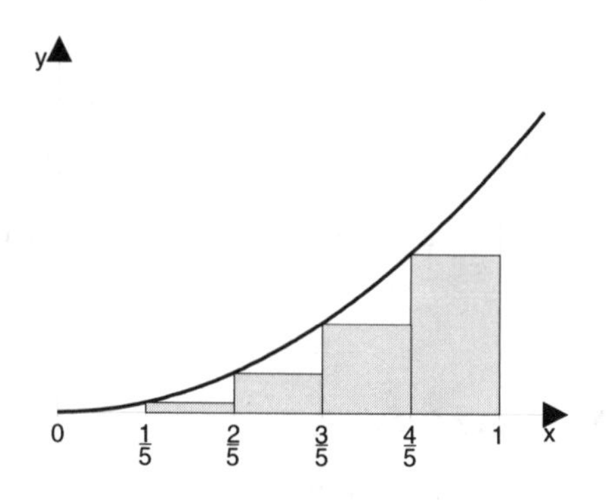

Figure 8.16

In general, when the x-axis between $x = 0$ and $x = 1$ is divided into n equal parts, the area under the curve may be approximated by the total area, A, of $(n - 1)$ rectangles each of width $\frac{1}{n}$. Given that

$$\sum_{r=1}^{n} r^2 = \tfrac{1}{6}n(n+1)(2n+1)$$

show that $A = \dfrac{(n-1)(2n-1)}{6n^2}$ and explain briefly how the exact value of $\int_0^1 x^2\,dx$ may be deduced from this expression.

(UCLES)

2 It is given that:

$$f(x) = \frac{1}{\sqrt{(1 + \sqrt{x})}}$$

and the integral $\int_0^1 f(x)\,dx$ is denoted by I.

(i) Using the trapezium rule, with four trapezia of equal width, obtain an approximation I_1 to the value of I, giving three decimal places in your answer.

(ii) A sketch of the graph of $y = f(x)$ is given in Figure 8.17. Use this diagram to justify the inequality $I < I_1$.

(iii) Evaluate I_2, where:

$$I_2 = \frac{1}{3}\sum_{r=1}^{3} f\left(\tfrac{1}{3}r\right)$$

giving three decimal places in your answer, and use the diagram to justify the inequality $I > I_2$.

(iv) By means of the substitution $\sqrt{x} = u - 1$, show that the exact value of I is $\frac{4}{3}(2 - \sqrt{2})$.

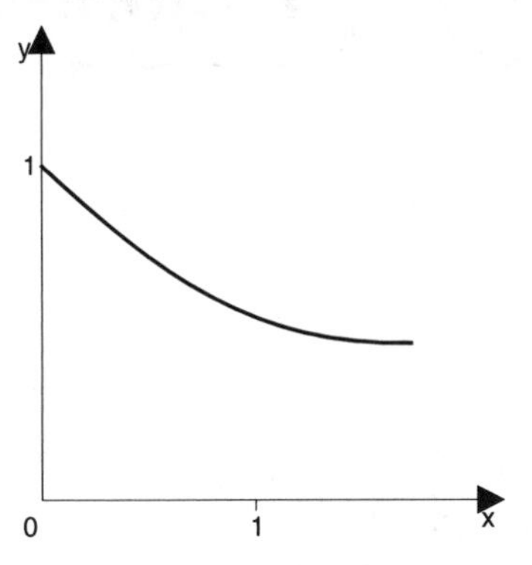

Figure 8.17

(UCLES)

9 Trigonometry

Most of the trigonometry that you will have encountered so far involves angles less than 90°, and right-angled triangles (you may have used the sine rule and cosine rule for an obtuse angled triangle). Many situations in real life, however, involve situations, particularly with machinery, where angles can be any size, positive or negative. This chapter helps you to master the skills needed to cope with this type of situation. You are advised to work through it systematically.

9.1 The four quadrants (sin x, cos x, tan x)

Once angles become greater than 90°, the traditional definition of sine, cosine and tangent using right-angled triangles become inappropriate. Instead, coordinates will be used.

Consider a point $P(x, y)$ rotating **anti-clockwise** in a circle of radius r (see Figure 9.1). The trigonometric ratios are defined as follows:

$$\sin\theta = \frac{y}{r} \quad \cos\theta = \frac{x}{r}$$

$$\tan\theta = \frac{y}{x} = \frac{y}{r} \div \frac{x}{r} = \frac{\sin\theta}{\cos\theta}$$

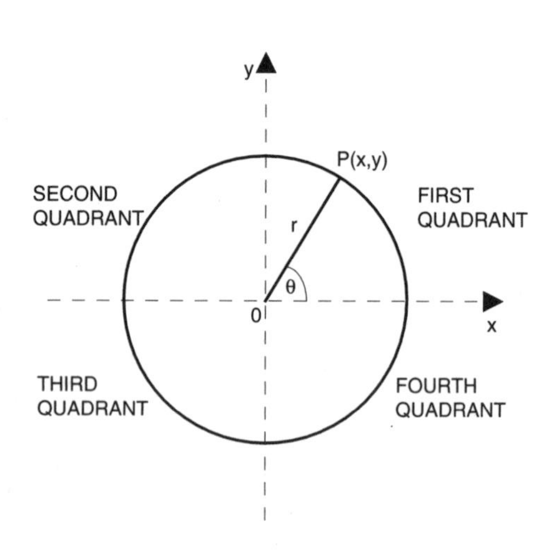

Figure 9.1

The radius of the circle r is always positive, and so the following situations arise:

1st quadrant $0 \leq \theta \leq 90°$	$\sin\theta$ is positive $\cos\theta$ is positive $\tan\theta$ is positive
2nd quadrant $90° \leq \theta \leq 180°$	$\sin\theta$ is positive $\cos\theta$ is negative $\tan\theta$ is negative
3rd quadrant $180° \leq \theta \leq 270°$	$\sin\theta$ is negative $\cos\theta$ is negative $\tan\theta$ is positive
4th quadrant $270° \leq \theta \leq 360°$	$\sin\theta$ is negative $\cos\theta$ is positive $\tan\theta$ is negative

The calculator will, of course, work out these values for you. If you plot graphs of the different trigonometric functions, then these look as shown in Figure 9.2.

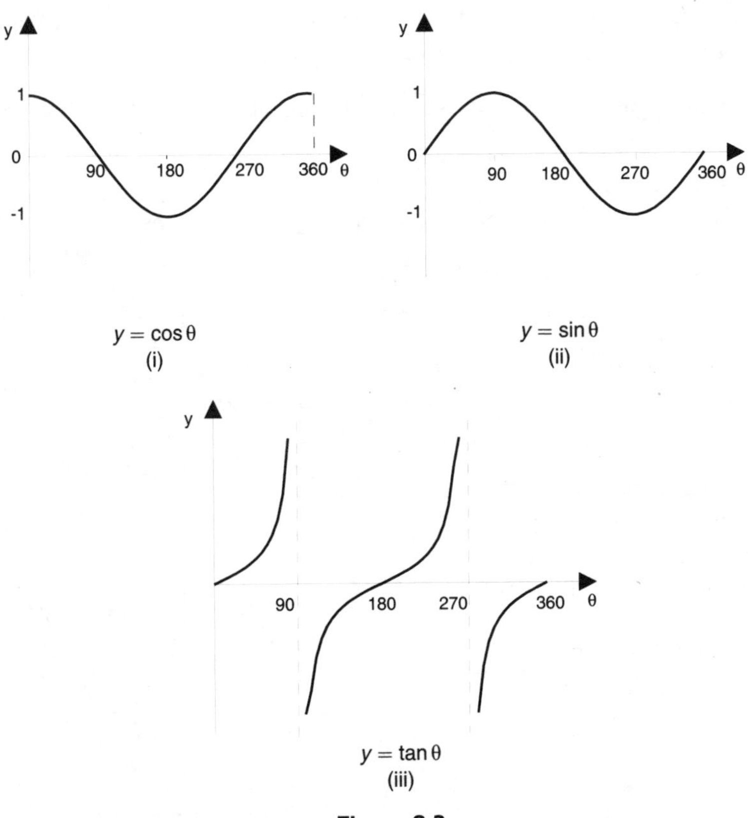

Figure 9.2

These graphs can be continued to angles greater than 360° (more than one complete revolution) or less than 0° (rotation in a *clockwise* direction).

Now look at how this all works.

(i) $90° \le \theta \le 180°$ (Subtract the angle from 180° and apply the appropriate sign.)

$$\sin 140° = \sin(180° - 140°) = \sin 40°$$
$$\cos 140° = -\cos(180° - 140°) - \cos 40°$$
$$\tan 140° = -\tan(180° - 140°) = -\tan 40°$$

(ii) $180° \le \theta \le 270°$ (Subtract 180° from the angle and apply the appropriate sign.)

$$\sin 220° = -\sin(220° - 180°) = -\sin 40°$$
$$\cos 220° = -\cos(220° - 180°) = -\cos 40°$$
$$\tan 220° = \tan(220° - 180°) = \tan 40°$$

(iii) $270° \le \theta \le 360°$ (Subtract the angle from 360° and apply the appropriate sign.)

$$\sin 300° = -\sin(360° - 300°) = -\sin 60°$$
$$\cos 300° = \cos(360° - 300°) = \cos 60°$$
$$\tan 300° = -\tan(360° - 300°) = -\tan 60°$$

(iv) For angles greater than 360°, reduce the angle to one less than 360° first, by subtracting multiples of 360°.

So: (a) $\sin 480° = \sin(480° - 360°)$
$= \sin 120° = \sin(180° - 120°) = \sin 60°$

(b) $\cos 865° = \cos(865° - 720°)$
$= \cos 145° = \cos(180° - 145°) = -\cos 35°$

(c) $\tan 1200° = \tan(1200° - 1080°)$
$= \tan 120° = -\tan(180° - 120°)$
$= -\tan 60°$

(v) For negative angles, the following rules will apply, which you can see by extending the graphs (see Figures 9.3a and 9.3b)

(a) $\cos\theta$ is symmetrical about the y-axis, and so $\cos(-\theta) = \cos\theta$ ($\cos\theta$ is an even function).

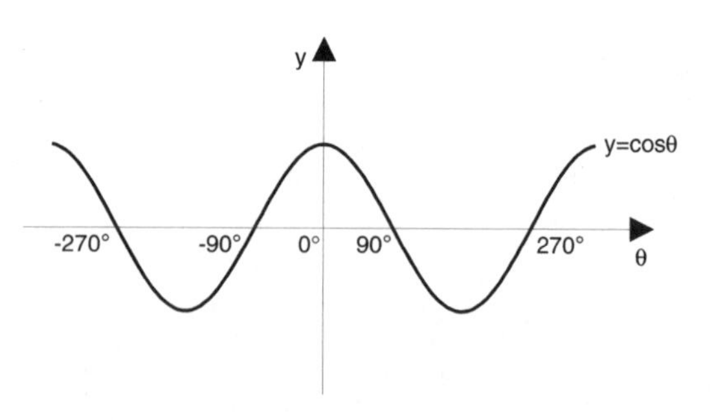

Figure 9.3a

(b) $\sin\theta$ is opposite in sign for negative angles, hence $\sin(-\theta) = -\sin\theta$ ($\sin\theta$ is an odd function).

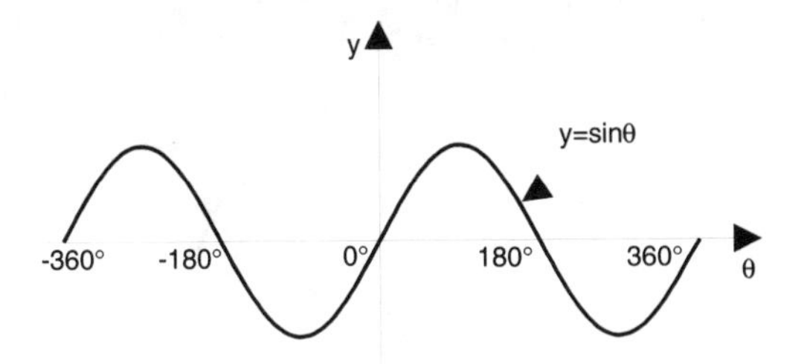

Figure 9.3b

(c) since $\tan\theta = \dfrac{\sin\theta}{\cos\theta}$, then $\tan(-\theta) = \dfrac{\sin(-\theta)}{\cos(-\theta)} = \dfrac{-\sin\theta}{\cos\theta} = -\tan\theta$

($\tan\theta$ is an odd function).

Hence, for example:

$$\begin{aligned}\sin(-120^\circ) &= -\sin 120^\circ = -\sin(180^\circ - 120^\circ)\\ &= -\sin 60^\circ\end{aligned}$$

$$\begin{aligned}\cos(-400^\circ) &= \cos 400^\circ = \cos(400^\circ - 360^\circ)\\ &= \cos 40^\circ\end{aligned}$$

$$\begin{aligned}\tan(-840^\circ) &= -\tan 840^\circ\\ &= -\tan(840^\circ - 720^\circ)\\ &= -\tan 120^\circ\\ &= --\tan(180^\circ - 120^\circ) = \tan 60^\circ\end{aligned}$$

Check that you have understood this section by trying the following questions.

Exercise 9(a)

Express the following trigonometric ratios of angles, as the ratio of an angle in the first quadrant, with the appropriate sign.

1 $\sin 140^\circ$ **2** $\cos 200^\circ$ **3** $\tan 135^\circ$
4 $\tan 600^\circ$ **5** $\sin(-80^\circ)$ **6** $\cos(-140^\circ)$
7 $\tan(-600^\circ)$ **8** $\tan 179^\circ$ **9** $\sin 600^\circ$
10 $\cos(-730^\circ)$ **11** $\cos 800^\circ$ **12** $\tan(-650^\circ)$
13 $\tan 380^\circ$ **14** $\cos 401^\circ$ **15** $\cos(-196^\circ)$
16 $\tan(-187^\circ)$ **17** $\sin(-636^\circ)$ **18** $\cos 1200^\circ$
19 $\sin 1080$ **20** $\cos 987$

ACTIVITY 7

The graph of $y = \sin 2x$ is as follows:

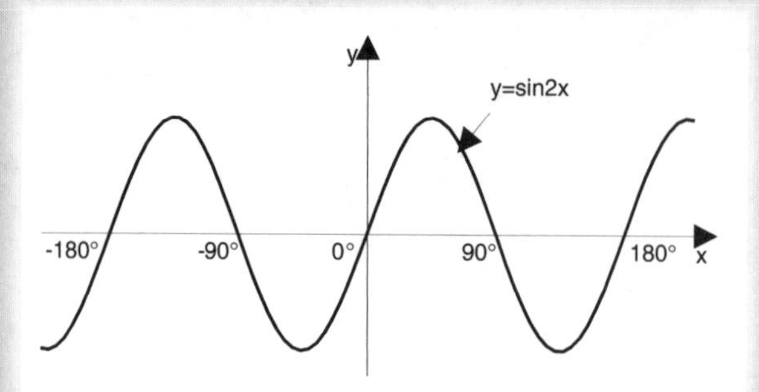

Figure 9.4

This is the graph of $y = \sin 3x$:

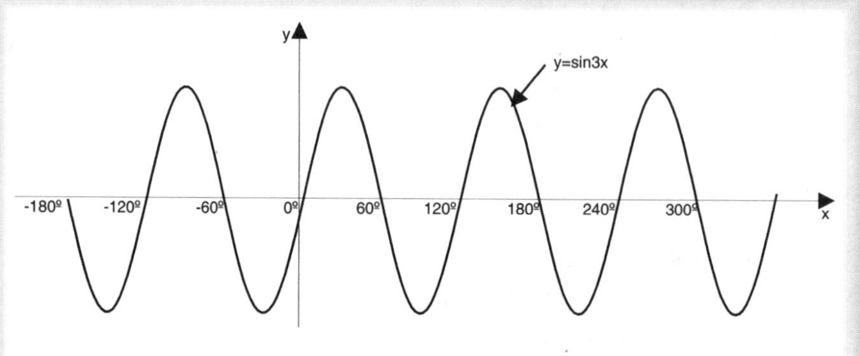

Figure 9.5

Investigate the graphs of $y = \sin nx$ and $y = \cos nx$ where n is an integer.
What does the graph of $y = 4\sin 2x + 3\sin 3x$ look like?
Investigate the curves $y = a\sin 2x + b\sin 3x$.

Note: This type of investigation is much easier if you have access to a graphics calculator.

9.2 Sec x, cosec x, cot x

It is often convenient to replace the reciprocals of $\cos x$, $\sin x$ and $\tan x$ in a calculation, and so three new trigonometric ratios are defined:

$$\sec x = \frac{1}{\cos x}; \quad \operatorname{cosec} x = \frac{1}{\sin x}; \quad \text{and } \cot x = \frac{1}{\tan x}$$

You will not usually find a button for these on the calculator.

In terms of quadrants, $\sec x$ behaves like $\cos x$, $\operatorname{cosec} x$ like $\sin x$, and $\cot x$ like $\tan x$.

So, for example,

$$\begin{aligned}\sec 280° &= -\sec(280° - 270°)\\ &= -\sec 10°\end{aligned}$$

$$\begin{aligned}\cot(-200°) &= -\cot 200°\\ &= -\cot(200° - 180°)\\ &= -\cot 20°\end{aligned}$$

$$\begin{aligned}\operatorname{cosec} 650° &= \operatorname{cosec}(650° - 360°)\\ &= -\operatorname{cosec}(360° - 290°)\\ &= -\operatorname{cosec} 70°\end{aligned}$$

Exercise 9(b)

Express the following trigonometric ratios of angles as the ratio of an angle in the first quadrant, with the appropriate sign.

1 $\sec 260°$ **2** $\operatorname{cosec} 400°$ **3** $\cot 240°$

4 $\sec 300°$ **5** $\cot 290°$ **6** $\operatorname{cosec}(-100°)$

7 $\sec(-200°)$ **8** $\cot(-350°)$ **9** $\sec 700°$

10 $\cot 900°$

9.3 Radian measure

The reason that angles are measured in degrees with 360° in a complete revolution dates back to the time of the Babylonians. It is a purely arbitrary number, although one cannot argue that 360 is a very useful number with many factors. However, when using the calculus, it turns out that a different way of measuring angles is necessary to make calculations easier.

We define

$$\begin{aligned}1 \text{ radian} &= \frac{360}{2\pi} = \frac{180}{\pi} \text{ degrees}\\ &= 57.3 \text{ degrees}\end{aligned}$$

hence 2π radians $= 360$ degrees

or π radians $= 180$ degrees

The abbreviations for radian are rad, c or even no unit at all.

There are a number of straightforward angles worth learning:

$90° = \frac{\pi}{2}$ $\qquad$ $60° = \frac{\pi}{3}$

$45° = \frac{\pi}{4}$ $\qquad$ $270° = \frac{3\pi}{2}$

$30° = \frac{\pi}{6}$ $\qquad$ and so on

If the angle is not any of these, use the following conversion rules:

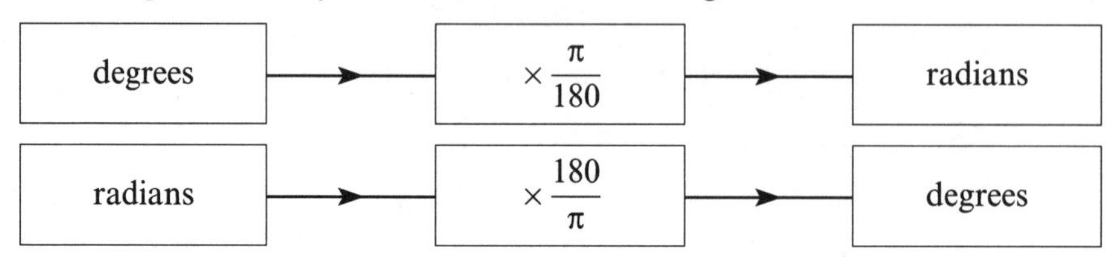

Exercise 9(c)

1 Convert to degrees the following angles given in radians:

(i) $\frac{\pi}{8}$ (ii) 3π (iii) $\frac{2\pi}{3}$ (iv) 1

(v) $\frac{5\pi}{12}$ (vi) 2.6 (vii) $\frac{7\pi}{4}$ (viii) $\frac{3\pi}{2}$

(ix) 6π (x) 8.42

2 Convert the following angles to radians (leave π in your answer where appropriate):

(i) 15° (ii) 50° (iii) 19.5° (iv) 420°

(v) −135° (vi) 600° (vii) $67\frac{1}{2}°$ (viii) 8°

(ix) 900° (x) 106.5°

9.4 Sectors of a circle

The two formulae most directly affected by measuring angles in radians are those concerning a sector of a circle of radius r, angle at the centre θ rad or $x°$ (see Figure 9.6).

In degrees,

$$\text{the area } = \pi r^2 \frac{x}{360}$$

$$\text{and arc length } L = 2\pi r \times \frac{x}{360}$$

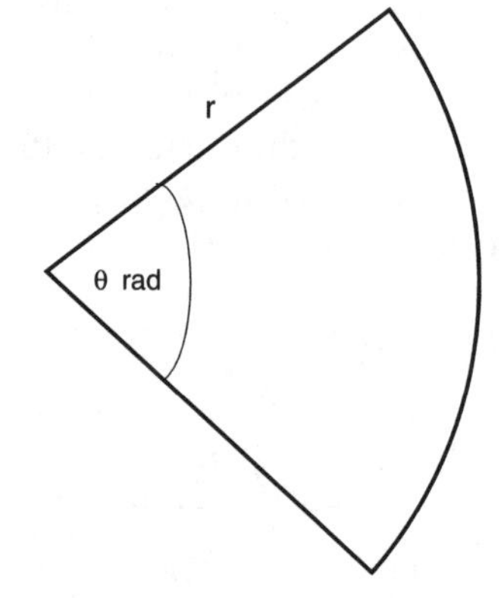

Figure 9.6

but if the angle $x^\circ = \theta$ rad, then since $360^\circ = 2\pi$ rad

$$\text{the area } = \pi r^2 \times \frac{\theta}{2\pi}$$

that is, area $= \frac{1}{2}r^2\theta$ (θ in radians) (SR1)

Hence π disappears from the formula.

Also arc length $= 2\pi r \times \dfrac{\theta}{2\pi}$

hence arc length $= r\theta$ (θ in radians) (SR2)

Example 9.1

Referring to Figure 9.7, O is the centre of the circle, and the angle AOB is 1.5 radians. If the shaded area of the circle is 20 cm², find the radius of the circle, and also the length of the larger arc of the circle joining A and B.

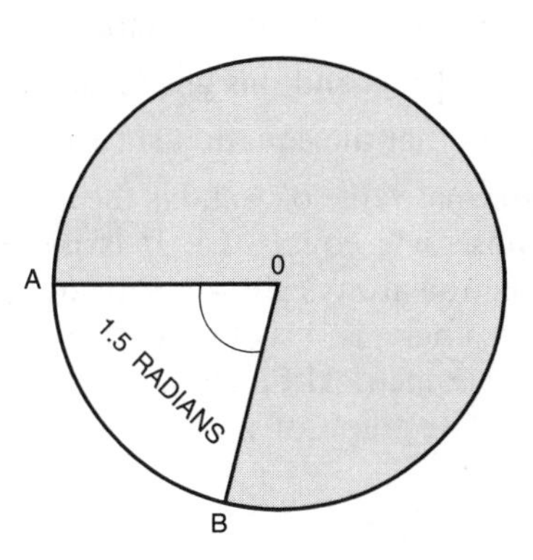

Figure 9.7

Solution

The radian formula for the area of a sector is:

area $= \frac{1}{2}r^2\theta$ (where r is the radius)

The required angle, however, is the reflex angle AOB.

This is $2\pi - 1.5 = 4.783$ radians

Hence: $20 = \frac{1}{2}r^2 \times 4.783$

$\therefore \quad r^2 = \dfrac{40}{4.783}$

Hence $r = 2.89$ cm

The longer arc AB is $r\theta = 2.89 \times 4.783 = 13.8$ cm

Exercise 9(d)

1 In Figure 9.8, O is the centre of a circle of radius 5 cm. Find the shaded area if angle $AOB = 0.8$ rad.

2 Referring to Figure 9.8, find the perimeter of the shaded area.

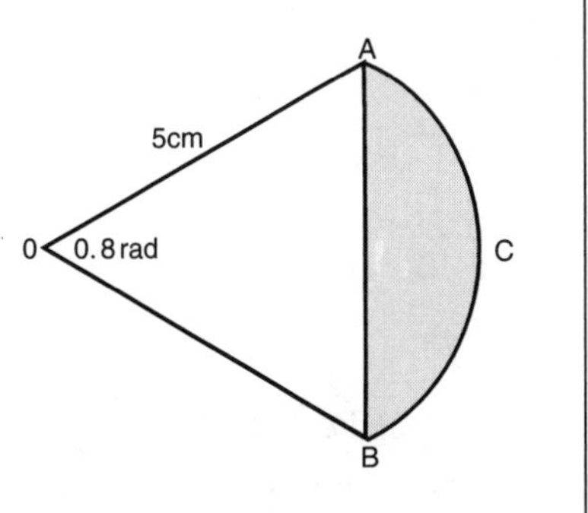

Figure 9.8

3 A sector of a circle of radius 10 cm has an area of 8 cm^2. Find the angle of the sector.

4 A and B are two points on a circle. The length of the longest arc between A and B is three times the length of the shortest arc. Find the angle subtended by the chord AB at the centre of the circle.

9.5 Inverse trigonometric functions

When you solve the equation $\sin x = 0.5$, using your calculator, you press the $\boxed{\sin^{-1}}$ key, and this gives a single value of $x = 30°$. However, we know from the graph that the equation $\sin x = 0.5$ has many solutions. The $\boxed{\sin^{-1}}$ key gives the **principal** value of x and is the inverse sine of 0.5. In other words, $\sin^{-1}$ is the **angle** whose sine equals 0.5. It is important that you do not forget the inverse sine function always gives you an angle. Because of the way inverse functions are used, this angle is usually written in radians. **Note:** The graphs of the inverse trigonometrical function are 1:1.

The graph of $y = \sin^{-1} x$ is shown in Figure 9.9a.

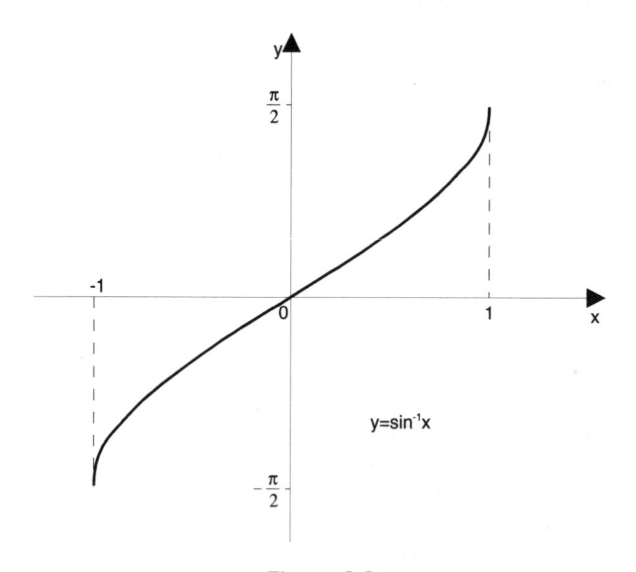

Figure 9.9a

Similarly, the keys for $\boxed{\cos^{-1}}$ and $\boxed{\tan^{-1}}$ give the principal values of the inverse cosine and inverse tangent functions. These two graphs are shown in Figures 9.9b and 9.9c.

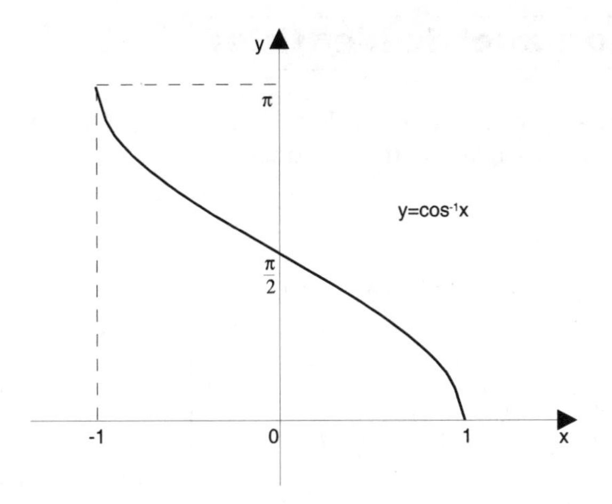

Figure 9.9b

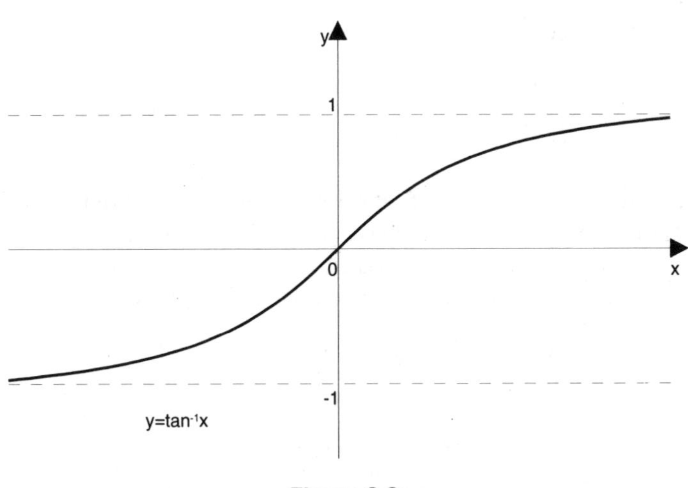

Figure 9.9c

A major use of these inverse trigonometric functions is in the field of integration (see Chapter 15).

Exercise 9(e)

Using the graphs of $\cot x$, $\operatorname{cosec} x$ and $\sec x$ (or otherwise), draw the graph of $y = \cot^{-1} x$, $y = \operatorname{cosec}^{-1} x$ and $y = \sec^{-1} x$.

9.6 Trigonometric identities

As with ordinary algebra, it is possible to simplify expressions in trigonometry using a wide range of trigonometric identities.

(i) Pythagorean identities

In a right-angled triangle (see Figure 9.10),

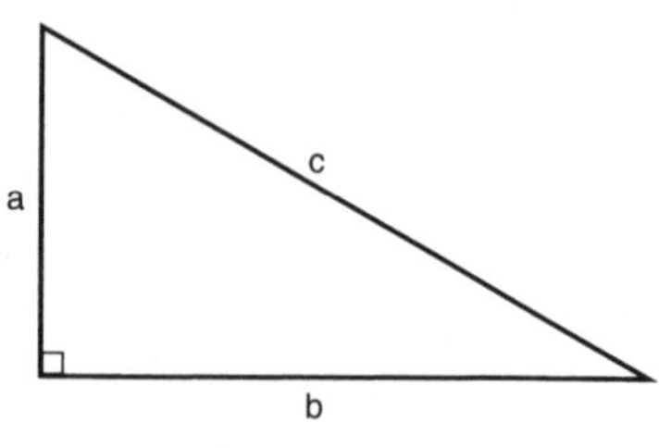

Figure 9.10

$$c^2 = a^2 + b^2$$

$$\div c^2 \qquad 1 = \frac{a^2}{c^2} + \frac{b^2}{c^2} = \left(\frac{a}{c}\right)^2 + \left(\frac{b}{c}\right)^2$$

that is, $1 = (\sin x)^2 + (\cos x)^2$

$(\sin x)^2$ is always written $\sin^2 x$ (sine squared x)

Hence $\sin^2 x + \cos^2 x \equiv 1$ (TI1)

or $\sin^2 x \equiv 1 - \cos^2 x$ (TI1A)

$\cos^2 x \equiv 1 - \sin^2 x$ (TI1B)

Using the definitions given in Section 9.1, you can show that this identity is true for all angles. (**Note** the equivalence sign $\equiv$, which states that it is always true. You do not *have* to use this sign, but it emphasises the fact that it is a different way of writing an expression, rather than solving an equation.

$\div$ TI1 by $\sin^2 x$ $\qquad 1 + \dfrac{\cos^2 x}{\sin^2 x} = \dfrac{1}{\sin^2 x}$

that is, $\qquad 1 + \cot^2 x = \operatorname{cosec}^2 x$

usually written:

$\operatorname{cosec}^2 x = 1 + \cot^2 x$ (TI2)

$\div$ TI1 by $\cos^2 x$ and you obtain:

$\sec^2 x = 1 + \tan^2 x$ (TI3)

> **MEMORY JOGGER**
>
> It is important to remember that the letter x in a given identity can be replaced by any algebraic or numerical expression.

Hence in TI1, if $x = 2A$, you get:

$$\sin^2 2A + \cos^2 2A = 1$$

In TI2, if $x = 45°$, you get:

$$\text{cosec}^2 45° = 1 + \cot^2 45°$$

In TI3, if $x = \theta - \frac{\pi}{4}$, you get:

$$\sec^2\left(\theta - \frac{\pi}{4}\right) = 1 + \tan^2\left(\theta - \frac{\pi}{4}\right)$$

The identities TI1–3 can be referred to as the Pythagorean identities, because they originate from Pythagoras's theorem.

Example 9.2

Simplify:

(i) $\dfrac{\sqrt{1 - \sin^2 x}}{\cos x}$ (ii) $\dfrac{1 + \tan^2 x}{\sec^3 x}$ (iii) $\sqrt{\dfrac{1 - \cos x}{1 + \cos x}}$

Solution

(i) $\dfrac{\sqrt{1 - \sin^2 x}}{\cos x} = \dfrac{\sqrt{\cos^2 x}}{\cos x} = \dfrac{\cos x}{\cos x} = 1$

(ii) $\dfrac{1 + \tan^2 x}{\sec^3 x} = \dfrac{\sec^2 x}{\sec^3 x} = \dfrac{1}{\sec x} = \cos x$

(iii) This is not easy to see. Multiply top and bottom of the fraction inside the square root sign by $1 - \cos x$:

$$\sqrt{\frac{(1 - \cos x)(1 - \cos x)}{(1 + \cos x)(1 - \cos x)}} = \sqrt{\frac{(1 - \cos x)^2}{1 - \cos^2 x}}$$

$$= \frac{1 - \cos x}{\sqrt{\sin^2 x}} = \frac{1 - \cos x}{\sin x}$$

$$\text{or} \quad = \text{cosec}\, x - \cot x$$

Exercise 9(f)

Simplify the following expressions as much as possible:

1 $\sqrt{1 + \tan^2 x}$ **2** $\dfrac{1}{1 - \sin^2 x}$ **3** $\dfrac{\cos x}{\sqrt{1 - \sin^2 x}}$

4 $(\text{cosec}^2 x - 1)^2$ **5** $\sqrt{\dfrac{1 + \cos x}{1 - \cos x}}$ **6** $\dfrac{\sec^3 x}{\sqrt{1 + \tan^2 x}}$

9.7 Solving trigonometric equations

If you have started reading at this point in the text, make sure you know the following:

DO YOU KNOW?

(i) The sign of the trigonometrical ratios in the four quadrants? Figure 9.11 shows which ratios are positive.

(ii) What radian measure is?

$360° = 2\pi$ radians

So $1° = \dfrac{\pi}{180}$ radians

or 1 radian $= \dfrac{180°}{\pi}$

(iii) In the sequence $a, a + 360, a + 720, \ldots$, the formula for predicting the $(n + 1)$th term is $a + 360n$.

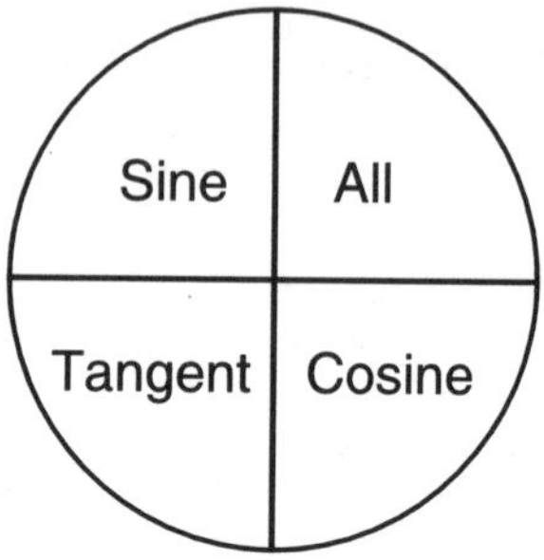

Sine | All
Tangent | Cosine

Figure 9.11

Consider the statement: $\sin x = 0.5$ How can we find what x is equal to?

Using the $\boxed{\sin^{-1}}$ button on a calculator, you would get $x = 30°$. However, is this the only solution? If you look at the graph of $y = \sin x$ over a wide range of values of x (Figure 9.12), it shows that in fact there are an infinite number of possible values for x for which $y = 0.5$.

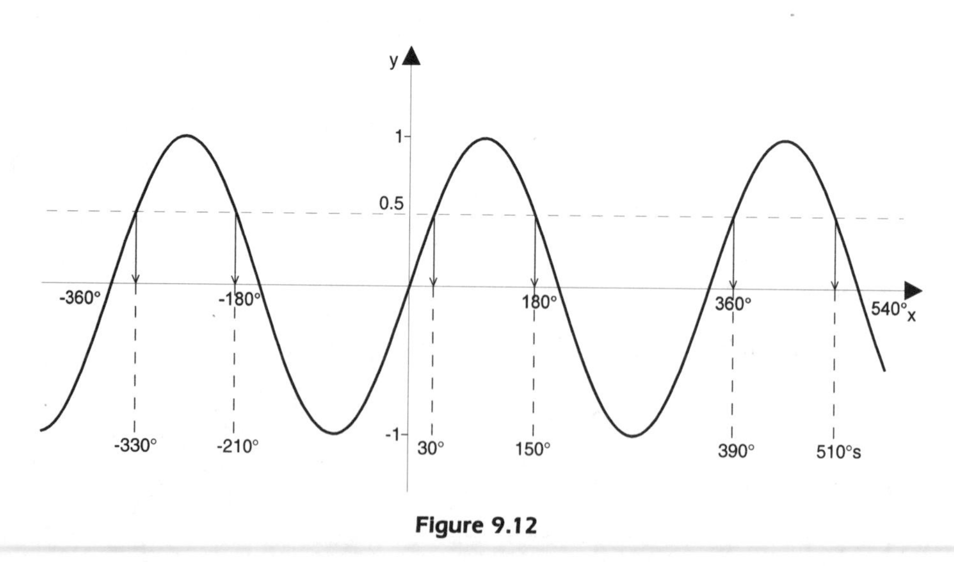

Figure 9.12

These are $x = \ldots -300°, -210°, 30°, 150°, 390°, 510°, \ldots$

If you group these alternately, as shown, you can see that in each group the angles increase by 360°. This can be summarised in the following way:

$$x = 30° + 360°n$$

or $x = 150° + 360°n$

This is called the **general solution**

where n is a positive or negative integer, or zero.

Very often, we require the solution to an equation within a given range. However, it is nearly always easier to find the general solution first and then select the answers you require.

Example 9.3

Look at the equation: $\tan 3x = -0.4$ with $180° \leq x \leq 180°$.

Solution

The tangent is negative in the **second** and **fourth** quadrants.

Use $+0.4$ [$\tan^{-1}$] on the calculator to give an angle of 21.8°. Using this angle, we can get an angle in the second quadrant of $180° - 21.8° = 158.2°$ and in the fourth quadrant of $360° - 21.8° = 338.2°$.

Hence: $3x = 158.2° + 360°n$

or: $3x = 338.2° + 360°n$

If we now divide by 3, we get:

$$x = 52.7° + 120°n \quad \text{(i)}$$

or: $x = 112.7° + 120°n$ (ii)

MEMORY JOGGER

Remember that n can take any integer value, positive, negative or zero

Using Equation (i):

$n = -1 \quad x = 52.7° - 120° = -67.3°$

$n = 0 \quad x = 52.7°$

$n = 1 \quad x = 52.7° + 120° = 172.7°$

These are the only three values that you can get in the range $-180° \leq x \leq 180°$.

Using Equation (ii):

with $n = -2, -1, 0$, you get angles $-127.3°, -7.3°, 112.7°$

Combining all these answers, we find the solution of the equation:

$\tan 3x = -0.4$ is $-127.3°, -67.3°, -7.3°, 52.7°, 112.7°, 172.7°$

On many occasions, when you solve a trigonometric equation, you want the answers in radians. The following example shows you that, provided you remember to convert to the radian mode on your calculator, the method of solution is the same.

Example 9.4

Solve the equation $\cos\left(2x - \frac{\pi}{4}\right) = -0.2 \quad 0 \leq x \leq 2\pi$.

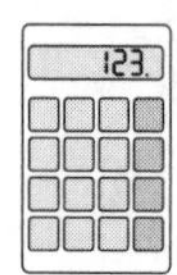

> **MEMORY JOGGER**
> The condition $0 \leq x \leq 2\pi$ tells you that a *radian* solution is required. Remember to put the calculator into *radian mode*.

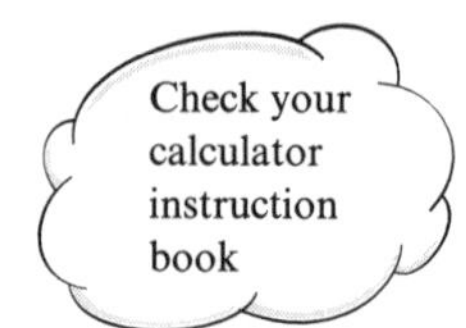

Solution

Find 0.2 $\boxed{\cos^{-1}}$ to give 1.369 radians. But a cosine equal to −0.2 means that we require an angle in the **second** quadrant $(\pi - 1.369)$, and in the **third** quadrant $(\pi + 1.369)$.

Hence: $2x - \frac{\pi}{4} = (\pi - 1.369) + 2n\pi$

or: $2x - \frac{\pi}{4} = (\pi + 1.369) + 2n\pi$

Hence: $2x = 2.558 + 2n\pi$

or: $2x = 5.296 + 2n\pi$

Dividing by 2 gives:

$x = 1.279 + n\pi$

or: $x = 2.648 + n\pi$

Remember that $0 \leq x \leq 6.28$, hence using the general solution with $n = 0$, 1 in each line, we get:

$x = 1.279, 2.648, 4.421, 5.790$

9.8 Solving more difficult trigonometric equations

However different a trigonometric equation looks, it can always be reduced to the basic type shown in the previous section. This is illustrated in the next three examples.

Example 9.5

Solve the equation $\sin^2 x = \frac{1}{2}$

> **MEMORY JOGGER**
> Remember $\sin^2 x$ means $(\sin x)^2$.

Solution

Take the square root of this equation, and you get two equations:

$$\sin x = \pm\frac{1}{\sqrt{2}}$$

If $\sin x = \frac{1}{\sqrt{2}}$, $\quad x = 45°, 135°$

If $\sin x = -\frac{1}{\sqrt{2}}$ $\quad x = 225°, 315°$

The solution is $x = 45°, 135°, 225°, 315°$

Example 9.6

Solve the equation $\sec 2x = 3$ if $-180° \leq x \leq 180°$.

Solution

Since $\sec\theta = \frac{1}{\cos\theta}$, the equation can be rewritten:

$$\cos 2x = \tfrac{1}{3}$$

It can then be solved in the usual way:

$$\begin{aligned} 2x &= 70.5° + 360°n \\ \text{or:}\quad 2x &= 289.5° + 360°n \\ \therefore\quad x &= 35.25° + 180°n \\ \text{or:}\quad x &= 144.75° + 180°n \end{aligned}$$

in the required range:

$$x = -144.75°, \ -35.25°, 35.25°, 144.75°$$

Example 9.7

Find the general solution in radians of the equation $3\sin^2\theta - 14\sin\theta - 4 = 0$.

Solution

This is a quadratic equation in $\sin\theta$. It does not factorise, and so you must use the quadratic equation formula:

$$x = \frac{-b \pm \sqrt{b^2 - 4ac}}{2a}$$

Hence: $\sin\theta = \dfrac{+14 \pm \sqrt{14^2 - 4 \times 3 \times -4}}{6}$

$= 4.937$ or -0.270

$\sin\theta$ cannot be greater than 1, and so the solution $\sin\theta = 4.937$ can be ignored.

Hence: $\sin\theta = -0.27$

$\therefore$ $\theta = (\pi + 0.273) + 2n\pi$

or: $\theta = (2\pi - 0.273) + 2n\pi$

that is, $\theta = 3.415 + 2n\pi$

or: $\theta = 6.010 + 2n\pi$

Exercise 9(g)

1 Solve the following equations in the range $0 \leq x \leq 360°$:

(i) $\sin x = 0.6$ (ii) $\cos x = -0.7$

(iii) $\tan 2x = \dfrac{1}{\sqrt{2}}$ (iv) $\cos(2x - 30°) = -\frac{1}{2}$

(v) $\cos 2x = 0.6$ (vi) $\cot x = 4$

2 Solve the following equations in the range $-180° \leq x \leq 180°$:

(i) $\tan 2x = -1$ (ii) $\sec 2x = 4$

(iii) $\sin 3x = -\dfrac{1}{\sqrt{2}}$ (iv) $\cos(30° - x) = 0.6$

3 Find the following solutions in radians in the range $0 \leq x \leq 2\pi$:

(i) $\tan 3x = \sqrt{3}$ (ii) $\cos\left(2x - \dfrac{\pi}{2}\right) = 0.6$

(iii) $\operatorname{cosec} 2x = 2.3$ (iv) $\sin\left(x + \dfrac{\pi}{4}\right) = -1$

(v) $\sin\left(3x - \dfrac{\pi}{6}\right) = \dfrac{\sqrt{3}}{2}$

4 Find the general solution of the following equations:

(i) $\sin 3x° = 1$ (ii) $\tan 4x° = -1$

(iii) $\cot x° = 2$ (iv) $\sin\left(2x - \dfrac{\pi}{4}\right) = 0.6$

5 Solve the following equations in the range $0 \leq x \leq 360°$:

(i) $\tan^2 x = 2$ (ii) $\sin^2 3x = 0.6$

(iii) $4\sin^2 x - 8\sin x + 3 = 0$ (iv) $\sin^2 x + \sin x - 1 = 0$

9.9 Identities 2 (the addition formulae, or compound angle formulae)

It would be very tempting to write an expression such as $\cos(A+B)$ as $\cos A + \cos B$. However, it is not difficult to prove this is not allowed.

If $A = 60°$ and $B = 60°$, then $\cos(60+60) = \cos 120° = -0.5$

but $\quad \cos 60° + \cos 60° = 0.5 + 0.5 = 1$

Hence: $\quad \cos(60° + 60°) \neq \cos 60° + \cos 60°$

Look at Figure 9.13.

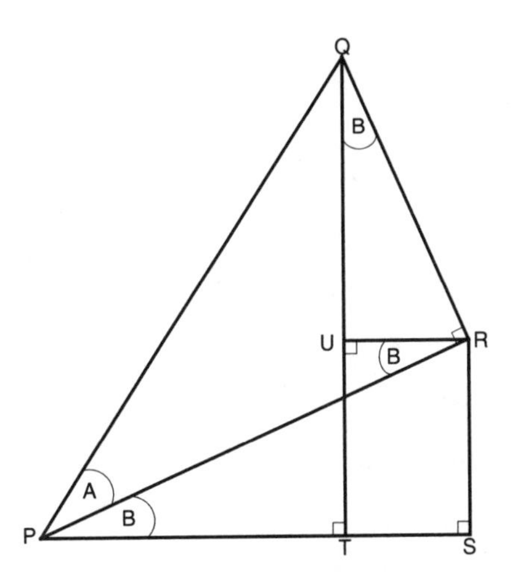

Figure 9.13

$$\begin{aligned}
\cos(A+B) &= \frac{PT}{PQ} \\
&= \frac{PS - TS}{PQ} \\
&= \frac{PS}{PQ} - \frac{UR}{PQ} \quad \text{(since } TS = UR\text{)} \\
&= \frac{PR}{PQ}\cos B - \frac{RQ}{PQ}\sin B \\
&= \cos A \cos B - \sin A \sin B \qquad \text{(TI4)}
\end{aligned}$$

In a similar way, it can be shown that:

$$\cos(A-B) \equiv \cos A \cos B + \sin A \sin B \qquad \text{(TI5)}$$
$$\sin(A+B) \equiv \sin A \cos B + \cos A \sin B \qquad \text{(TI6)}$$
$$\sin(A-B) \equiv \sin A \cos B - \cos A \sin B \qquad \text{(TI7)}$$

Since $\tan(A+B) = \dfrac{\sin(A+B)}{\cos(A+B)} = \dfrac{\sin A \cos B + \cos A \sin B}{\cos A \cos B - \sin A \sin B}$,

$\div$ top and bottom lines by $\cos A \cos B$, you get:

$$\tan(A+B) \equiv \frac{\tan A + \tan B}{1 - \tan A \tan B} \qquad \text{(TI8)}$$

Also $$\tan(A-B) \equiv \frac{\tan A - \tan B}{1 + \tan A \tan B} \qquad \text{(TI9)}$$

If you replace B by A in Equation (TI4)

$$\cos(A+A) = \cos 2A = \cos A \cos A - \sin A \sin A$$

that is, $$\cos 2A = \cos^2 A - \sin^2 A \qquad \text{(TI10)}$$

or: $$= \cos^2 A - (1 - \cos^2 A)$$

$$= 2\cos^2 A - 1 \qquad \text{(TI10a)}$$

or: $$= 2(1 - \sin^2 A) - 1$$

$$= 1 - 2\sin^2 A \qquad \text{(TI10b)}$$

The last two of these are often needed in integration when rearranged, to give:

$$\cos^2 A = \tfrac{1}{2}(1 + \cos 2A)$$

or: $$\sin^2 A = \tfrac{1}{2}(1 - \cos 2A)$$

Replace B by A in Equation (TI6):

$$\sin(A+A) = \sin A \cos A + \cos A \sin A$$

that is, $$\sin 2A = 2 \sin A \cos A \qquad \text{(TI11)}$$

Replace B by A in Equation (TI8):

$$\tan(A+A) = \frac{\tan A + \tan A}{1 - \tan A \tan A}$$

that is, $$\tan 2A = \frac{2\tan A}{1 - \tan^2 A} \qquad \text{(TI12)}$$

Formulae given in Equations (TI10), (TI11) and (TI12) are often called the **double angle** formulae.

The following examples should be studied carefully. They cover a wide range of types of question.

Then try Exercise 9(h).

Example 9.8

Find, without using a calculator:

(i) $\cos 15^\circ$ (ii) $\tan 75^\circ$

(iii) $\frac{1}{2}\cos 15^\circ - \frac{\sqrt{3}}{2}\sin 15^\circ$ (iv) $\cot 22\frac{1}{2}^\circ$

Solution

The simplest right-angled triangles with angles of 45° and 60° are shown in Figure 9.14. From these, you can get the trig ratios without using a calculator for 30°, 45° and 60°.

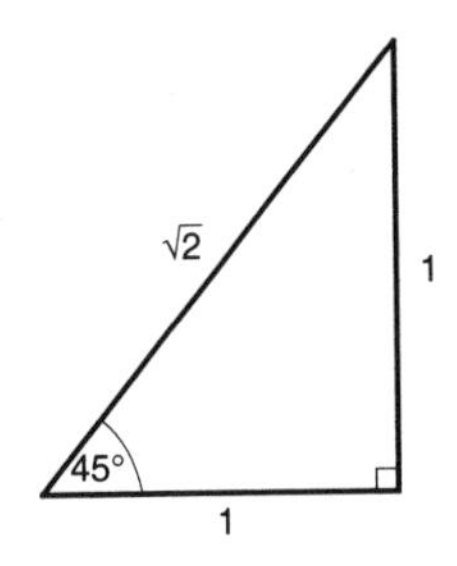

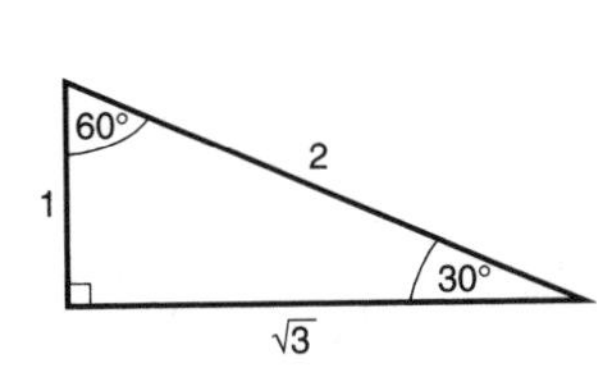

Figure 9.14

(i) So $\cos 15^\circ = \cos(45^\circ - 30^\circ)$

$$= \cos 45^\circ \cos 30^\circ + \sin 45^\circ \sin 30^\circ \qquad \text{(using Equation (TI6))}$$

$$= \frac{1}{\sqrt{2}} \times \frac{\sqrt{3}}{2} + \frac{1}{\sqrt{2}} \times \frac{1}{2} = \frac{\sqrt{3}+1}{2\sqrt{2}}$$

or: $= \dfrac{\sqrt{6}+\sqrt{2}}{4}$ if you rationalise the denominator

(ii) $\tan 75^\circ = \tan(45^\circ + 30^\circ)$

$$= \frac{\tan 45^\circ + \tan 30^\circ}{1 - \tan 45^\circ \tan 30^\circ} \qquad \text{(using Equation (TI8))}$$

$$= \frac{1 + \frac{1}{\sqrt{3}}}{1 - \frac{1}{\sqrt{3}}} = \frac{\sqrt{3}+1}{\sqrt{3}-1}$$

or: $= 2 + \sqrt{3}$ (after rationalising the denominator)

(iii) Since $\cos 60^\circ = \frac{1}{2}$ and $\sin 60^\circ = \frac{\sqrt{3}}{2}$,

$$\tfrac{1}{2}\cos 15^\circ - \tfrac{\sqrt{3}}{2}\sin 15^\circ = \cos 60^\circ \cos 15^\circ - \sin 60^\circ \sin 15^\circ$$

$$= \cos(60^\circ + 15^\circ)$$

$$= \cos 75^\circ$$

or:

$$= \sin 30^\circ \cos 15^\circ - \cos 30^\circ \sin 15^\circ$$

$$= \sin(30^\circ - 15^\circ)$$

$$= \sin 15^\circ$$

(iv) At first sight, this might seem different to the previous question. But if you use Equation (TI12) with $A = 22\frac{1}{2}°$,

$$\text{then:} \qquad \tan 45° = \frac{2\tan 22\frac{1}{2}°}{1 - \tan^2 22\frac{1}{2}°}$$

$$\text{that is,} \quad 1 = \frac{2\tan 22\frac{1}{2}°}{1 - \tan^2 22\frac{1}{2}°}$$

This leads to the quadratic equation:

$$\tan^2 22\tfrac{1}{2}° + 2\tan 22\tfrac{1}{2}° - 1 = 0$$

If you solve this, using the quadratic formula:

$$\tan 22\tfrac{1}{2}° = \frac{-2 \pm \sqrt{8}}{2}$$

$$\text{but} \quad \tan 22\tfrac{1}{2}° > 0 \qquad \therefore \quad \tan 22\tfrac{1}{2}° = \sqrt{2} - 1$$

$$\text{but} \quad \cot x = \frac{1}{\tan x}$$

$$\therefore \quad \cot 22\tfrac{1}{2}° = \frac{1}{\sqrt{2} - 1} = \sqrt{2} + 1$$

Example 9.9

If A is obtuse, and B acute, and also $\sin A = \frac{4}{5}$ and $\tan B = \frac{5}{12}$, find, without using a calculator:

(i) $\sin(A + B)$; (ii) $\tan(A - B)$; (iii) $\sec(A + B)$.

Solution

You can use Pythagoras's theorem to complete two right-angled triangles, one for each angle as shown. Remember, however, that A is in fact an **obtuse** angle.

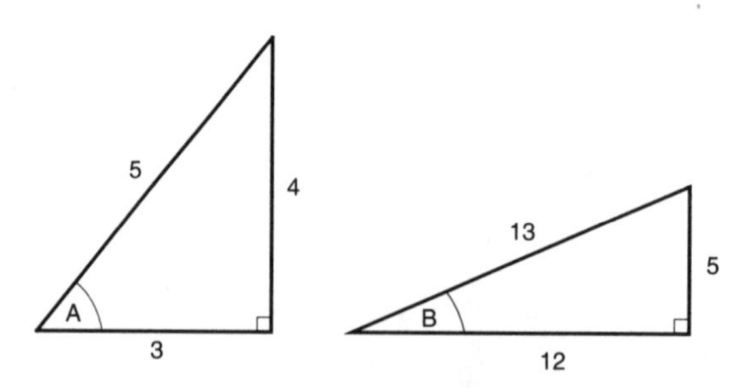

Figure 9.15

(i) $\sin(A+B) = \sin A \cos B + \cos A \sin B$

$$= \left(\tfrac{4}{5}\right) \times \left(\tfrac{12}{13}\right) + \left(-\tfrac{3}{5}\right) \times \left(\tfrac{5}{13}\right) = \tfrac{33}{65}$$

↑

negative because
A is obtuse

negative because
A is obtuse

↓

(ii) $$\tan(A-B) = \frac{\tan A - \tan B}{1 + \tan A \tan B} = \frac{-\frac{4}{3} - \frac{5}{12}}{1 + \left(-\frac{4}{3}\right) \times \left(\frac{5}{12}\right)}$$

$$= \tfrac{63}{16}$$

(iii) Find $\cos(A+B)$ first

$$\cos(A+B) = \cos A \cos B - \sin A \sin B$$

$$= \left(-\tfrac{3}{5}\right) \times \left(\tfrac{12}{13}\right) - \left(\tfrac{4}{5}\right) \times \left(\tfrac{5}{13}\right) = -\tfrac{56}{65}$$

$$\therefore \quad \sec(A+B) = -\tfrac{65}{56}$$

Example 9.10

Solve the equation $\sin x = 2\sin(x - 30°)$ for $0 \leq x \leq 360°$.

Solution

$$\sin x = 2\sin(x - 30°)$$

$$= 2[\sin x \cos 30 - \cos x \sin 30]$$

$$= 2(\sin x) \times 0.866 - (2\cos x) \times 0.5$$

$$\therefore \quad \cos x = 1.732 \sin x - \sin x = 0.732 \sin x$$

$$\therefore \quad \frac{1}{0.732} = \frac{\sin x}{\cos x} = \tan x$$

that is, $\tan x = 1.366$

$$\therefore \quad x = 53.8° \quad \text{or} \quad 233.8°$$

Example 9.11

Prove the following identities:

(i) $\dfrac{\cos(X+Y)}{\cos X \cos Y} \equiv 1 - \tan X \tan Y$

(ii) $\dfrac{\cos A}{\sin A} + \dfrac{\sin A}{\cos A} \equiv 2\operatorname{cosec} 2A$

(iii) $\frac{1}{2}(\tan\theta + \cot\theta) \equiv \operatorname{cosec} 2\theta$

(iv) $\cos 3A \equiv 4\cos^3 A - 3\cos A$

> **MEMORY JOGGER**
> When proving a trig identity try starting with the more complicated side and rearrange it to equal the other side.

Solution

(i) L.H.S. $= \dfrac{\cos(X+Y)}{\cos X \cos Y} = \dfrac{\cos X \cos Y - \sin X \sin Y}{\cos X \cos Y}$

$= \dfrac{\cos X \cos Y}{\cos X \cos Y} - \dfrac{\sin X \sin Y}{\cos X \cos Y}$

$= 1 - \tan X \tan Y =$ R.H.S.

(ii) L.H.S. $= \dfrac{\cos^2 A + \sin^2 A}{\sin A \cos A} = \dfrac{1}{\sin A \cos A}$

$= \dfrac{2}{2 \sin A \cos A} = \dfrac{2}{\sin 2A} = 2 \operatorname{cosec} 2A$

$=$ R.H.S.

(iii) L.H.S. $= \frac{1}{2}(\tan\theta + \cot\theta) = \frac{1}{2}\left(\tan\theta + \dfrac{1}{\tan\theta}\right)$

$= \frac{1}{2}\left(\dfrac{\tan^2\theta + 1}{\tan\theta}\right) = \dfrac{1 + \tan^2\theta}{2\tan\theta}$

$= \dfrac{\sec^2\theta}{2\tan\theta} = \frac{1}{2} \times \dfrac{1}{\cos^2\theta} \times \dfrac{\cos\theta}{\sin\theta} = \dfrac{1}{2\cos\theta\sin\theta}$

$= \dfrac{1}{\sin 2\theta} = \operatorname{cosec} 2\theta$

(iv)

> **MEMORY JOGGER**
> Identities containing triple angles usually have to be worked out in two stages.

$$
\begin{aligned}
\cos 3A &= \cos(2A + A) = \cos 2A \cos A - \sin 2A \sin A \\
&= (2\cos^2 A - 1)\cos A - 2\sin A \cos A \sin A \\
&= 2\cos^3 A - \cos A - 2\sin^2 A \cos A \\
&= 2\cos^3 A - \cos A - 2(1 - \cos^2 A)\cos A \\
&= 4\cos^3 A - 3\cos A
\end{aligned}
$$

Exercise 9(h)

1 Prove the identities:

(i) $\sin(A + 45°) \equiv \frac{1}{\sqrt{2}}(\sin A + \cos A)$

(ii) $\frac{1 - \cos\theta}{1 + \cos\theta} \equiv \tan^2 \frac{1}{2}\theta$

(iii) $\frac{1}{1 + \cos 2A} \equiv \frac{\tan 2A}{\tan 2A + 1}$

(iv) $\frac{\cos\theta - \sin\theta}{\cos\theta + \sin\theta} \equiv \frac{\cos 2\theta}{1 + \sin\theta}$

(v) $\frac{\sin X}{\sin Y} + \frac{\cos X}{\cos Y} \equiv 2\sin(X + Y)\operatorname{cosec} 2Y$

(vi) $\frac{\cos 4\theta}{1 + \sin 4\theta} + \frac{1 + \sin 4\theta}{\cos 4\theta} \equiv \frac{2}{\cos 4\theta}$

2 Solve the equations for $0° \leq x \leq 360°$:

(i) $\sin^2 x - \cos^2 x = 0.5$
(ii) $\tan 2x + \tan x = 0$
(iii) $\sin(x + 30°) + \cos(x - 60°) = \cos(x + 10°)$

9.10 $a\cos\theta + b\sin\theta$

The expression $a\cos\theta + b\sin\theta$ is very common in mathematics. You need to remember that it comes indirectly from the addition formulae in Equations (TI4–7):

Let $R = \sqrt{a^2 + b^2}$

then $a\cos\theta + b\sin\theta = R\left[\frac{a}{R}\cos\theta + \frac{b}{R}\sin\theta\right]$

Check by multiplying out the R.H.S.

Draw a right-angled triangle, base angle α as shown in Figure 9.16.

Then $\frac{a}{R} = \cos\alpha$ and $\frac{b}{R} = \sin\theta$

$$\therefore \quad a\cos\theta + b\sin\theta = R[\cos\alpha\cos\theta + \sin\alpha\sin\theta]$$
$$= R\cos(\theta - \alpha)$$

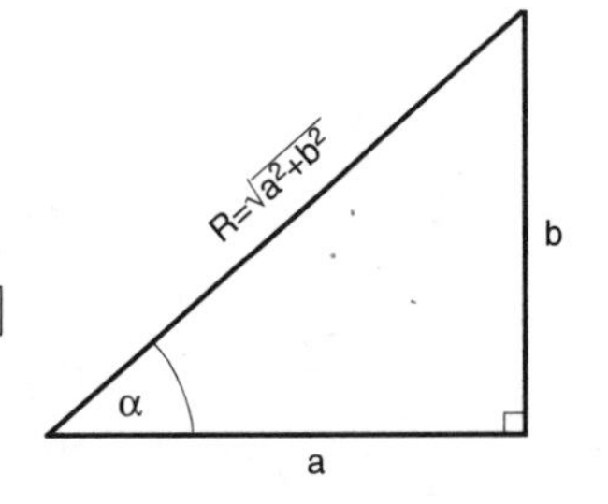

Figure 9.16

For example, consider:

(i) $3\cos\theta + 4\sin\theta$

$$a = 3,\ b = 4,\ \text{so } R = \sqrt{3^2 + 4^2} = 5$$

$$\cos\alpha = \tfrac{3}{5},\ \sin\alpha = \tfrac{4}{5} \qquad \therefore \quad \alpha = 53.1^\circ$$

$$\therefore \quad 3\cos\theta + 4\sin\theta = 5\cos(\theta - 53.1^\circ)$$

(ii) $5\cos x - 4\sin x$

(Take care with the negative sign.)

$$a = 5,\ b = -4 \quad R = \sqrt{5^2 + (-4)^2} = \sqrt{41}$$

$$\cos\alpha = \frac{5}{\sqrt{41}} \quad \sin\alpha = \frac{-4}{\sqrt{41}}$$

(**Note:** Since $\cos\alpha > 0$ and $\sin\alpha < 0$, α must be in the fourth quadrant, so $\alpha = 321.3^\circ$.

$$\therefore \quad 5\cos x - 4\sin x = \cos(x - 321.3^\circ)$$

We are now in a position to solve equations of the type $a\cos\theta + b\sin\theta = c$.

Example 9.12

Solve the equation $4\sin x - 2\cos x = 3$ for $0 \leq x \leq 360^\circ$.

Solution

Here, note that to get the equation in the required form, it needs to be rewritten:

$$-2\cos x + 4\sin x = 3$$

$$\therefore \quad R = \sqrt{(-2)^2 + 4^2} = \sqrt{20}, \qquad a = -2,\ b = 4$$

$$\therefore \quad \cos\alpha = \frac{-2}{\sqrt{20}} \quad \text{and} \quad \sin\alpha = \frac{4}{\sqrt{20}}$$

Since $\cos\alpha < 0$ and $\sin\alpha > 0$, α is in the second quadrant,

$$\therefore \quad \sqrt{20}\cos(x - 116.6^\circ) = 3$$

$$\therefore \quad \cos(x - 116.6^\circ) = \frac{3}{\sqrt{20}}$$

So: $x - 116.6^\circ = 47.9^\circ + 360^\circ n$

or $312.1^\circ + 360^\circ n$

$\therefore \quad x = 164.5^\circ + 360^\circ n$

or $428.7^\circ + 360^\circ n$

in the required range $x = 68.7^\circ,\ 164.5^\circ$.

Exercise 9(I)

1 Express the following in the form $R\cos(x - \alpha)$:

(i) $2\sin x + 3\cos x$ (ii) $5\sin x - 12\cos x$

(iii) $\sin x - \cos x$ (iv) $3\cos x - 2\sin x$

2 Solve the following equations for $0^\circ \leq x \leq 360^\circ$:

(i) $2\sin x + 3\cos x = 1$ (ii) $\sin x + \cos x = 0.5$

(iii) $2\sin x - 13\cos x = -0.5$ (iv) $4\sin x - 3\cos x = -1$

9.11 Further identities

(i) Since $\sin(A + B) = \sin A\cos B + \cos A\sin B$ (i)

and $\sin(A - B) = \sin A\cos B - \cos A\sin B$ (ii)

(i) + (ii) gives: $\sin(A + B) + \sin(A - B) = 2\sin A\cos B$

Replace $A + B$ with X and $A - B$ with Y, to get:

$$\sin X + \sin Y = 2\sin\frac{X + Y}{2}\cos\frac{X + Y}{2} \qquad \text{(TI13)}$$

similarly $$\sin X - \sin Y = 2\cos\frac{X + Y}{2}\cos\frac{X + Y}{2} \qquad \text{(TI14)}$$

$$\cos X + \cos Y = 2\cos\frac{X + Y}{2}\cos\frac{X - Y}{2} \qquad \text{(TI15)}$$

$$\cos X - \cos Y = -2\sin\frac{X + Y}{2}\sin\frac{X - Y}{2} \qquad \text{(TI16)}$$

These identities are particularly useful in integration.

(ii) Using $\cos 2A = \cos^2 A - \sin^2 A$ and $\cos^2 A + \sin^2 A = 1$,

then $$\cos 2A = \frac{\cos^2 A - \sin^2 A}{\cos^2 A + \sin^2 A} \quad (\div \text{ top and bottom by } \cos^2 A)$$

$$= \frac{1 - \tan^2 A}{1 + \tan^2 A} \quad \left(\text{often written } \frac{1 - t^2}{1 + t^2}\right) \qquad \text{(TI17)}$$

also $$\sin 2A = \frac{2\tan A}{1 + \tan^2 A} \quad \text{written } \frac{2t}{1 + t^2} \qquad \text{(TI18)}$$

and $$\tan 2A = \frac{2\tan A}{1 - \tan^2 A} \quad \text{written } \frac{2t}{1 - t^2} \qquad \text{(TI19)}$$

These are often called the t formulae. They enable you to write any trig function in terms of the same quantity, t.

1 If $\tan A = -\frac{3}{4}$, and $\sin B = \frac{5}{13}$, where A and B are obtuse, find:

(i) $\tan(A+B)$ (ii) $\sin(A-B)$ (iii) $\cos(A+B)$

2 Solve the equation $3\sin x - 2\cos x = 1$ for $0 \le x \le 2\pi$.

3 Find an expression for $\tan 3x$ in terms of $\tan x$.

4 Without using the calculus, find the maximum value of the expression $y = 4\cos x + 6\sin x$. Sketch the graph for $0° \le x \le 360°$.

5 Find the general solution of the equation $4\cos(2x - 30°) = 1$.

6 Prove the identities.

(i) $\dfrac{\sin\theta}{1-\cos\theta} \equiv \cot\frac{1}{2}\theta$ (ii) $\dfrac{\sin A + \sin B}{\cos A + \cos B} \equiv \tan\frac{1}{2}(A+B)$

7 Show that $\sin\theta + \sin 2\theta + \sin 3\theta \equiv \sin 2\theta(1 + 2\cos\theta)$, and hence solve the equation $\sin\theta + \sin 2\theta + \sin 3\theta = 0$.

8 If $\sin(\theta - \alpha) = 4\cos(\theta + \alpha)$, find $\tan\alpha$.

Revision Problems 9

1 The diagram shows part of a trace on an oscilloscope. The equation of the curve is: $y = 4\sin 2x$.

At point A, $y = -0.6$. What is the largest negative value that x could be (measured in radians)?

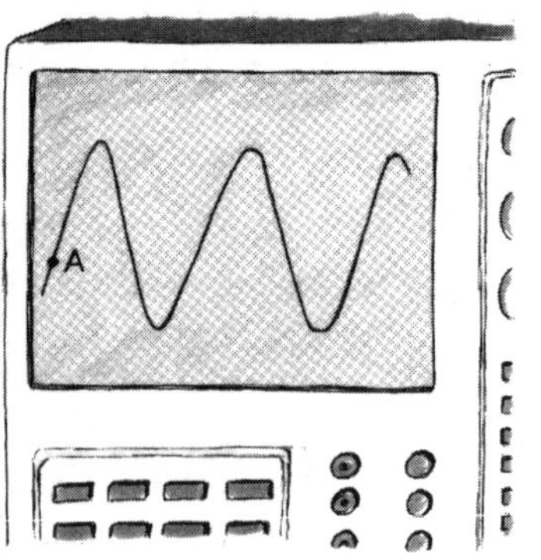

Figure 9.17

2 A cork bobs up and down in water (see Figure 9.18), so that the distance x cm of the top of the cork below a fixed line is given by:

$$x = 4\sin\left(2t + \frac{\pi}{4}\right)$$

where t is the time in seconds.

(i) Find the earliest time for which $x = 2$ cm.

(ii) Find x, when $t = 2.5$ secs. Interpret your answer clearly.

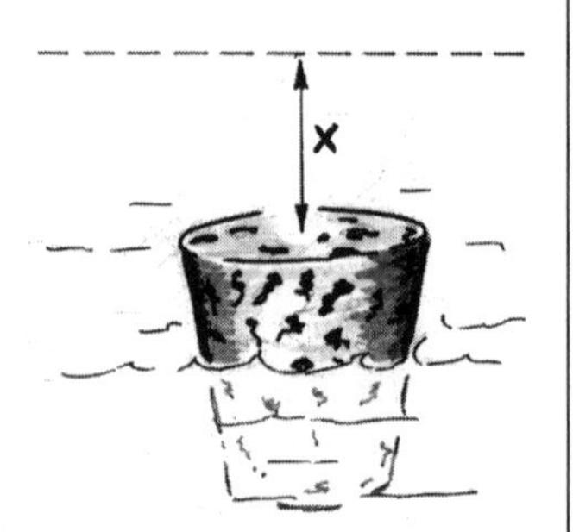

Figure 9.18

3 Find the values of x (in degrees) at the points where the two curves $y = 4\sin^2 x$, and $2y = 3\sin x + 1$ meet in the range $-180° \le x \le 180°$.

4 The amount x that a car suspension spring compresses after t seconds as it travels along a bumpy road is given by the equation:

$$x = 1.2\cos\left(4t + \frac{\pi}{2}\right)$$

How many times does $x = 0.6$ for values of t between 0 and 10 inclusive?

5 A spider is making a web. Twelve equally spaced radial threads, all lying in the same plane, have been established from the centre O of the web. The spider is on one ray at A, distance h from O, and moves in a straight path to B on an adjacent ray.

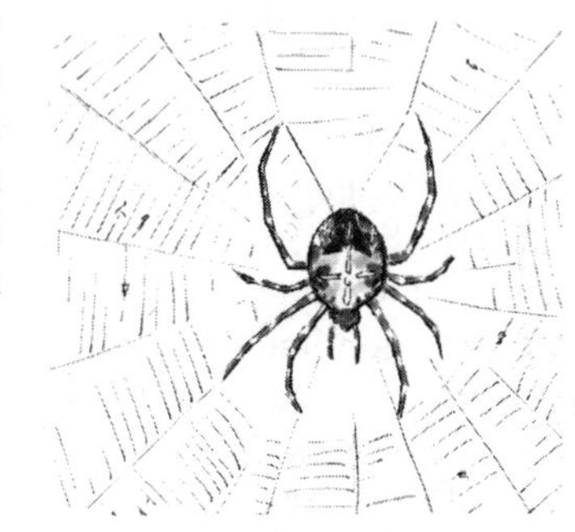

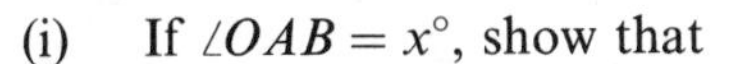

(i) If $\angle OAB = x°$, show that

$$OB = \frac{h \sin x°}{\sin(150 - x)°}$$

For what angle x is OB equal to h?

(ii) Expand $\sin(150 - x)°$, and hence, by writing the equation $\sin x° = \sin(150 - x)°$ as an equation in $\tan x°$, deduce an **exact** expression for the tangent of the angle in the final part of your answer to (i) above.

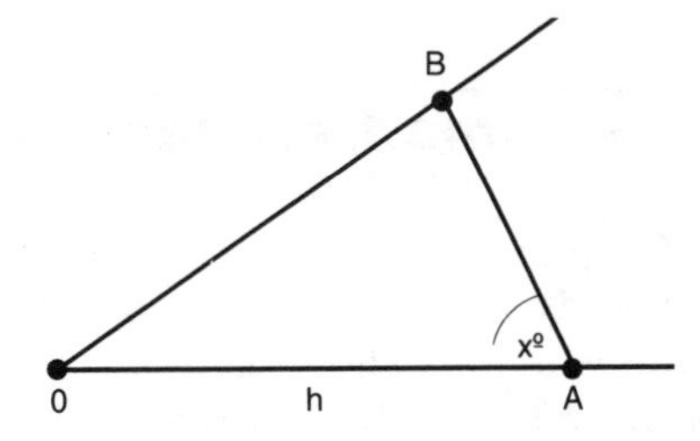

Figure 9.19

(Oxford & Cambridge)

10 Calculus applied to trig functions

We have looked at the calculus applied to situations where the functions are only algebraic. However, in Chapter 9 we looked at a whole new range of trigonometric functions which we can now work with. In order to differentiate them, we need to look more closely at why it is necessary to measure angles using radians.

10.1 Small angles

In Chapter 9, we came across the idea of measuring an angle using radians. This idea leads to some extremely useful results if the angles are small.

Figure 10.1 shows a sector of a circle OAB centre O of radius r. OB is extended to C where CA is a tangent to the circle at A. Angle $AOB = \theta$ radians.

It is not difficult to see that

area of triangle $AOB <$ area of sector $OAB <$ area of triangle OAC

$\therefore \quad \frac{1}{2} \times r \times r \times \sin\theta < \frac{1}{2}r^2\theta < \frac{1}{2} \times OA \times AC$

Hence: $\quad \frac{1}{2}r^2 \sin\theta < \frac{1}{2}r^2\theta < \frac{1}{2}r \times r\tan\theta$

$\div\frac{1}{2}r^2 \quad \sin\theta < \theta < \tan\theta$

If θ is zero, the areas are all *equal* to zero, and so the inequality can be written:

$$\sin\theta \leq \theta \leq \tan\theta$$

$$\div \sin\theta \quad 1 \leq \frac{\theta}{\sin\theta} \leq \frac{1}{\cos\theta}$$

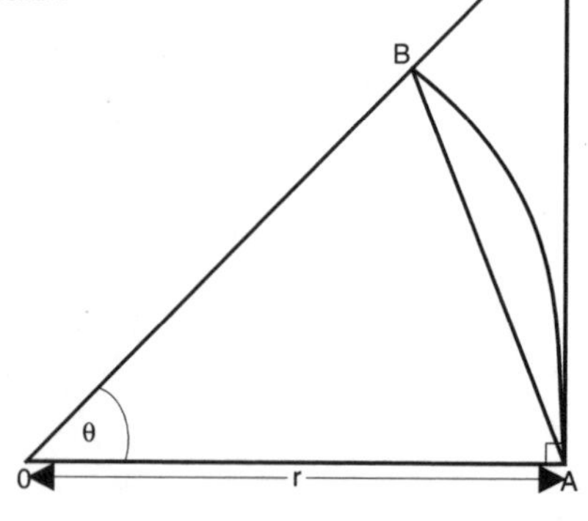

Figure 10.1

Now look at a few values of θ (measured in radians).

$$\text{If}\quad \theta = 1 \qquad 1 \leq \frac{\theta}{\sin\theta} \leq 1.85$$

$$\theta = 0.1 \qquad 1 \leq \frac{\theta}{\sin\theta} \leq 1.005$$

$$\theta = 0.01 \qquad 1 \leq \frac{\theta}{\sin\theta} \leq 1.000045$$

You can see that as $\theta \to 0$, $\frac{\theta}{\sin\theta} \approx 1$. In other words, if θ is small and measured in radians:

$$\sin\theta \approx \theta$$

Note: 0.1 radians = 5.7° and so this result is true for a reasonably large angle.

$$\begin{aligned} \text{Now}\quad \cos\theta &= \sqrt{1-\sin^2\theta} \\ &= \sqrt{1-\theta^2} = \left(1-\theta^2\right)^{\frac{1}{2}} && \text{for small angles} \\ &= 1 - \tfrac{1}{2}\theta^2 && \text{(using the binomial theorem and ignoring the } \theta^4 \text{ term and beyond).} \end{aligned}$$

$$\begin{aligned} \text{Also}\quad \tan\theta &= \frac{\sin\theta}{\cos\theta} = \frac{\theta}{1-\frac{1}{2}\theta^2} = \theta(1-\tfrac{1}{2}\theta^2)^{-1} \\ &= \theta(1+\theta^2) && \text{(using the binomial theorem and neglecting small terms)} \\ &= \theta + \tfrac{1}{2}\theta^3 \\ &= \theta && \text{(neglecting small terms.)} \end{aligned}$$

Hence we can summarise this by saying that if θ is measured in radians, and θ is about 0.1 or less:

$$\sin\theta \approx \theta \qquad \text{(SA1)}$$

$$\cos\theta \approx 1 - \tfrac{1}{2}\theta^2 \qquad \text{(SA2)}$$

$$\tan\theta \approx \theta \qquad \text{(SA3)}$$

Example 10.1

If θ is small and measured in radians, show that $\frac{\sin 2\theta}{1+\tan\theta} \approx 2\theta - 2\theta^2$.

Solution

If θ is small, $\sin 2\theta = 2\theta$,

$$\begin{aligned} \therefore \quad \frac{\sin 2\theta}{1+\tan\theta} &= \frac{2\theta}{1+\theta} = 2\theta(1+\theta)^{-1} \\ &= 2\theta(1-\theta+\ldots) \\ &= 2\theta - 2\theta^2 \end{aligned}$$

Using the binomial theorem as far as the term in θ

Example 10.2

The points with coordinates $A(\alpha, \cos\alpha)$ and $B(\beta, \cos\beta)$ lie on the line $y = \cos x$. Find the gradient of the straight line joining A and B. If $\beta - \alpha$ is very small, use your result to find the gradient of the tangent at A.

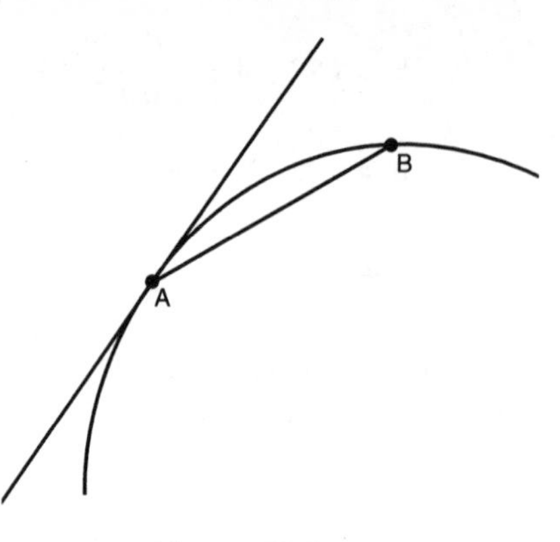

Figure 10.2

Solution

The gradient of AB is

$$\frac{\cos\beta - \cos\alpha}{\beta - \alpha} = \frac{-2\sin\dfrac{\beta+\alpha}{2}\sin\dfrac{\beta-\alpha}{2}}{\beta - \alpha}$$

(Using Equation (TI16))

If $\dfrac{\beta-\alpha}{2} = \theta$, then if $\beta - \alpha$ is small, θ is small.

$$\text{Hence the gradient } = \frac{-2\sin\dfrac{\beta+\alpha}{2}\sin\theta}{2\theta} = -\sin\frac{\beta+\alpha}{2} \times \frac{\sin\theta}{\theta}$$

Now $\dfrac{\sin\theta}{\theta} \to 1$ as $\theta \to 0$, and if $\beta - \alpha$ is small, then $\beta \approx \alpha$. Hence:

$$\frac{\beta+\alpha}{2} = \frac{\alpha+\alpha}{2} = \alpha$$

Hence the gradient of $AB = -\sin\alpha$.

Now if B and A are very close together, the gradient of AB is approximately the same as the gradient of the tangent at A. Hence the gradient of the tangent at $A = -\sin\alpha$. (This result is the same as differentiating $y = \cos x$; see page 217.)

Exercise 10(a)

1 In the following examples, θ is assumed to be measured in radians, and small enough so that θ^3 can be ignored. Obtain the answer in the form $a + b\theta + c\theta^2$.

(i) $\sin 4\theta$ (ii) $1 + \cos 3\theta$

(iii) $1 + \sin^2\theta$ (iv) $\dfrac{\sin 2\theta}{1 + \cos 2\theta}$

(v) $\sin 3\theta(1 + \cos 2\theta)$ (vi) $\tan 4\theta$

2 In triangle PQR, angle Q is $\frac{1}{3}\pi$ radians, and angle P is $\frac{\pi}{2} + x$ radians where x is small. Prove that

$$\frac{QR}{PR} \approx \frac{2}{\sqrt{3}} - \frac{1}{\sqrt{3}}x^2 \qquad \text{(Hint: use the sine rule.)}$$

10.2 Differentiation of sin x, cos x and tan x

(i) If $y = \sin x$, then if we increase x by an amount δx, y changes by an amount δy.

Hence: $y + \delta y = \sin(x + \delta x)$ (i)

also $y = \sin x$ (ii)

(i) – (ii) $\delta y = \sin(x + \delta x) - \sin x$

$= \sin x \cos \delta x + \cos x \sin \delta x - \sin x$

(Using Equation (TI4))

$x + \delta x$ must be all inside the sine function

Now if δx is very small, $\cos \delta x \approx 1$ and $\sin \delta x \approx \delta x$,

$$\therefore \quad \delta y \approx \sin x + \delta x \cos x - \sin x$$

$$\approx \delta x \cos x$$

$$\therefore \quad \frac{\delta y}{\delta x} \approx \cos x$$

In other words, as $\delta x \to 0$,

$$\frac{\mathrm{d}y}{\mathrm{d}x} = \cos x$$

(ii) Similarly, if $y = \cos x$ (i)

$y + \delta y = \cos(x + \delta x)$ (ii)

(ii) – (i) $\delta y = \cos(x + \delta x) - \cos x$

$\delta y = \cos x \cos \delta x - \sin x \sin \delta x - \cos x$

if $\delta x \to 0$,

$$\delta y \approx \cos x - \delta x \sin x - \cos x$$

$$\approx -\delta x \sin x$$

$$\therefore \quad \frac{\delta y}{\delta x} \approx -\sin x$$

hence as $\delta x \to 0$ $\frac{\mathrm{d}y}{\mathrm{d}x} = -\sin x$.

(iii) If $y = \tan x$, it is easier to write this as a quotient:

So $y = \frac{\sin x}{\cos x}$

Using Equation (D5) for differentiation of a quotient, we get:

$$\frac{\mathrm{d}y}{\mathrm{d}x} = \frac{(\cos x) \times (\cos x) - (\sin x) \times (-\sin x)}{(\cos x)^2}$$

$$= \frac{\cos^2 x + \sin^2 x}{\cos^2 x} = \frac{1}{\cos^2 x}$$

that is, $\frac{\mathrm{d}y}{\mathrm{d}x} = \sec^2 x$

SUMMARY

$$\frac{d}{dx}(\sin x) = \cos x \qquad \text{(D7)}$$

$$\frac{d}{dx}(\cos x) = -\sin x \qquad \text{(D8)}$$

$$\frac{d}{dx}(\tan x) = \sec^2 x \qquad \text{(D9)}$$

10.3 Function of a function (chain rule), etc.

As with the examples covered in Chapter 5, it is better to write these examples out in full until you have really grasped the method. The following example gives a wide range of applications of the chain rule to trigonometric type functions.

Example 10.2

Differentiate the following functions with respect to x. Simplify your answers if possible.

(i) $\sin(2x+5)$ (ii) $\sin x^2$ (iii) $\cos^2 x$

(iv) $\tan(3x+1)$ (v) $\tan^2(2x+\frac{\pi}{4})$ (vi) $\sin^2(4x+1)$

(vii) $x \sin^2 x$ (viii) $\dfrac{2x}{1+x\cos x}$

Solution

(i) $y = \sin(2x+5)$: if $u = 2x+5$

$$y = \sin u \quad \frac{dy}{du} = \cos u \quad \text{and} \quad \frac{du}{dx} = 2$$

So: $$\frac{dy}{dx} = \cos u \times 2 = 2\cos(2x+5).$$

(ii) $y = \sin x^2$: let $u = x^2$

$$\therefore \quad y = \sin u \quad \frac{dy}{du} = \cos u \quad \text{and} \quad \frac{du}{dx} = 2x$$

Hence: $$\frac{dy}{dx} = \cos u \times 2x$$

$$= 2x \cos x^2$$

(iii) $y = \cos^2 x$

$y = (\cos x)^2$

let $u = \cos x \quad \therefore y = u^2$,

Note: $\cos^2 x$ means $(\cos x) \times (\cos x)$ and can be written $(\cos x)^2$

$$\therefore \quad \frac{dy}{du} = 2u, \; \frac{du}{dx} = -\sin x$$

$$\frac{dy}{dx} = 2u \times -\sin x$$

$$= -2\cos x \sin x$$

$$= -\sin 2x \qquad \text{(Using Equation (TI11))}$$

(iv) $y = \tan(3x+1)$, let $u = 3x+1$

$$\therefore \quad y = \tan u, \quad \frac{dy}{du} = \sec^2 u \quad \text{and} \quad \frac{du}{dx} = 3$$

$$\therefore \quad \frac{dy}{dx} = \sec^2 u \times 3$$

$$= 3\sec^2(3x+1)$$

(v) This example requires use of the chain rule twice, since the expression is a function of a function of a function.

So: if $u = 2x + \frac{\pi}{4}$, $\quad y = \tan^2 u = (\tan u)^2$

If $\quad v = \tan u, \quad$ then $y = v^2$

$$\text{So:} \quad \frac{dy}{dv} = 2v, \quad \frac{dv}{du} = \sec^2 u \quad \text{and} \quad \frac{du}{dx} = 2$$

$$\therefore \quad \frac{dy}{dx} = \frac{dy}{dv} \times \frac{dv}{du} \times \frac{du}{dx}$$

$$= 2v \times \sec^2 u \times 2$$

$$= 2\tan u \times \sec^2(2x + \tfrac{\pi}{4}) \times 2$$

$$= 4\tan(2x + \tfrac{\pi}{4})\sec^2(2x + \tfrac{\pi}{4})$$

(vi) This is similar to the last part:

if $\quad u = 4x + 1, \quad$ and $\quad v = \sin u, \quad$ then $\quad y = v^2$

$$\text{So} \quad \frac{dy}{dv} = 2v, \quad \frac{dv}{du} = \cos u \quad \text{and} \quad \frac{du}{dx} = 4$$

$$\text{Hence:} \quad \frac{dy}{dx} = \frac{dy}{dv} \times \frac{dv}{du} \times \frac{du}{dx}$$

$$= 2v \times \cos u \times 4$$

$$= 8\sin u \cos u$$

$$= 4\sin 2u \qquad \text{(Using Equation (TI11))}$$

$$= 4\sin(8x+2)$$

(vii) We now combine algebraic and trigonometric functions. This is a product:

$$\frac{dy}{dx} = x \times 2\sin x \cos x + \sin^2 x \times 1$$
$$= x \sin 2x + \sin^2 x$$

(viii) This is a quotient:

$$\frac{dy}{dx} = \frac{(1 + x\cos x) \times 2 - 2x \times (x \times -\sin x + \cos x \times 1)}{(1 + x\cos x)^2}$$
$$= \frac{2 + 2x\cos x + 2x^2 \sin x - 2x\cos x}{(1 + x\cos x)^2}$$
$$= \frac{2 + 2x^2 \sin x}{(1 + x\cos x)^2}$$
$$= \frac{2(1 + x^2 \sin x)}{(1 + x\cos x)^2}$$

It is worth expanding our previous summary to give:

$$\frac{d}{dx}(\sin(ax + b)) = a\cos(ax + b)$$
$$\frac{d}{dx}(\cos(ax + b)) = -a\sin(ax + b)$$
$$\frac{d}{dx}(\tan(ax + b)) = a\sec^2(ax + b)$$

Now try the following exercise.

Exercise 10(b)

Differentiate the following expressions with respect to x; simplify your answers if possible.

1	$\sin 4x$	**2**	$\cos\left(x + \frac{\pi}{4}\right)$	**3**	$\tan^2 x$
4	$4\cos 2x$	**5**	$3\tan 4x$	**6**	$5\cos^2 3x$
7	$\tan\left(2x - \frac{\pi}{4}\right)$	**8**	$\tan^3 x$	**9**	$\sin^2 2x$
10	$x\sin x$	**11**	$x\cos^3 x$	**12**	$\frac{x}{1 + \sin x}$
13	$x\sin\left(2x + \frac{\pi}{4}\right)$	**14**	$x^2 \sin 3x$	**15**	$\frac{(1 + x)^2}{\sin x}$
16	$(1 + \sin^2 2x)^2$	**17**	$\frac{x - \sin x}{x + \sin x}$	**18**	$x^3 \tan 2x$
19	$x^2 \tan^2 x$	**20**	$4\sin^3(x^3 + 1)$		

10.4 Differentiating sec x, cosec x, cot x

(i) If $y = \sec x$, then $y = (\cos x)^{-1}$.

So $\dfrac{dy}{dx} = -(\cos x)^{-2} \times -\sin x$ (Using the chain rule)

$$= \frac{\sin x}{\cos^2 x}$$

This is not a particularly useful form for the answer.

Better $\dfrac{dy}{dx} = \dfrac{\sin x}{\cos x} \times \dfrac{1}{\cos x} = \sec x \tan x$

(ii) If $y = \operatorname{cosec} x$, then $y = (\sin x)^{-1}$

So $\dfrac{dy}{dx} = -(\sin x)^{-2} \times \cos x$ (Using the chain rule)

$$= -\frac{\cos x}{\sin^2 x} = \frac{-\cos x}{\sin x} \times \frac{1}{\sin x}$$

$$= -\operatorname{cosec} x \cot x$$

(iii) If $y = \cot x$, then $y = \dfrac{\cos x}{\sin x}$,

$\therefore \dfrac{dy}{dx} = \dfrac{(\sin x) \times (-\sin x) - (\cos x) \times (\cos x)}{\sin^2 x}$ (Using the quotient rule)

$$= \frac{-(\sin^2 x + \cos^2 x)}{\sin^2 x} = \frac{-1}{\sin^2 x}$$

$$= -\operatorname{cosec}^2 x$$

In a similar fashion to the last section, these results can be slightly extended to the following general results:

$$\frac{d}{dx}(\sec(ax+b)) = a\sec(ax+b)\tan(ax+b) \quad \text{(D10)}$$

$$\frac{d}{dx}(\operatorname{cosec}(ax+b)) = -a\operatorname{cosec}(ax+b)\cot(ax+b) \quad \text{(D11)}$$

$$\frac{d}{dx}(\cot(ax+b)) = -\operatorname{cosec}^2(ax+b) \quad \text{(D12)}$$

Example 10.3

Differentiate the following examples with respect to x, simplifying your answers where possible.

(i) $\sec\left(3x + \frac{\pi}{4}\right)$ (ii) $\operatorname{cosec} x^2$ (iii) $\cot^2 x$

(iv) $\operatorname{cosec}^2(3x+4)$ (v) $\dfrac{x^2}{1 + \sec x}$ (vi) $\dfrac{\tan 2x}{1 + \cot 2x}$

Solution

(i) If $y = \sec(3x + \frac{\pi}{4})$, let $u = 3x + \frac{\pi}{4}$

Hence: $y = \sec u \quad \dfrac{dy}{du} = \sec u \tan u$ and $\dfrac{du}{dx} = 3$

Hence: $$\frac{dy}{dx} = \sec u \tan u \times 3$$

$$= 3\sec\left(3x + \frac{\pi}{4}\right)\tan\left(3x + \frac{\pi}{4}\right)$$

(ii) If $u = x^2$, $y = \operatorname{cosec} u$

$$\frac{dy}{du} = -\operatorname{cosec} u \cot u \quad \text{and} \quad \frac{du}{dx} = 2x$$

$$\therefore \quad \frac{dy}{dx} = -\operatorname{cosec} u \cot u \times 2x$$

$$= -2x \operatorname{cosec} x^2 \cot x^2$$

(iii) If $u = \cot x$: $y = u^2$,

then $$\frac{dy}{du} = 2u \quad \text{and} \quad \frac{du}{dx} = -\operatorname{cosec}^2 x$$

$$\therefore \quad \frac{dy}{dx} = 2u \times -\operatorname{cosec}^2 x$$

$$= -2\cot x \operatorname{cosec}^2 x$$

(iv) This is an example where the chain rule is used twice.
If $u = 3x + 4$, and $v = \operatorname{cosec} u$, then $y = v^2$,

$$\frac{dy}{dv} = 2v, \quad \frac{dv}{du} = -\operatorname{cosec} u \cot u \quad \text{and} \quad \frac{du}{dx} = 3$$

Hence: $$\frac{dy}{dx} = 2v \times -\operatorname{cosec} u \cot u \times 3$$

$$= 6\operatorname{cosec} u \times -\operatorname{cosec} u \cot u$$

$$= -6\operatorname{cosec}^2 u \cot u$$

$$= -6\operatorname{cosec}^2(3x + 4)\cot(3x + 4)$$

(v) The quotient rule applies here.

So: $$\frac{dy}{dx} = \frac{(1 + \sec x) \times 2x - x^2 \sec x \tan x}{(1 + \sec x)^2}$$

$$= \frac{2x + 2x\sec x - x^2 \sec x \tan x}{(1 + \sec x)^2}$$

$$= \frac{x(2 + 2\sec x - x\sec x \tan x)}{(1 + \sec x)^2}$$

(vi) Again the quotient rule applies:

$$\frac{dy}{dx} = \frac{(1 + \cot 2x) \times \sec^2 x \times 2 - \tan 2x \times - \operatorname{cosec}^2 2x \times 2}{(1 + \cot 2x)^2}$$

$$= \frac{2[\sec^2 2x + \cot 2x \sec^2 2x + \tan 2x \operatorname{cosec}^2 2x]}{(1 + \cot 2x)^2}$$

It is possible (although unlikely) that this could be simplified further. However, unless you were trying to solve, say, $\frac{dy}{dx} = 0$, it is really not worth the effort.

Exercise 10(c)

Differentiate the following expressions with respect to x, simplifying your answers where possible.

1	$\sec 2x$	**2**	$x \cot x$	**3**	$\operatorname{cosec}^3 2x$
4	$\operatorname{cosec}\left(x + \frac{\pi}{4}\right)$	**5**	$\sec\left(2x - \frac{\pi}{3}\right)$	**6**	$x^3 \sec 3x$
7	$\frac{1 + \sec x}{1 - \sec x}$	**8**	$\sin x \sec^2 x$	**9**	$(1 + \cot 2x)^3$
10	$\operatorname{cosec}\left(\frac{1 + x}{1 - x}\right)$				

10.5 Integrating trigonometric functions

We can now look at integrating trigonometric functions.

Now if $y = \sin x$, $\frac{dy}{dx} = \cos x$

Hence: $\int \cos x \, dx = \sin x + c$

if $y = \cos x$, $\frac{dy}{dx} = -\sin x$, and so $\int \sin x \, dx = -\cos x + c$

You need to be careful with + and − signs in the integrating of trigonometric functions.

Also, if $y = \sin nx$, $\frac{dy}{dx} = n \cos nx$ (Using the chain rule)

Hence: $\int \cos nx \, dx = \frac{1}{n} \sin nx + c$ (I12)

Similarly, $\int \sin nx \, dx = -\frac{1}{n} \cos nx + c$ (I13)

There are several other types of trigonometric function that can be integrated easily, by making use of some of the identities.

So: $\int \cos^2 x \, dx$ can be evaluated by using the identity $\cos 2x = 2\cos^2 x - 1$, which can be rearranged as $\cos^2 x = \frac{1}{2}(\cos 2x + 1)$.

$$\text{Hence} \quad \int \cos^2 x \, dx = \int \left(\tfrac{1}{2}\cos 2x + \tfrac{1}{2}\right) dx$$
$$= \tfrac{1}{4}\sin 2x + \tfrac{1}{2}x + c$$

$$\text{Similarly,} \quad \int \sin^2 x \, dx = \int \tfrac{1}{2}(1 - \cos 2x) \, dx$$
$$= \tfrac{1}{2}x - \tfrac{1}{4}\sin 2x + c$$

$$\text{and} \quad \int \tan^2 x \, dx = \int (\sec^2 x - 1) \, dx$$
$$= \tan x - x + c$$

Note: $\frac{d}{dx}(\tan x) = \sec^2 x$

Since we are working with the calculus, all angles must be measured in radians. Look at the following examples:

Example 10.4

Find:

(i) $\displaystyle\int_0^{\frac{\pi}{4}} \cos^2 4x \, dx$ (ii) $\displaystyle\int_{-\frac{\pi}{3}}^{\frac{\pi}{3}} \tan^2 x \, dx$

giving answers to 3 significant figures.

Solution

(i)
$$\int_0^{\frac{\pi}{4}} \cos^2 4x \, dx = \int_0^{\frac{\pi}{4}} \tfrac{1}{2}(\cos 8x + 1) \, dx$$
$$= \left[\tfrac{1}{16}\sin 8x + \tfrac{1}{2}x\right]_0^{\frac{\pi}{4}}$$
$$= \frac{1}{16}\sin 2\pi + \frac{\pi}{8} = \frac{\pi}{8}$$
$$= 0.393$$

(ii)
$$\int_{-\frac{\pi}{3}}^{\frac{\pi}{3}} \tan^2 x \, dx = \int_{-\frac{\pi}{3}}^{\frac{\pi}{3}} (\sec^2 x - 1) \, dx = \Big[\tan x - x\Big]_{-\frac{\pi}{3}}^{\frac{\pi}{3}}$$
$$= \left[\tan\tfrac{\pi}{3} - \tfrac{\pi}{3}\right] - \left[\tan\left(-\tfrac{\pi}{3}\right) - -\tfrac{\pi}{3}\right]$$
$$= 1.37$$

Example 10.5

Find the area enclosed between the graph of $y = \sin x$, $0 \leq x \leq \pi$, and the line $y = 0.5$.

Solution

The region required is shown in Figure 10.3.

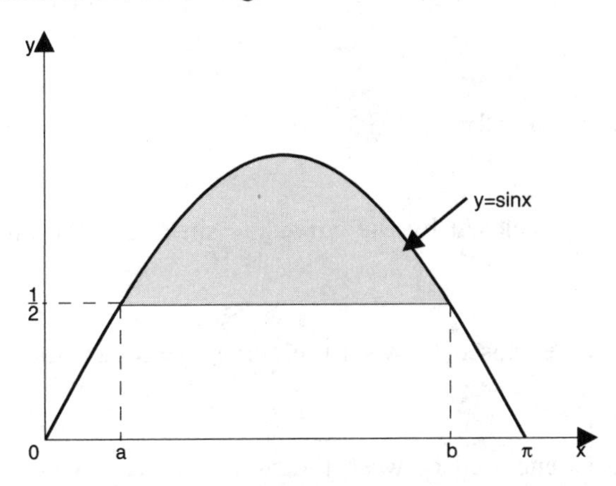

Figure 10.3

Before evaluating the required integral, the limits $x = a$ and $x = b$ need to be calculated. These values will be the solutions of the equation $\sin x = 0.5$,

$$\text{that is,} \quad x = \frac{\pi}{6} \quad \text{or} \quad \frac{5\pi}{6}$$

$$\text{Hence the area } = \int_a^b (y_2 - y_1)\, dx$$

$$= \int_{\frac{\pi}{6}}^{\frac{5\pi}{6}} (\sin x - 0.5)\, dx$$

$$= \Big[-\cos x - 0.5x\Big]_{\frac{\pi}{6}}^{\frac{5\pi}{6}}$$

$$= \left[-\cos\frac{5\pi}{6} - \frac{5\pi}{12}\right] - \left[-\cos\frac{\pi}{6} - \frac{\pi}{12}\right] = 0.685$$

Exercise 10(d)

1 Evaluate the integrals of the following functions with respect to x.

(i) $\cos 4x$ (ii) $3 \sin 2x$ (iii) $\cos^2 4x$

(iv) $\tan^2 3x$ (v) $\sec 2x \tan 2x$ (vi) $\sec^2 2x$

(vii) $-\text{cosec}\, 2x \cot 2x$

2 Find:

(i) $\int_0^{\frac{\pi}{6}} \cos 2x \, dx$ (ii) $\int_0^{\frac{\pi}{2}} 3 \sin 2x \, dx$

(iii) $\int_0^{\frac{\pi}{8}} \sec^2 2t \, dt$ (iv) $\int_{\frac{\pi}{4}}^{\frac{\pi}{2}} \operatorname{cosec}^2 x \, dx$

(v) $\int_{\frac{\pi}{6}}^{\frac{\pi}{3}} \operatorname{cosec} x \cot x \, dx$

3 Find the area enclosed by the curve $y = \sin x$, and the lines $x = \frac{\pi}{2}$ and $y = \frac{\sqrt{3}}{2}$.

4 Find the area enclosed between the curve $y = \cos 2x$ and the x-axis for $-\frac{\pi}{4} \leq x \leq \frac{\pi}{4}$.

5 Find the area enclosed between the curve $y = \sec^2 x$, the x-axis and the lines $x = 0$, $x = \frac{\pi}{4}$.

10.6 Applications

The problems using the calculus, developed in Chapter 5, can now be expanded to include trigonometric functions. The following examples should be studied carefully.

Example 10.6

Find the equation of the tangent to the curve $y = 4 \sin 2x + 5x$ at the point where $x = \frac{\pi}{6}$.

Solution

$$\frac{dy}{dx} = 8 \cos 2x + 5$$

$$\text{at } x = \frac{\pi}{6} \quad \frac{dy}{dx} = 8 \cos \frac{2\pi}{6} + 5 = 9$$

$$\text{Also, if } x = \frac{\pi}{6}, \; y = 4 \sin \frac{2\pi}{6} + \frac{5\pi}{6} = 6.08 \quad \text{(3 sig. figs)}$$

The equation is:

$$(y - 6.08) = 9\left(x - \frac{\pi}{6}\right) = 9x - 4.71$$

Hence: $$y = 9x + 1.37$$

Example 10.7

Find where the first minimum occurs after $x = 0$ on the curve with the equation $y = \sin\left(3x + \frac{\pi}{6}\right)$.

Solution

$$\frac{dy}{dx} = 3\cos\left(3x + \frac{\pi}{6}\right) = 0$$

if $\cos\left(3x + \frac{\pi}{6}\right) = 0$

that is, $3x + \frac{\pi}{6} = \frac{\pi}{2} + 2n\pi$

or $\frac{3\pi}{2} + 2n\pi$

Hence: $x = \frac{\pi}{9} + \frac{2n\pi}{3}$ or $x = \frac{4\pi}{9} + \frac{2n\pi}{3}$

You cannot yet be certain which value of x to take.

Now $\frac{d^2y}{dx^2} = -9\sin\left(3x + \frac{\pi}{6}\right)$

$x = \frac{\pi}{9}$ $\frac{d^2y}{dx^2} = -9\sin\left(\frac{\pi}{3} + \frac{\pi}{6}\right) = -9 < 0$

Hence this point is a maximum.

$x = \frac{4\pi}{9}$ $\frac{d^2y}{dx^2} = -9\sin\left(\frac{4\pi}{3} + \frac{\pi}{6}\right) = 9 > 0$

Hence $x = \frac{4\pi}{9}$ is the first minimum after $x = 0$.

Example 10.8

The end of a piston moves along a straight rod. At $t = 0$, it is at a point A on the rod (see Figure 10.4). After t secs, its distance x from A is given by the formula:

$$x = 1 - \cos^3 2t$$

Find: (i) the speed of the piston after 2 seconds; (ii) the acceleration of the piston and the direction it is moving in after 4 seconds.

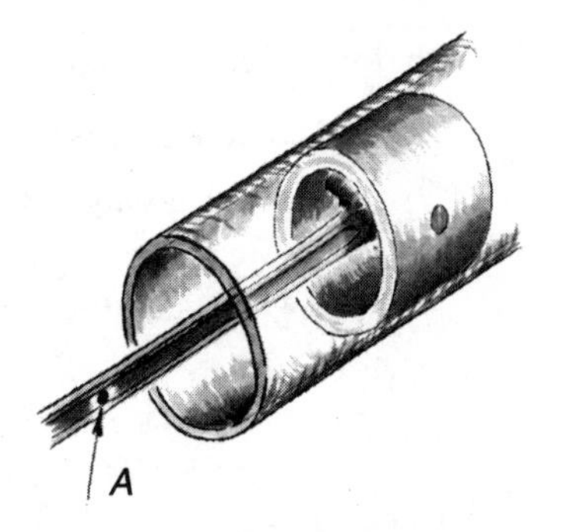

Figure 10.4

Solution

(i) Velocity $= \dfrac{dx}{dt} = -3\cos^2 2t \times -\sin 2t \times 2$

$$= 6\cos^2 2t \sin 2t$$

If $t = 2$ $\quad \dfrac{dx}{dt} = 6\cos^2 4 \sin 4$

$$= -1.94$$

Remember these angles are in radians

Hence the speed $= 1.94\ \text{ms}^{-1}$.

(ii) Acceleration $= \dfrac{d^2x}{dt^2} = 6\cos^2 2t \times 2\cos 2t + \sin 2t \times 12\cos 2t \times -\sin 2t \times 2$

$$= 12\cos^3 2t - 24\sin^2 2t \cos 2t$$

when $t = 4$: $\quad \dfrac{d^2x}{dt^2} = 12\cos^3 8 - 24\sin^2 8\cos 8 = 3.38$

The acceleration is therefore $3.38\ \text{ms}^{-2}$

To find the direction, you need to find the velocity, using the formula from (i).

$$\frac{dx}{dt} = 6\cos^2 8 \sin 8 = 0.126$$

Hence the piston is moving away from A and accelerating.

Example 10.9

Find the turning points on the curve $y = \sin 2x - 2\sin x$ for $0 \le x \le 2\pi$ and sketch the curve.

Solution

$$\frac{dy}{dx} = 2\cos 2x - 2\cos x = 0$$

if $\quad 2(2\cos^2 x - 1) - 2\cos x = 0$

$$4\cos^2 x - 2\cos x - 2 = 0$$

that is, $\quad 2\cos^2 x - \cos x - 1 = 0$

$$(2\cos x + 1)(\cos x - 1) = 0$$

$\therefore \quad \cos x = 1$ giving $x = 0,\ 2\pi$

or $\quad \cos x = -\frac{1}{2}$ giving $x = \dfrac{2\pi}{3}, \dfrac{4\pi}{3}$

$$\frac{d^2y}{dx^2} = -4\sin 2x + 2\sin x$$

Hence at $x = 0,\ 2\pi$, $\quad \dfrac{d^2y}{dx^2} = 0$

This suggests these are points of inflexion.

$$\text{At } x = \frac{2\pi}{3} \quad \frac{d^2y}{dx^2} = -4\sin\frac{4\pi}{3} + 2\sin\frac{2\pi}{3} = 5.2 > 0$$

therefore it is a minimum point.

$$\text{At } x = \frac{4\pi}{3} \quad \frac{d^2y}{dx^2} = -4\sin\frac{8\pi}{3} + 2\sin 4\pi 3 = -5.2 < 0$$

therefore it is a maximum point.

$$\text{At } x = \frac{2\pi}{3},\ y = -2.6$$

$$\text{At } x = \frac{4\pi}{3},\ y = 2.6$$

Also, at $x = 0,\ \pi,\ 2\pi,\ y = 0$

The graph can be sketched as shown in Figure 10.5.

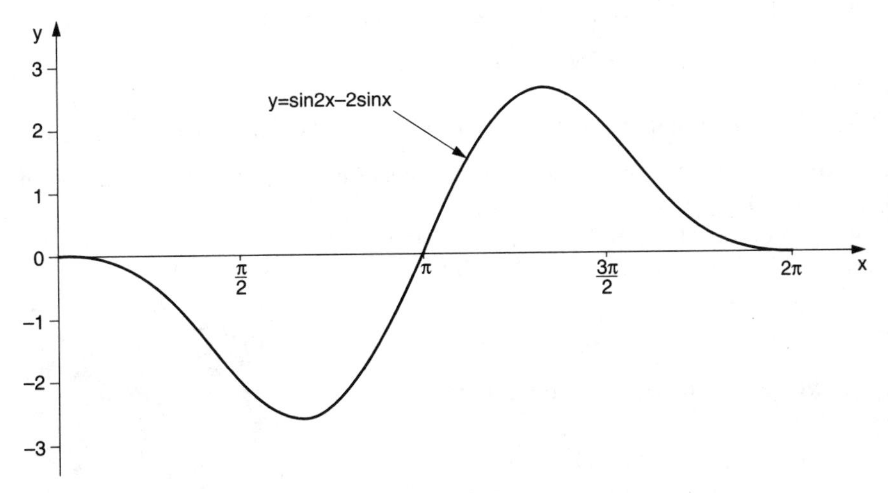

Figure 10.5

Miscellaneous Examples 10

1 If θ is small, find in the form $a + b\theta + c\theta^2$, an expression for:

$$\frac{\sin 3\theta + 1}{1 + \cos^2\theta}$$

2 Differentiate with respect to x:

(i) $x\sin^2 x$ (ii) $\sec x \tan^2 x$ (iii) $\dfrac{1+\cos x}{1-\cos x}$

(iv) $x\sec x$ (v) $\operatorname{cosec}^2 x$ (vi) $\dfrac{\cot 4x}{1-\tan 4x}$

3 Find the equation of the tangent and normal to the curve $y = x\sin x$, at the point $\left(\frac{\pi}{4}, \frac{\pi}{4\sqrt{2}}\right)$.

4 Find

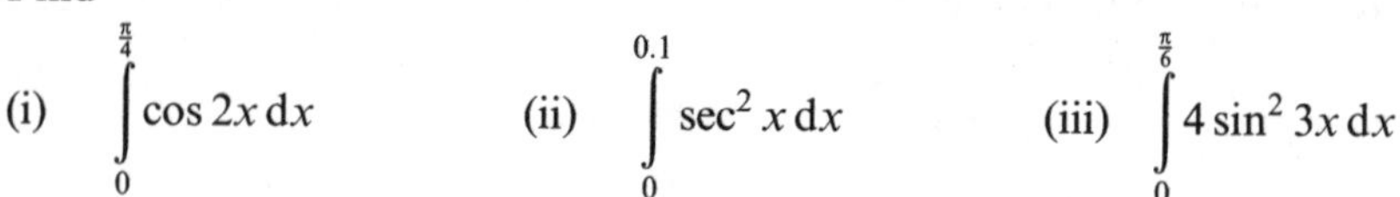

(i) $\int_0^{\frac{\pi}{4}} \cos 2x \, dx$ (ii) $\int_0^{0.1} \sec^2 x \, dx$ (iii) $\int_0^{\frac{\pi}{6}} 4\sin^2 3x \, dx$

5 Find the area enclosed between the curve $y = \sin x$, and the line $y = 0.5$, where $0 \leq x \leq \pi$.

Revision Problems 10

1 Find the area enclosed between the graph of $y = \cos 2x$, and the x and y axes.

2 Find the volume generated when the curve $y = \sin 3x$ between $x = 0$ and $x = \frac{\pi}{6}$ is rotated by 360° about the x-axis.

3 Find the point on the curve $y = 2x + \sin 2x$ for which $\frac{dy}{dx} = 0$, and $0 \leq x \leq \pi$. Determine the nature of this point. Hence sketch the graph of y for $0 \leq x \leq \pi$.

4 The depth of water in metres, h, at the end of a pier is modelled by the equation:

$$h = 2.3 + 2\cos(30t)^\circ$$

where t is the time in hours after high water.

(i) What is $(30t)^\circ$ in radians?

(ii) Find the value of t when the depth of water at the end of the pier is first 3.1 m.

(iii) What is the minimum depth of water at the end of the pier? Find the value of t when this minimum depth first occurs.

(iv) Find $\frac{dh}{dt}$ and deduce the rate in mh^{-1} at which the water is rising or falling.

(a) 4 hours after high water;

(b) 8 hours after high water.

(v) By considering your expression for $\frac{dh}{dt}$, find the value of t at which the water is next rising most quickly.

(Oxford & Cambridge)

11 Vectors

The use of vectors in mathematics and science provides a method of generalising results from two dimensions into three dimensions, which might otherwise be difficult to visualise. The use of vectors is often thought to be difficult, but if you follow the rules they are very straightforward.

11.1 Basic ideas

Figure 11.1 shows a map with a sailing course $A..B..C$ marked on it. The position of a point is fixed by axes running north–south (NS) and east–west (EW). Each square represents 1km.

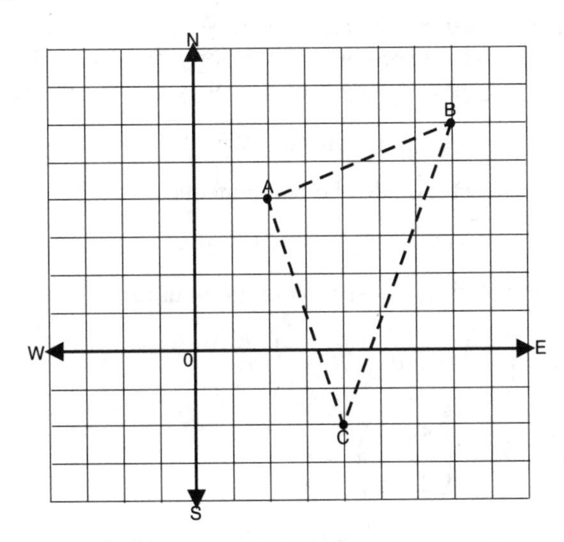

Figure 11.1

Hence A is $(2,4)$, B $(7,6)$ and C $(4,-2)$. If you start at A, the instruction for travelling to B might be to sail for 5.4 km on a bearing of 068.2°. This instruction is a **vector**, because it has size (or length) *and* direction. It will be denoted by $\overrightarrow{AB}$. Without the arrow, AB just equals 5.4, the length from A to B. So $\overrightarrow{AB} = 5.4$ km on a bearing 068.2°. Although this instruction is quite clear, it leads to problems as we develop the theory.

Another way of travelling from A to B is to go 5 km east, then 2 km north. This could be written 5E + 2N or $\begin{pmatrix} 5 \\ 2 \end{pmatrix}$.

$$\text{Hence} \quad \overrightarrow{AB} = 5\text{E} + 2\text{N} \quad \text{or} \quad \overrightarrow{AB} = \begin{pmatrix} 5 \\ 2 \end{pmatrix}$$

To travel from B to C, you would go to 3 km west and 8 km south. This could be written 3W + 8S, but is better written −3E − 8N (**Note:** W can be written −E, and S can be written −N.)

$$\text{So} \qquad \overrightarrow{BC} = -3\text{E} - 8\text{N} \quad \text{or} \quad \overrightarrow{BC} = \begin{pmatrix} -3 \\ -8 \end{pmatrix}$$

$$\begin{aligned} \text{Now} \quad \overrightarrow{AB} + \overrightarrow{BC} &= 5\text{E} + 2\text{N} - 3\text{E} - 8\text{N} \\ &= 2\text{E} - 6\text{N} \end{aligned}$$

If you start at A and go 2E − 6N, you end up at C. This could be written $\overrightarrow{AC}$. But 2E − 6N would take us from A to C.

$$\text{Hence} \quad \overrightarrow{AB} + \overrightarrow{BC} = \overrightarrow{AC}$$

This demonstrates how vectors are added, using the triangle rule of addition.

The E and N parts of the vector are referred to as the **components** of the vector.

Since vectors are used in a variety of ways, mathematicians use the letter **i** to denote 1 unit in the x-axis direction, and the letter **j** to denote 1 unit in the y-axis direction. Because they have a length of one, they are called **unit vectors.**

$$\text{Hence} \quad \overrightarrow{AB} = 5\mathbf{i} + 2\mathbf{j} \quad \text{and} \quad \overrightarrow{BC} = -3\mathbf{i} - 8\mathbf{j}$$

To add together two vectors, add the components:

$$\text{hence,} \quad \overrightarrow{AB} + \overrightarrow{BC} = (5 - 3)\mathbf{i} + (2 - 8)\mathbf{j} = 2\mathbf{i} - 6\mathbf{j}$$

The column vector notation is just as commonly used:

$$\text{hence} \quad \overrightarrow{AB} = \begin{pmatrix} 5 \\ 2 \end{pmatrix} \quad \text{and} \quad \overrightarrow{BC} = \begin{pmatrix} -3 \\ -8 \end{pmatrix}$$

$$\text{So} \quad \overrightarrow{AB} + \overrightarrow{BC} = \begin{pmatrix} 5 - 3 \\ 2 - 8 \end{pmatrix} = \begin{pmatrix} +2 \\ -6 \end{pmatrix}$$

Subtraction of vectors is perhaps slightly more difficult to understand.

What do we mean by $\overrightarrow{AB} - \overrightarrow{BC}$? This is best understood by realising that $-\overrightarrow{BC}$ is the same as $\overrightarrow{CB}$ (in other words, the direction is reversed).

$$\text{So} \quad \overrightarrow{AB} - \overrightarrow{BC} = \overrightarrow{AB} + \overrightarrow{CB} = 5\mathbf{i} + 2\mathbf{j} + 3\mathbf{i} + 8\mathbf{j} = 8\mathbf{i} + 10\mathbf{j}$$

However, in effect you are just subtracting the components,

$$\text{since} \quad \overrightarrow{AB} - \overrightarrow{BC} = (5 - -3)\mathbf{i} + (2 - -8)\mathbf{j} = 8\mathbf{i} + 10\mathbf{j}$$

To multiply a vector by an ordinary number (scalar) is very straightforward. $3\overrightarrow{AB}$ means $\overrightarrow{AB}+\overrightarrow{AB}+\overrightarrow{AB}$ and so each component is multiplied by 3. Since $\overrightarrow{AB} = 5\mathbf{i} + 2\mathbf{j}$, it follows that

$$3\overrightarrow{AB} = 15\mathbf{i} + 6\mathbf{j}$$

As was stated at the beginning of the chapter, the real value in using vectors is when they are extended to three dimensions.

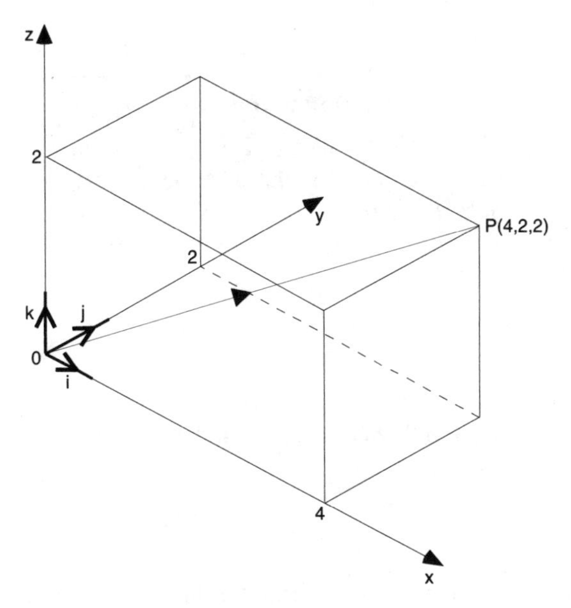

Figure 11.2

In Figure 11.2, the point P has coordinates $(4, 2, 2)$. The vector can be written:

$$\overrightarrow{OP} = 4\mathbf{i} + 2\mathbf{j} + 2\mathbf{k}$$

where $\mathbf{k}$ is now a unit vector in the direction of the z-axis.

To find the length of a vector (its magnitude or modulus), you use the formula:

$$\left|\overrightarrow{OP}\right| = \sqrt{x^2 + y^2 + z^2} \qquad \text{(V1)}$$

where $\overrightarrow{OP} = x\mathbf{i} + y\mathbf{j} + z\mathbf{k}$ and $\left|\overrightarrow{OP}\right|$ stands for the length of the vector.

If $z = 0$, you can see this simply reduces to Pythagoras's theorem. Think of this formula as being a three-dimensional version of Pythagoras's theorem. You can see from this that you need to be able to get the vector that joins two points easily. It is worthwhile summarising this by formula.

Consider two points $A(x_1, y_1, z_1)$ and $B(x_2, y_2, z_2)$. The vector joining A to B, namely $\overrightarrow{AB}$, is given by:

$$\overrightarrow{AB} = (x_2 - x_1)\mathbf{i} + (y_2 - y_1)\mathbf{j} + (z_2 - z_1)\mathbf{k} \qquad \text{(V2)}$$

If you think in terms of **position vectors**, the position vector of B is given by the vector $\overrightarrow{OB}$, where O is the origin. This is often denoted by **b**. Also, the position vector of A is given by $\overrightarrow{OA} = \mathbf{a}$.

$$\text{Hence} \quad \overrightarrow{OA} = x_1\mathbf{i} + y_1\mathbf{j} + z_1\mathbf{k} = \mathbf{a}$$

$$\text{and} \quad \overrightarrow{OB} = x_2\mathbf{i} + y_2\mathbf{j} + z_2\mathbf{k} = \mathbf{b}$$

$$\text{Hence} \quad \overrightarrow{AB} = \overrightarrow{OB} - \overrightarrow{OA} = \mathbf{b} - \mathbf{a}$$

MEMORY JOGGER

The vector from A to B is the position vector of B minus the position vector of A.

Example 11.1

$A(4, 1, 4)$, $B(2, 0, -1)$ and $C(3, -1, -2)$ are three points in three-dimensional space. Find the length of AB, BC and CA, and hence find the area of triangle ABC.

Solution

This question can be answered easily without a three-dimensional diagram, if you think 'vectors'.

$$\overrightarrow{AB} = (2-4)\mathbf{i} + (0-1)\mathbf{j} + (-1-4)\mathbf{k}$$

$$= -2\mathbf{i} - \mathbf{j} - 5\mathbf{k}$$

$$\therefore \quad \left|\overrightarrow{AB}\right| \sqrt{(-2)^2 + (-1)^2 + (-5)^2} = 5.48$$

$$\overrightarrow{BC} = (3-2)\mathbf{i} + (-1-0)\mathbf{j} + (-2 - -1)\mathbf{k}$$

$$= \mathbf{i} - \mathbf{j} - \mathbf{k}$$

$$\therefore \quad \left|\overrightarrow{BC}\right| = \sqrt{1^2 + (-1)^2 + (-1)^2} = 1.73$$

$$\overrightarrow{CA} = (4-3)\mathbf{i} + (1 - -1)\mathbf{j} + (4 - -2)\mathbf{k}$$

$$= \mathbf{i} + 2\mathbf{j} + 6\mathbf{k}$$

$$\therefore \quad \left|\overrightarrow{CA}\right| = \sqrt{1^2 + 2^2 + 6^2} = 6.40$$

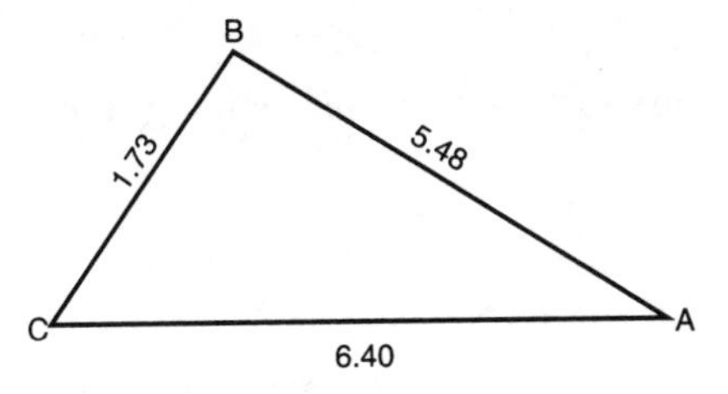

Figure 11.3

The sketch in Figure 11.3 shows what we have found. To find the area of the triangle, use the formula:

$$\text{Area} = \sqrt{s(s-a)(s-b)(s-c)}$$

where $s = \frac{1}{2}(a+b+c) = 6.81$

$$\begin{aligned}\text{Area} &= \sqrt{6.81 \times (6.81-1.73)(6.81-6.40)(6.81-5.48)}\\ &= 4.14\end{aligned}$$

Exercise 11(a)

1 Find the vectors joining the following pairs of points.
   (i) $(2,3)$ and $(1,-2)$
   (ii) $(1,0,1)$ and $(3,-3,0)$
   (iii) $(3,1,2)$ and $(1,1,0)$
   (iv) $(2,-1,2)$ and $(1,3,-2)$
   (v) $(1,t,3)$ and $(2t,1,-t)$

2 Find the lengths of the vectors found in Question 1.

3 Find the area of the triangle formed by the points $A(1,1,2)$, $B(2,1,0)$ and $C(3,0,-1)$.

4 Find the volume of the pyramid formed by the points O, $A(2,0,0)$, $B(0,-3,0)$ and $C(0,0,4)$.

11.2 Equation of a line

You know that in Cartesian coordinates, the equation of a line is given by $y = mx + c$ in two dimensions. In three dimensions, to find the equation of a line in Cartesian coordinates, the formula:

$$\frac{x-a}{d_1} = \frac{x-b}{d_2} = \frac{x-c}{d_3}$$

is used, where (a,b,c) is a point on the line, and $d_1\mathbf{i} + d_2\mathbf{j} + d_3\mathbf{k}$ represents the direction of the line.

Example 11.2

Find the Cartesian equation of the line joining the points $A(1,2,2)$ and $B(-1,3,4)$.

Solution

$$\text{Here the direction } \overrightarrow{AB} = (-1-1)\mathbf{i} + (3-2)\mathbf{j} + (4-2)\mathbf{k}$$
$$= -2\mathbf{i} + \mathbf{j} + 2\mathbf{k}$$

Hence $d_1 = -2$, $d_2 = 1$ and $d_3 = 2$.

The equation is $\dfrac{x-1}{-2} = \dfrac{y-2}{1} = \dfrac{z-2}{2}$

You could also use the coordinates of B, to give $\dfrac{x+1}{-2} = \dfrac{y-3}{1} = \dfrac{z-4}{2}$

The Cartesian form of the equation is not easy to manipulate, and a vector version is preferable.

If we let: $\dfrac{x-a}{d_1} = \dfrac{y-b}{d_2} = \dfrac{z-c}{d_3} = \lambda$ (a constant)

Then, looking at each part separately, we have:

$$x - a = \lambda d_1, \; y - b = \lambda d_2 \text{ and } z - c = \lambda d_2$$
$$\therefore \quad x = a + \lambda d_1$$
$$y = b + \lambda d_2$$
$$z = c + \lambda d_3$$

that is, $\begin{pmatrix} x \\ y \\ z \end{pmatrix} = \begin{pmatrix} a \\ b \\ c \end{pmatrix} + \begin{pmatrix} \lambda d_1 \\ \lambda d_2 \\ \lambda d_3 \end{pmatrix} = \begin{pmatrix} a \\ b \\ c \end{pmatrix} + \lambda \begin{pmatrix} d_1 \\ d_2 \\ d_3 \end{pmatrix}.$

$\begin{pmatrix} a \\ b \\ c \end{pmatrix}$ or $a\mathbf{i} + b\mathbf{j} + c\mathbf{k}$ represents the position of a point on the line.

$\begin{pmatrix} d_1 \\ d_2 \\ d_3 \end{pmatrix}$ is the direction of the line.

$\mathbf{r} = \begin{pmatrix} x \\ y \\ z \end{pmatrix}$ gives the **position vector** of any point on the straight line.

We can summarise this as follows:

> The equation of a line through A with position vector $\mathbf{a}$ in the direction $\mathbf{d}$ is
>
> $$\mathbf{r} = \mathbf{a} + \lambda\mathbf{d} \qquad \text{(V3)}$$

λ is a variable parameter, and its value determines which point on the line you are considering.

Example 11.3

Find the vector equation of the line through the point $A(1, 2, -1)$ in the direction $3\mathbf{i} + 2\mathbf{j} - \mathbf{k}$.

Solution

Using $\mathbf{r} = \mathbf{a} + \lambda\mathbf{d}$, we have:

$$\mathbf{r} = \begin{pmatrix} 1 \\ 2 \\ -1 \end{pmatrix} + \lambda \begin{pmatrix} 3 \\ 2 \\ -1 \end{pmatrix}$$

It is often more convenient to write this as:

$$\mathbf{r} = \begin{pmatrix} 1 + 3\lambda \\ 2 + 2\lambda \\ -1 - \lambda \end{pmatrix}$$

If you prefer the **i**, **j**, **k** notation:

Then: $\mathbf{r} = \mathbf{i} + 2\mathbf{j} - \mathbf{k} + \lambda(3\mathbf{i} + 2\mathbf{j} - \mathbf{k})$

or: $\mathbf{r} = (1 + 3\lambda)\mathbf{i} + (2 + 2\lambda)\mathbf{j} + (-1 - \lambda)\mathbf{k}$

or: $\mathbf{r} = (1 + 3\lambda)\mathbf{i} + (2 + 2\lambda)\mathbf{j} - (1 + \lambda)\mathbf{k}$

Example 11.4

Find the vector equation of the line joining the points $U(1, 4, -3)$ and $V(2, 2, 0)$.

Solution

You will need to find the direction of the line first. This will be given by the vector $\overrightarrow{UV}$.

So: $\overrightarrow{UV} = (2 - 1)\mathbf{i} + (2 - 4)\mathbf{j} + (0 - -3)\mathbf{k} = \mathbf{i} - 2\mathbf{j} + 3\mathbf{k}$

We can use either U or V as the point on the line. If we use U, then the equation becomes:

$$\mathbf{r} = \begin{pmatrix} 1 \\ 4 \\ -3 \end{pmatrix} + \lambda \begin{pmatrix} 1 \\ -2 \\ 3 \end{pmatrix}$$

Example 11.5

Decide whether the points $P(10, 3, 4)$ and $Q(1, 0, 3)$ lie on the line with equation:

$$\mathbf{r} = \begin{pmatrix} 4 \\ 1 \\ 2 \end{pmatrix} + \mu \begin{pmatrix} 3 \\ 1 \\ 1 \end{pmatrix}$$

Solution

If a point *does* lie on the line, there will be a particular value of the parameter that gives its coordinates.

(i) Does $\begin{pmatrix} 10 \\ 3 \\ 4 \end{pmatrix} = \begin{pmatrix} 4 \\ 1 \\ 2 \end{pmatrix} + \mu \begin{pmatrix} 3 \\ 1 \\ 1 \end{pmatrix}$ have a solution for μ?

There are, in fact, three equations here. Look at each component separately:

$$10 = 4 + 3\mu \quad \therefore \quad \mu = 2$$
$$3 = 1 + \mu \quad \therefore \quad \mu = 2$$
$$4 = 2 + \mu \quad \therefore \quad \mu = 2$$

In each case, $\mu = 2$, and so the point *does* lie on the line.

(ii) Similarly,

Does $\begin{pmatrix} 1 \\ 0 \\ 3 \end{pmatrix} = \begin{pmatrix} 4 \\ 1 \\ 2 \end{pmatrix} + \mu \begin{pmatrix} 3 \\ 1 \\ 1 \end{pmatrix}$ have a solution for μ?

$$1 = 4 + 3\mu \quad \therefore \quad \mu = -1$$
$$0 = 1 + \mu \quad \therefore \quad \mu = -1$$
$$3 = 2 + \mu \quad \therefore \quad \mu = 1$$

In this example, μ is not the same for each coordinate, so we must conclude that Q does *not* lie on the line.

Example 11.6

Find the point of intersection, of the straight lines:

$$\mathbf{r} = (2\mathbf{i} + \mathbf{j} + 3\mathbf{k}) + \lambda(\mathbf{i} + \mathbf{j} - \mathbf{k})$$
$$\text{and} \quad \mathbf{r} = (5\mathbf{i} + \mathbf{j} + 5\mathbf{k}) + \mu(2\mathbf{i} - \mathbf{j} + 3\mathbf{k})$$

Solution

If the lines intersect at a point, then **r** must be the same for each line.

So: $(2\mathbf{i} + \mathbf{j} + 3\mathbf{k}) + \lambda(\mathbf{i} + \mathbf{j} - \mathbf{k}) = (5\mathbf{i} + \mathbf{j} + 5\mathbf{k}) + \mu(2\mathbf{i} - \mathbf{j} + 3\mathbf{k})$

that is, $(2 + \lambda)\mathbf{i} + (1 + \lambda)\mathbf{j} + (3 - \lambda)\mathbf{k} = (5 + 2\mu)\mathbf{i} + (1 - \mu)\mathbf{j} + (5 + 3\mu)\mathbf{k}$

Equating coefficients of **i**, **j**, and **k** gives:

$$2+\lambda = 5+2\mu \quad \therefore \quad \lambda - 2\mu = 3 \qquad \text{(i)}$$
$$1+\lambda = 1-\mu \quad \therefore \quad \lambda + \mu = 0 \qquad \text{(ii)}$$
$$3-\lambda = 5+3\mu \quad \therefore \quad \lambda + 3\mu = -2 \qquad \text{(iii)}$$
$$\text{(i)} - \text{(ii)} \quad -3\mu = 3 \quad \therefore \quad \mu = -1$$

Substituting into (ii) gives $\lambda = 1$.

Check in (iii) $1 + 3x - 1 = -2$

Using $\lambda = 1$ in the first equation gives:

$$\mathbf{r} = (2\mathbf{i}+\mathbf{j}+3\mathbf{k}) + (\mathbf{i}+\mathbf{j}-\mathbf{k})$$
$$= 3\mathbf{i}+2\mathbf{j}+2\mathbf{k}$$

Note: You could also put $\mu = -1$ into the second equation.
Hence the lines intersect at the point $(3, 2, 2)$.

Exercise 11(b)

1 Find the equation of the following lines through a given point in the given direction.

(i) $(1, 2, 1)$; direction $3\mathbf{i} - 2\mathbf{j} + \mathbf{k}$
(ii) $(2, -1, -3)$; direction $4\mathbf{i} - \mathbf{j} - 2\mathbf{k}$
(iii) $(1, 0, 1)$; direction $3\mathbf{i} - 2\mathbf{k}$

2 Find the vector equation of the lines joining the following pairs of points:

(i) $(1, 3, 1)$ and $(2, 1, 2)$ (ii) $(4, -1, 0)$ and $(3, 0, 2)$
(iii) $(t, 2t, 3t)$ and $(4t, 5, 6t)$

3 Find the points of intersection of the following pairs of lines:

(i) $\mathbf{r} = \mathbf{i}+\mathbf{j}+\mathbf{k}+\lambda(\mathbf{i}-\mathbf{j}+\mathbf{k})$ and $\mathbf{r} = 2\mathbf{i}+\mathbf{j}-\mathbf{k}+\mu(\mathbf{i}-2\mathbf{j}+4\mathbf{k})$
(ii) $\mathbf{r} = \mathbf{i}+2\mathbf{k}+\lambda(\mathbf{i}-\mathbf{j})$ and $\mathbf{r} = 3\mathbf{i}+\mathbf{k}+\mu(3\mathbf{i}+\mathbf{j}-2\mathbf{k})$
(iii) $\mathbf{r} = \mathbf{i}+3\mathbf{j}-2\mathbf{k}+\lambda(\mathbf{i}+2\mathbf{j}-\mathbf{k})$ and $\mathbf{r} = \mathbf{i}-2\mathbf{j}-5\mathbf{k}+\mu(\mathbf{i}+7\mathbf{j}+2\mathbf{k})$

11.3 The scalar product

Although a vector quantity is not a single number, it is possible to define a way of multiplying two vectors together, which turns out to be extremely useful.

If $\mathbf{x} = a_1\mathbf{i} + a_2\mathbf{j} + a_3\mathbf{k}$
and $\mathbf{y} = b_1\mathbf{i} + b_2\mathbf{j} + b_3\mathbf{k}$

then the **scalar product** of **x** and **y** is written **x.y** and is defined by:

$$\mathbf{x}.\mathbf{y} = a_1b_1 + a_2b_2 + a_3b_3 \qquad \text{(V4)}$$

The answer is no longer a vector, hence the name scalar product.

$$\begin{aligned} \text{So,} \quad \text{if} \quad \mathbf{x} &= 4\mathbf{i} + 3\mathbf{j} - 2\mathbf{k} \\ \text{and} \quad \mathbf{y} &= 2\mathbf{i} - \mathbf{j} + 4\mathbf{k} \\ \text{then} \quad \mathbf{x}.\mathbf{y} &= 4 \times 2 + 3 \times -1 + -2 \times 4 \\ &= 8 - 3 - 8 = -3 \end{aligned}$$

In order to understand why this is useful, you need to look at the fact that it is possible to show (beyond the scope of this book), that **x.y** can also be proved geometrically to equal $|\mathbf{x}||\mathbf{y}|\cos\theta$, where θ is the angle between **x** and **y**.

In order to find θ, it is important that the two vectors both point **away** from a common point, as in Figure 11.4.

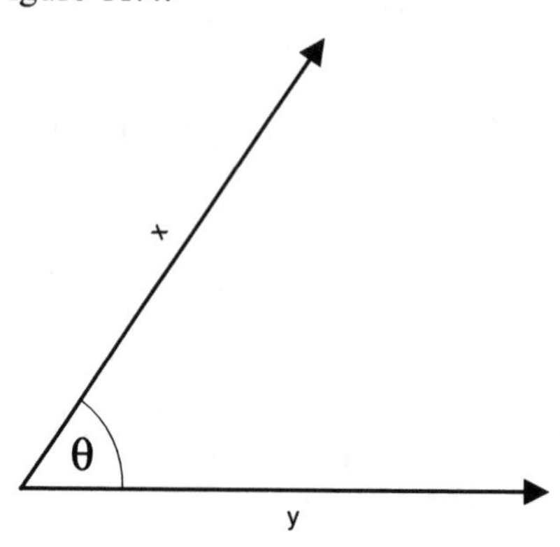

Figure 11.4

This means that:

$$\mathbf{x}.\mathbf{y} = |\mathbf{x}||\mathbf{y}|\cos\theta = a_1b_1 + a_2b_2 + a_3b_3$$

$$\text{that is,} \quad \cos\theta = \frac{a_1b_1 + a_2b_2 + a_3b_3}{|\mathbf{x}||\mathbf{y}|}$$

$$\text{So:} \quad \cos\theta = \frac{a_1b_1 + a_2b_2 + a_3b_3}{\sqrt{a_1^2 + a_2^2 + a_3^2}\sqrt{b_1^2 + b_2^2 + b_3^2}} \qquad \text{(V5)}$$

This formula is used wherever you want to find the angle between two vectors.

Using the example from earlier in this section:

$$|\mathbf{x}| = \sqrt{4^2 + 3^2 + (-2)^2} = \sqrt{29}$$

$$|\mathbf{y}| = \sqrt{2^2 + (-1)^2 + 4^2} = \sqrt{21}$$

$$\text{So:} \quad \cos\theta = \frac{\mathbf{x}.\mathbf{y}}{|\mathbf{x}||\mathbf{y}|} = \frac{-3}{\sqrt{29}\sqrt{21}}$$

$$\text{giving:} \quad \theta = 97^\circ$$

You can use this idea to solve problems in shapes that otherwise would prove quite difficult to visualise.

Example 11.7

A, B and C are three corners of a cuboid shown in Figure 11.5. The sides of the cuboid are of length 6 cm, 4 cm and 3 cm, as marked on the diagram. Find the angle ACB.

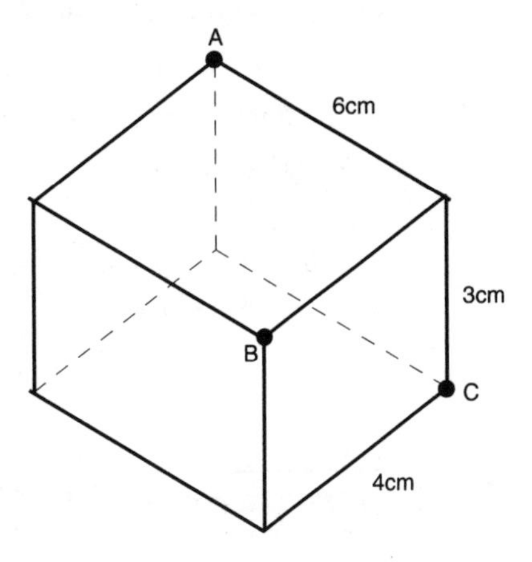

Figure 11.5

Solution

Although this could be solved using the ideas of conventional trigonometry, a vector approach is much more straightforward.

If you draw axes along the sides of the cuboid as shown in Figure 11.6, the coordinates of the points are: $A(0,4,3)$, $B(6,0,3)$ and $C(6,4,0)$.

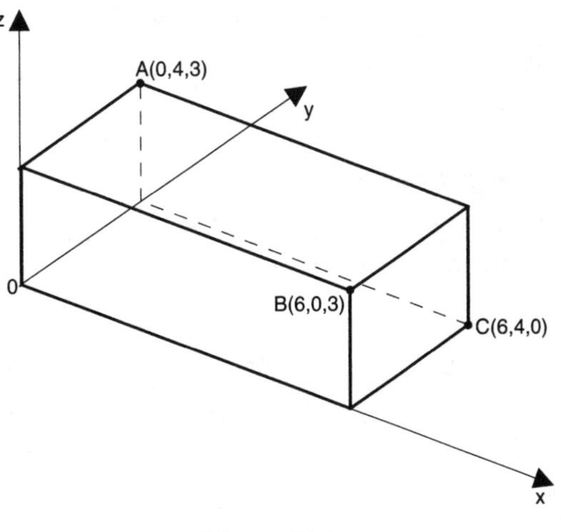

Figure 11.6

$$\vec{CA} = \mathbf{a} - \mathbf{c} = \begin{pmatrix} -6 \\ 0 \\ 3 \end{pmatrix}$$

$$\vec{CB} = \mathbf{b} - \mathbf{c} = \begin{pmatrix} 0 \\ -4 \\ 3 \end{pmatrix}$$

$$\therefore \quad \left|\vec{CA}\right| = \sqrt{6^2 + 3^2} = \sqrt{45}$$

$$\left|\vec{CB}\right| = \sqrt{4^2 + 3^2} = 5$$

$$\vec{CA}.\,\vec{CB} = (-6 \times 0) + (0 \times -4) + (3 \times 3) = 9$$

$$\therefore \quad \cos\theta = \frac{9}{\sqrt{45} \times 5} = 0.2683$$

$$\therefore \quad \theta = 74.4^\circ$$

that is, angle $ACB = 74.4^\circ$.

Exercise 11(c)

1 Find the angle between the following pairs of vectors:

(i) $2\mathbf{i}+3\mathbf{j}+\mathbf{k}$ and $3\mathbf{i}-\mathbf{j}-\mathbf{k}$ (ii) $2\mathbf{i}+\mathbf{j}$ and $\mathbf{i}-\mathbf{j}-2\mathbf{k}$

2 Find the angle between the straight lines:

(i) $\mathbf{r}=(\mathbf{i}+\mathbf{j}+2\mathbf{k})+\lambda(\mathbf{i}-\mathbf{j}-\mathbf{k})$ and $\mathbf{r}=(3\mathbf{i}-\mathbf{j}-\mathbf{k})+\mu(2\mathbf{i}-\mathbf{j}-\mathbf{k})$

(ii) $\mathbf{r}=\begin{pmatrix}2+\mu\\1-2\mu\\-1+\mu\end{pmatrix}$ and $\mathbf{r}=\begin{pmatrix}3-\lambda\\0\\2+2\lambda\end{pmatrix}$.

11.4 Perpendicular lines

If two vectors are perpendicular, then the angle between them is 90°. Since $\cos 90^\circ = 0$, it follows that:

$$\mathbf{x}.\mathbf{y}=0 \quad \text{means} \quad \mathbf{x} \text{ is perpendicular to } \mathbf{y} \qquad \text{(V6)}$$

We can use this fact whenever we want to construct a line that is perpendicular to another line.

Example 11.8

Find the equation of the line through the point $P(1,-4,-1)$ which is perpendicular to the line joining $A(1,1,2)$ and $B(-2,1,3)$.

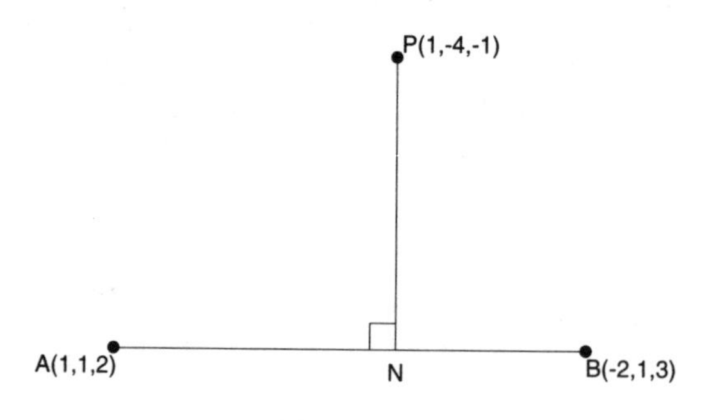

Figure 11.7

Solution

$$\overrightarrow{AB}=(-2-1)\mathbf{i}+(1-1)\mathbf{j}+(3-2)\mathbf{k}$$
$$=-3\mathbf{i}+\mathbf{k}$$

So the equation of the line AB is:

$$\mathbf{r}=(\mathbf{i}+\mathbf{j}+2\mathbf{k})+\lambda(-3\mathbf{i}+\mathbf{k})$$
$$=(1-3\lambda)\mathbf{i}+\mathbf{j}+(2+\lambda)\mathbf{k}$$

We next need the point N on the line AB so that PN is perpendicular to AB.

$$\text{Now} \quad \overrightarrow{PN} = \mathbf{n} - \mathbf{p} = (1 - 3\lambda)\mathbf{i} + \mathbf{j} + (2 + \lambda)\mathbf{k} - (\mathbf{i} - 4\mathbf{j} - \mathbf{k})$$
$$= -3\lambda\mathbf{i} + 5\mathbf{j} + (3 + \lambda)\mathbf{k}$$

Since $\overrightarrow{PN}$ is perpendicular to $\overrightarrow{AB}$,

$$\overrightarrow{PN} . \overrightarrow{AB} = 0$$

$\therefore \quad (-3\lambda\mathbf{i} + 5\mathbf{j} + (3 + \lambda)\mathbf{k}).(-3\mathbf{i} + \mathbf{k}) = 0$

that is, $\quad 9\lambda + 3 + \lambda = 0$

$\therefore \quad 3 = -10\lambda, \quad \lambda = -\frac{3}{10}$

Therefore: $\quad \overrightarrow{PN} = \frac{9}{10}\mathbf{i} + 5\mathbf{j} + \frac{27}{10}\mathbf{k}$

A better direction vector for the line PN would be:

$$10\,\overrightarrow{PN} = 9\mathbf{i} + 50\mathbf{j} + 27\mathbf{k}$$

The equation required is:

$$\mathbf{r} = \mathbf{i} + 4\mathbf{j} - \mathbf{k} + \mu(9\mathbf{i} + 50\mathbf{j} + 27\mathbf{k})$$

11.5 Cartesian equation of a plane

The equation $ax + by + cz = d$ (V7)

represents a plane in three dimensions.

For example, if $3x + 2y + 5z = 15$, then to locate its position, proceed as follows:

If $x = y = 0$, $\quad 5z = 15$ and $z = 3$
If $x = z = 0$, $\quad 2y = 15$ and $y = 7\frac{1}{2}$
If $y = z = 0$, $\quad 3x = 15$ and $x = 5$

A part of the plane is shown in Figure 11.8, indicating where the axes intersect the plane.

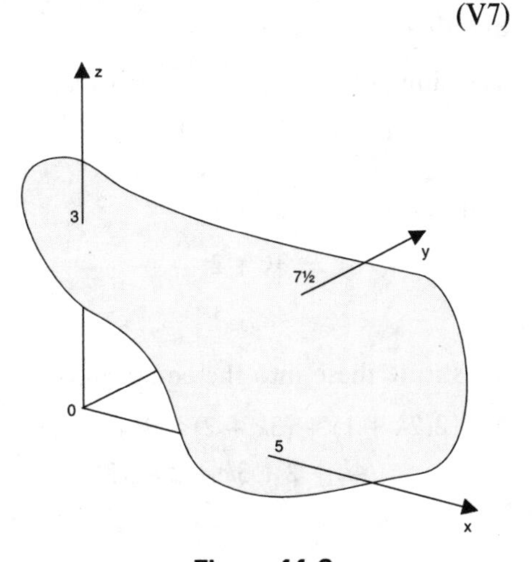

Figure 11.8

Example 11.9

Find the equation of the plane that passes through the points $(2, 1, 1)$, $(3, 1, -1)$ and $(2, 2, 1)$.

Solution

Let the equation of the plane be $ax + by + cz = 1$.

(**Note:** It is convenient to let $d = 1$ in the equation.)
Substitute the coordinates of the points into the equation.

so $\quad 2a + b + c = 1 \quad$ (i)

$3a + b - c = 1 \quad$ (ii)

$2a + 2b + c = 1 \quad$ (iii)

(i) + (ii) $\quad 5a + 2b = 2 \quad$ (iv)

(ii) + (iii) $\quad 5a + 3b = 2 \quad$ (v)

(iv) − (v) $\quad -b = 0$

$\therefore \quad b = 0$

hence $a = \frac{2}{5}$, $c = \frac{1}{5}$.

The equation of the plane is:

$$\tfrac{2}{5}x + \tfrac{1}{5}z = 1$$

or $\quad 2x + z = 5$

(Hence d ends up equal to 5.)

Example 11.10

Find where the line $\dfrac{x-1}{2} = \dfrac{y-2}{3} = \dfrac{z+1}{4}$ cuts the plane $2x + y - 3z = 5$.

Solution

The simplest way to do this is to let:

$$\frac{x-1}{2} = \frac{y-2}{3} = \frac{z+1}{4} = k$$

and hence $\quad x = 2k + 1$

$y = 3k + 2$

$z = 4k - 1$

Substitute these into the equation of the plane:

so: $\quad 2(2k + 1) + (3k + 2) - 3(4k - 1) = 5$

$4k + 2 + 3k + 2 - 12k + 3 = 5$

$-5k = -2$

$\therefore \quad k = \frac{2}{5}$

Hence: $x = 2 \times \frac{2}{5} + 1 = 1\frac{4}{5}$
$y = 3 \times \frac{2}{5} + 2 = 3\frac{1}{5}$
$z = 4 \times \frac{2}{5} - 1 = \frac{3}{5}$

Hence the point of intersection is $\left(1\frac{4}{5}, 3\frac{1}{5}, \frac{3}{5}\right)$

Example 11.11

Find the equation of the plane which passes through $(1, -4, -3)$ and is parallel to the plane $2x - y - 4z = 5$.

Solution

If the plane is parallel to $2x - y - 4z = 5$, its equation must be:

$$2x - y - 4z = d$$

It passes through $(1, -4, -3)$, hence:

$$2 \times 1 - -4 - 4 \times -3 = d = 18$$

The plane is $2x - y - 4z = 18$.

Example 11.12

Find the equation of the line of intersection of the planes $3x + 2y + z = 1$ and $2x + 3y - z = 4$.

Solution

If two planes are not parallel, they intersect along a line common to both planes. See Figure 11.9.

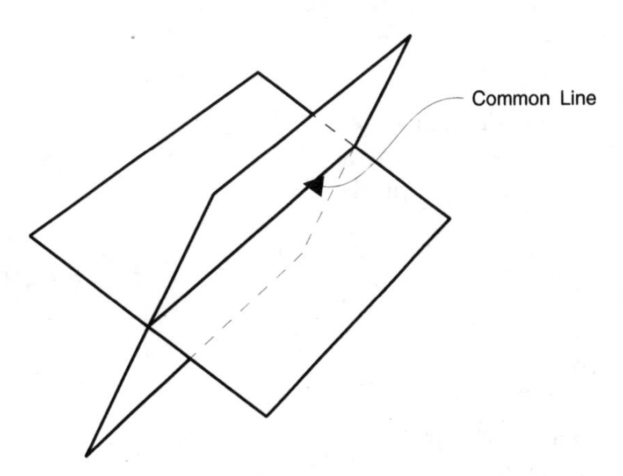

Figure 11.9

$$3x + 2y + z = 1$$
$$2x + 3y - z = 4$$

We have two equations with three unknowns. This means they will not have a unique solution. Let $z = t$ (a parameter).

So: $3x + 2y = 1 - t$ (i)

$2x + 3y = 4 + t$ (ii)

(i) × 2 $6x + 4y = 2 - 2t$ (iii)

(ii) × 3 $6x + 9y = 12 + 3t$ (iv)

(iv) – (iii) $5y = 10 + 5t$

$\therefore \quad y = 2 + t$

Substitute into (i) $3x = 1 - t - 2(2 + t)$

$= -3 - 3t$

$\therefore \quad x = -1 - t$

The solution can be written: $\begin{pmatrix} x \\ y \\ z \end{pmatrix} = \begin{pmatrix} -1-t \\ 2+t \\ t \end{pmatrix} = \begin{pmatrix} -1 \\ 2 \\ 0 \end{pmatrix} + t\begin{pmatrix} -1 \\ 1 \\ 1 \end{pmatrix}$

You should recognise this as the vector equation of a straight line.

11.6 Vector equation of a plane

If a plane is drawn through three points A, B, C, with position vectors **a**, **b**, and **c**, and if **r** is the vector position of any point P in the plane, then:

$$\overrightarrow{OP} = \overrightarrow{OA} + \overrightarrow{AP} \qquad \text{(i)}$$

Now $\overrightarrow{AP}$ is in the plane defined by A, B, C and so $\overrightarrow{AP}$ can be achieved by a multiple of $\overrightarrow{AB}$ followed by a multiple of $\overrightarrow{AC}$.

$$\text{So} \quad \overrightarrow{AP} = \lambda\,\overrightarrow{AB} + \mu\,\overrightarrow{AC} \qquad \text{(ii)}$$

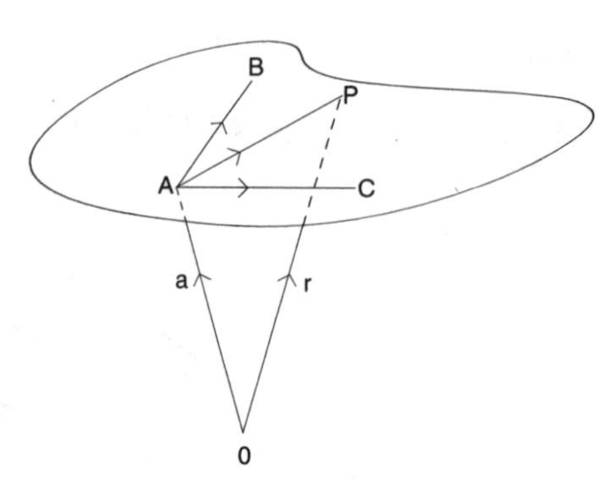

Figure 11.10

$\therefore$ from (i) and (ii)

$$\overrightarrow{OP} = \mathbf{a} + \lambda\,\overrightarrow{AB} + \mu\,\overrightarrow{AC}$$

that is, $\mathbf{r} = \mathbf{a} + \lambda(\mathbf{b} - \mathbf{a}) + \mu(\mathbf{c} - \mathbf{a})$ (V7)

This version is not often used, as it contains two parameters, λ and μ, making it cumbersome to work with. However, it is interesting to see how this leads to the standard Cartesian version.

If $\mathbf{a} = 2\mathbf{i} + 3\mathbf{j} + \mathbf{k}$, $\mathbf{b} = 3\mathbf{i} - \mathbf{j} - 2\mathbf{k}$ and $\mathbf{c} = \mathbf{i} + 2\mathbf{j} + 4\mathbf{k}$,

$$\text{then} \quad \mathbf{r} = \begin{pmatrix} 2 \\ 3 \\ 1 \end{pmatrix} + \lambda \begin{pmatrix} 1 \\ -4 \\ -3 \end{pmatrix} + \mu \begin{pmatrix} -1 \\ -1 \\ 3 \end{pmatrix}$$

Equating components,

$$x = 2 + \lambda - \mu \qquad \text{(i)}$$

$$y = 3 - 4\lambda - \mu \qquad \text{(ii)}$$

$$z = 1 - 3\lambda + 3\mu \qquad \text{(iii)}$$

$$\text{(i)} - \text{(ii)} \qquad x - y = -1 + 5\lambda \qquad \text{(iv)}$$

$$3\text{(ii)} + \text{(iii)} \qquad 3y + z = 10 - 15\lambda \qquad \text{(v)}$$

$$\text{(v)} + 3\text{(iv)} \qquad 3x + z = 7 \qquad \text{(vi)}$$

The fact that y does not appear simply means that the plane is parallel to the y-axis and can take any value.

Equation (vi) can be written:

$$\begin{pmatrix} 3 \\ 0 \\ 1 \end{pmatrix} . \begin{pmatrix} x \\ y \\ z \end{pmatrix} = 7$$

$$\text{that is,} \quad \mathbf{r}. \begin{pmatrix} 3 \\ 0 \\ 1 \end{pmatrix} = 7$$

This version will be written:

$$\mathbf{r}.\mathbf{n} = k$$

$\mathbf{n}$ is the **normal** to the plane.

If you divide both sides by the length of $\mathbf{n}$, that is, $|\mathbf{n}|$.

$$\text{Then} \quad \mathbf{r}.\frac{\mathbf{n}}{|\mathbf{n}|} = \frac{\mathbf{k}}{|\mathbf{n}|}$$

$$\text{now} \quad \frac{\mathbf{n}}{|\mathbf{n}|} = \hat{\mathbf{n}}$$

where $\hat{\mathbf{n}}$ is a unit vector (length one). It is the unit vector normal to the plane. $\dfrac{\mathbf{k}}{|\mathbf{n}|}$ will be written d.

$$\text{So} \quad \mathbf{r}.\hat{\mathbf{n}} = d \qquad \text{(V8)}$$

To understand the meaning of d, look at Figure 11.11. You are looking at the plane side on. ON is perpendicular to the plane, in the direction of $\hat{\mathbf{n}}$. So $\vec{ON} = ON\hat{\mathbf{n}}$.

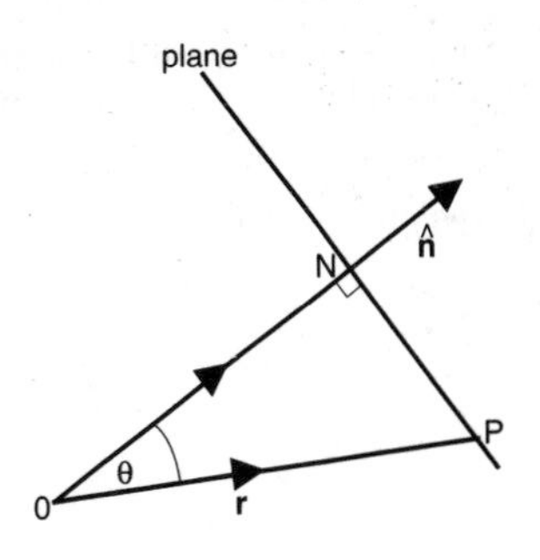

Figure 11.11

Now $ON = OP\cos\theta$

So $ON = |\mathbf{r}|\cos\theta$ (i)

But $\cos\theta = \dfrac{r.\hat{\mathbf{n}}}{|\mathbf{r}||\hat{\mathbf{n}}|} = \dfrac{\mathbf{r}.\hat{\mathbf{n}}}{|\mathbf{r}|}$ since $|\hat{\mathbf{n}}| = 1$

Substituting into (i):

$$ON = |\mathbf{r}|\frac{\mathbf{r}.\hat{\mathbf{n}}}{|\mathbf{r}|} = \mathbf{r}.\hat{\mathbf{n}} = d \qquad \text{(from (V8))}$$

Hence d is the perpendicular distance of the plane from the origin.

Example 11.13

Find the perpendicular distance of the plane $3x + 4y - 2z = 5$ from the origin.

Solution

$3x + 4y - 2z$ can be written $\mathbf{r}.\begin{pmatrix} 3 \\ 4 \\ -2 \end{pmatrix}$, hence $\mathbf{r}.\begin{pmatrix} 3 \\ 4 \\ -2 \end{pmatrix} = 5$

$$\therefore \quad \mathbf{n} = \begin{pmatrix} 3 \\ 4 \\ -2 \end{pmatrix} \qquad \therefore \quad |\mathbf{n}| = \sqrt{3^2 + 4^2 + (-2)^2} = \sqrt{29}$$

$$\therefore \quad \mathbf{r}.\hat{\mathbf{n}} = \frac{5}{\sqrt{29}}$$

The perpendicular distance $d = \dfrac{5}{\sqrt{29}}$.

Example 11.14

Find the angle between the planes $2x + 3y + z = 1$ and $4x - 2y - 3z = 2$.

Solution

This is not easy to illustrate, but if you look at the planes sideways, as in Figure 11.12, the angle between the planes θ is the same as the angle between the normals.

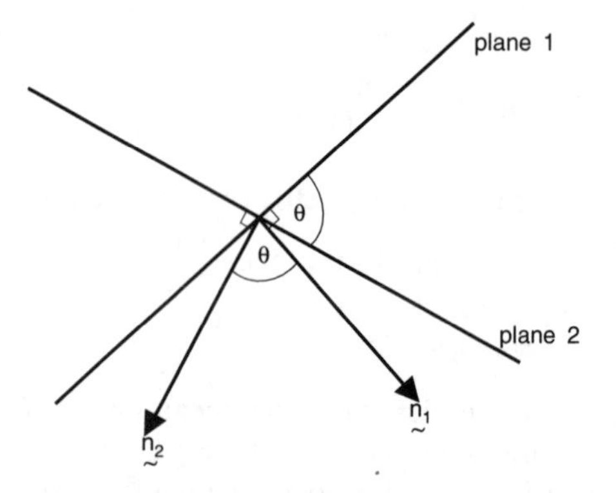

Figure 11.12

$$\mathbf{n}_1 = \begin{pmatrix} 2 \\ 3 \\ 1 \end{pmatrix} \text{ for the plane } 2x + 3y + z = 1$$

$$\mathbf{n}_2 = \begin{pmatrix} 4 \\ -2 \\ -3 \end{pmatrix} \text{ for the plane } 4x - 2y - 3z = 2$$

$$\mathbf{n}_1.\mathbf{n}_2 = 8 - 6 - 3 = -1$$

$$|\mathbf{n}_1| = \sqrt{14} \quad \text{and} \quad |\mathbf{n}_2| = \sqrt{29}$$

$$\text{Hence} \quad \cos\theta = \frac{-1}{\sqrt{14}\sqrt{29}}$$

$$\therefore \qquad \theta = 92.8°$$

Example 11.15

Find the perpendicular distance from the point $P(1, 4, -2)$ to the plane π: $3x - 2y - 7z = 4$.

Solution

First construct a plane through $(1, 4, -2)$ parallel to $3x - 2y - 7z = 4$. This will be $3x - 2y - 7z = d$. It passes through $(1, 4, -2)$, so these coordinates can be substituted into the equation:

$$\therefore \quad 3 - 8 + 14 = d = 9$$

The perpendicular distance of $3x - 2y - 7z = 4$ from the origin is:

$$\frac{4}{\sqrt{3^2 + 2^2 + 7^2}} = \frac{4}{\sqrt{62}}$$

The perpendicular distance of $3x - 2y - 7z = 9$ from the origin is:

$$\frac{9}{\sqrt{62}}$$

The perpendicular distance of P to the plane π is:

$$\frac{9}{\sqrt{62}} - \frac{4}{\sqrt{62}} = \frac{5}{\sqrt{62}}$$

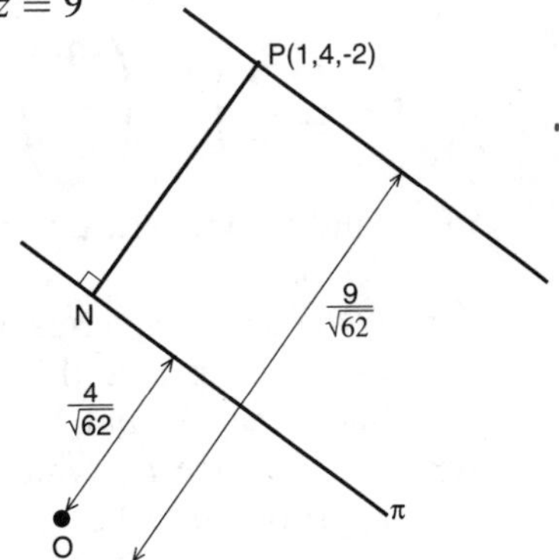

Figure 11.13

Exercise 11(d)

1 Find the equation of the plane through the point $(1, 3, 1)$ parallel to the plane $2x + 4y - 5z = 2$.

2 Find the equation of the line of intersection of the planes $2x + 3y - 5z = 6$ and $x + 2y + 3z = 2$.

3 Find the angle between the planes $3x + 2y = 8$ and $x + 2y - z = 4$.

4 Find the perpendicular distance of the points $(3, 1, -2)$ from the plane $\mathbf{r}.\begin{pmatrix} 2 \\ 1 \\ 1 \end{pmatrix} = 6$.

Miscellaneous Examples 11

1 Find the angle between the vectors $2\mathbf{i} + 3\mathbf{j} - \mathbf{k}$, and $4\mathbf{i} + 6\mathbf{j} + \mathbf{k}$.

2 Find the angle between the planes $\mathbf{r}.(2\mathbf{i} + 6\mathbf{j} - \mathbf{k}) = 5$ and $\mathbf{r}.(3\mathbf{i} - 6\mathbf{j} + 2\mathbf{k}) = 1$.

3 Find the equation of the plane that passes through the points $(1, 1, 2)$, $(3, 1, -1)$ and $(2, 2, -1)$.

4 Find the perpendicular distance from the point $P(1, -1, 2)$ to the plane π: $3x + 2y - z = 1$.

5 Find where the line $\dfrac{x-1}{2} = \dfrac{y-3}{4} = \dfrac{z+1}{1}$ meets the plane: $\mathbf{r}.\begin{pmatrix} 1 \\ 2 \\ -4 \end{pmatrix} = 2$.

6 Find the area of the triangle formed by the points $A(1, 1, 2)$, $B(1, 3, 1)$ and $C(-1, -1, 0)$.

Revision Problems 11

1 The equation of a line L_1, which passes through the point $A(2, 3, -5)$ is:

$$\mathbf{r} = \begin{pmatrix} 2 \\ 3 \\ -5 \end{pmatrix} + t\begin{pmatrix} 0 \\ -3 \\ 4 \end{pmatrix}$$

(a) Find the numerical values of p and q for which the point $B(p, q, -1)$ lies on L_1.

(b) Write down, in terms of a parameter s, a vector equation of the line L_2 which passes through the points $C(2, -1, -3)$ and $D(5, 1, 3)$. Show that L_1 and L_2 are skew lines.

(c) Another line L_3, which is parallel to L_1, intersects L_2. By using an appropriate scalar product, show that the lines L_2 and L_3 intersect at an angle $\cos^{-1}(18/35)$.

(NEAB)

2 A pyramid has a rectangular base $OABC$ and vertex D. The position vectors of A, B, C and D with reference to the fixed origin O are $\mathbf{a} = 8\mathbf{i}$, $\mathbf{b} = 8\mathbf{i} + 4\mathbf{j}$, $\mathbf{c} = 4\mathbf{j}$, $d = 4\mathbf{i} + 2\mathbf{j} + 6\mathbf{k}$ respectively.

(a) Express the vector $\overrightarrow{AD}$ in terms of **i**, **j** and **k**.

(b) Find the cosine of the angle ODA and hence show that the triangular face ODA has area $8\sqrt{10}$.

(c) By using the result from (b), or otherwise, find the perpendicular distance from O to the line AD.

(d) Determine a vector equation of the straight line through the points B and D.

(e) By considering an appropriate scalar product, or otherwise, find the position vector of the point on the line BD which is closest to O.

(AEB 94)

12 Numerical methods

A major effect that computers have had in the field of mathematics is to enable a whole range of numerical techniques to come easily within the range of everybody.

Equations can be solved now that would have proved impossible a few decades ago. Nevertheless, when using a numerical technique, it is important that you are aware of any errors occurring in the calculation, or limitations being imposed in the accuracy of the calculation.

MEMORY JOGGER

Do not approximate too early in a calculation. Store as much information in the calculator memory as you can.

12.1 Error bounds

(i) Absolute errors

Suppose that H is given by the formula:

$$H = \frac{4t}{p+q}$$

where it is known that t, p and q can be calculated to an accuracy of 0.05 (called the **absolute error**).

If $t = 4.3$, $p = 6.5$ and $q = 2.1$, what can you say about H?

At first sight, you can find H from:

$$H = \frac{4 \times 4.3}{(6.5 + 2.1)} = 2$$

But $4.25 \leq t \leq 4.35$; $6.45 \leq p \leq 6.55$ and $2.05 \leq q \leq 2.15$.
If you take the upper extremes of t, and the lower extremes of p and q:

$$H = \frac{4 \times 4.35}{(6.45 + 2.05)} = 2.05$$

Similarly, taking the lower extreme of t, and the upper extremes of p and q:

$$H = \frac{4 \times 4.25}{6.55 + 2.15} = 1.95$$

We can confidently say that:

$$1.95 \leq H \leq 2.05$$

(ii) Relative errors

Example 12.1

While measuring the surface area of the cone-shaped front of a jet aircraft, the formula:

$$A = \pi r \ell$$

was used.

It was later discovered there was a 5% error in the value of r, and a 2% error in the value of ℓ. Find the percentage relative error in A.

Solution

$$\text{The relative error} = \frac{\text{error}}{\text{correct value}} (\times 100\%)$$

This type of problem often causes difficulty because there are no numerical values apart from the percentages.

If r is in error by 5%, it must lie in the range $0.95r$ to $1.05r$

Similarly, ℓ lies in the range 0.98ℓ to 1.02ℓ

$\therefore$ A lies in the range $\pi \times 0.95r \times 0.98\ell$ to $\pi \times 1.05r \times 1.02\ell$

$= 0.931\pi r\ell$ to $1.071\pi r\ell$

Since $\pi r\ell$ was the original answer, $1.071\pi r\ell$ is a 7.1% error too high, and $0.931\pi r\ell$ is a 6.9% error too low. Since 7.1% is the greater of these, we are forced to conclude that the percentage error in A could be 7.1%.

Exercise 12(a)

In Questions 1–4, you should assume that all variables can only be measured to an absolute accuracy of 0.05. In each case, evaluate the subject of the formula and state the limits within which it can lie.

1 $H = 4x + 3y$ $x = 1.6$, $y = 12.3$ **2** $T = \dfrac{2t}{1-t}$ $t = 0.4$

3 $T = 2\pi\sqrt{\dfrac{L}{g}}$ $L = 84.9$, $g = 9.8$ **4** $\dfrac{1}{f} = \dfrac{1}{u} + \dfrac{1}{v}$ $u = 69.3$, $v = 0.4$

5 Using the formula $X = \dfrac{2y}{t+y}$, it was found that a 5% error was made in the measurement of t and y. What is the likely percentage relative error made in calculating X?

6 When measuring the dimensions of a cylinder, the lengths were stated to the nearest mm. If the height was given as 7.6 cm, and the radius as 2.3 cm, find the possible range in the ratio $\dfrac{v}{s}$, where v is the volume of the cylinder, and s its total surface area.

12.2 Solving equations numerically

One of the most common areas of mathematics where errors occur is in the solution of an equation. In other words, in the solution of $f(x) = 0$. We shall now look at several techniques for dealing with minimising the error and hence increasing the accuracy of the solution.

DO YOU KNOW

Trial and improvement methods can be used to solve $f(x) = 0$.

So to solve $x^3 + x = 6$,

write $f(x) = x^3 + x - 6 = 0$

$f(1) = -4 < 0$

$f(2) = 4 > 0$

The function has *changed sign*, hence: a solution lies between $x = 1$ and $x = 2$.

Try $f(1.5) = -1.125$ (too low)

try $f(1.6) = -0.304$ (too low)

and so on

In fact, $x = 1.63$ (3 sig. figs)

12.3 Linear interpolation

The method of linear interpolation is often useful as a first step in getting closer to the solution of an equation.

Suppose that we are trying to solve $f(x) = 0$. If $f(2) = -3$ and $f(3) = 12$, the graph of $f(x)$ might look like the one shown in Figure 12.1. If a straight line is drawn between the points $P_1(2, -3)$ and $P_2(3, 12)$ it will cut the x-axis at T, hopefully close to the solution of $f(x) = 0$.

Now P_1N_1T and P_2N_2T are similar triangles.

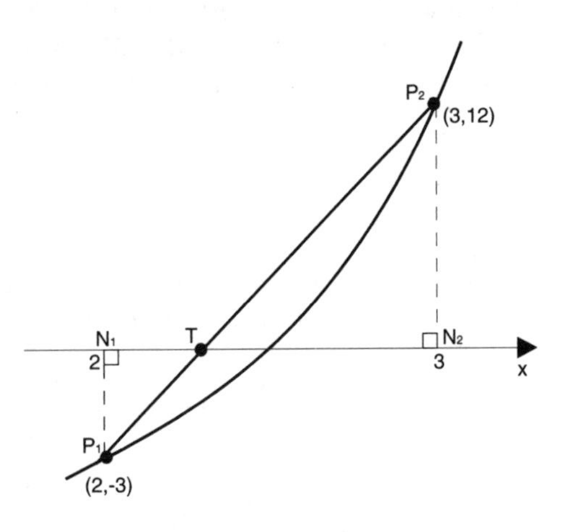

Figure 12.1

$$\text{So} \quad \frac{N_1T}{N_2T} = \frac{N_1P_1}{N_2P_2} = \frac{3}{12} = \frac{1}{4}$$

$\therefore \quad T$ divides N_1N_2 in the ratio $1 : 4$

The x coordinate of T is given by: $2 + \dfrac{1}{1+4} = 2.2$

We say that $x = 2.2$ would be a better approximation to the solution of $f(x) = 0$.

In general, if we are trying to solve the equation $f(x) = 0$, and $f(a) = y_1$, $f(b) = y_2$ where y_1 and y_2 are opposite signs with $a < b$, then the linear interpolation formula for a closer approximation to the solution is given by:

$$\text{closer solution} = a + \frac{|f(a)|(b-a)}{|f(a)| + |f(b)|} \qquad \text{(N1)}$$

The modulus signs are needed because it is only the length of the lines that are needed.

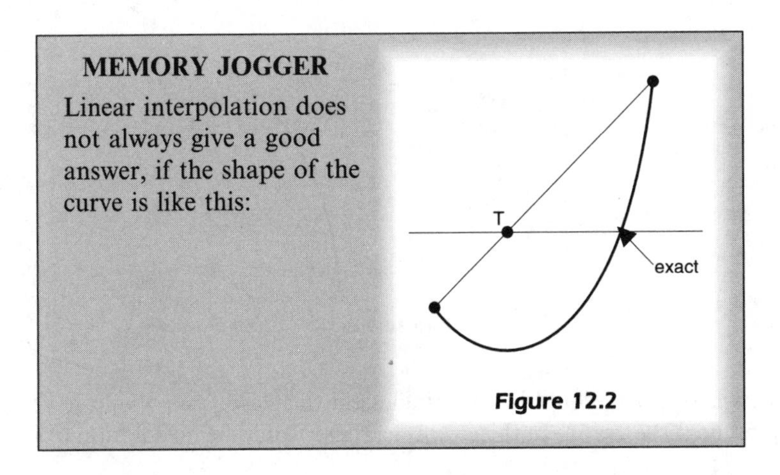

Figure 12.2

Example 12.2

It is known that for a given $f(x)$, $f(2) = 6$, and $f(5) = -8$. Use linear interpolation to find an approximation to the solution of $f(x) = 0$. Draw a sketch to show how, in fact, it is possible that there is no solution to the equation $f(x) = 0$ for $2 \leq x \leq 5$.

Solution

This simple question illustrates a major problem of solving equations.

Using Equation (N1), with $a = 2$, $b = 5$

$$\text{The closer approximation} = 2 + \frac{|6| \times (5-2)}{(|6| + |-8|)}$$

$$= 2 + \frac{6}{14} \times 3 = 3.3$$

If you look at the curve in Figure 12.3, the graph satisfies all of the conditions of the question, but clearly there is no solution to $f(x) = 0$ for $2 \leq x \leq 5$.

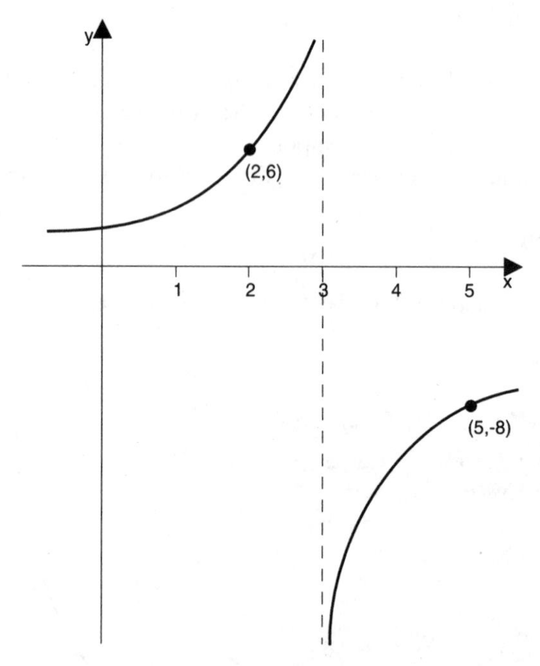

Figure 12.3

The answer, 3.3, would be meaningless in this case. You would have to look more closely at the graph. In this example, there would be no solution to $f(x) = 0$.

Exercise 12(b)

1. It is given that $f(1) = 2$, and $f(2) = -3$. Use linear interpolation to find an approximation to the solution of $f(x) = 0$.
2. If $g(2) = 4$ and $g(6) = -7$, use linear interpolation to find an approximate solution to the equation $g(x) = 0$. Draw a sketch to show that in fact $g(x) = 0$ may have several solutions for $2 \leq x \leq 6$.
3. If $f(2) = -1.2$ and $f(2.5) = 1.3$, use linear interpolation to find an approximate solution to the equation $f(x) = 0$.
4. Show that a solution to the equation $x^4 - 2\sqrt{x} - 8 = 0$ lies between 1 and 2. Use linear interpolation to find an approximation to the solution.
5. Find an approximation to 1 decimal place to the solution of the equation $2\ln x = 4 - x$.

Having established a good approximation to the solution of an equation, you now have to refine the solution to give accuracy to any necessary degree. Two main areas will be considered, algebraic and iterative.

12.4 Algebraic approximation

There are certain types of equation which respond to a method which one might refer to as that of small changes. In this case, if you know the solution is approximately $x = a$, then by substituting $x = a + h$ into the equation and ignoring powers of h above a certain level (because h is small), you can get quite a simple equation to solve, containing h. This solution can also be further refined if necessary.

Example 12.3

The equation $7x^3 - 5x^2 + 11x - 60 = 0$ has a solution approximately equal to 2. By substituting $x = 2 + h$, and ignoring the term in h^3, find the solution correct to two decimal places. Compare your answer with ignoring the term in h^2.

Solution

If $x = 2 + h$

$$7(2+h)^3 - 5(2+h)^2 + 11(2+h) - 60 = 0$$

$$7(8 + 12h + 6h^2 + h^3) - 5(4 + 4h + h^2) + 22 + 11h - 60 = 0$$

that is, $$56 + 84h + 42h^2 - 20 - 20h - 5h^2 + 22 + 11h - 60 = 0$$

if we ignore the term in h^3 (it is very small).

This simplifies to:

$$37h^2 + 75h - 2 = 0 \quad \text{(i)}$$

Using the quadratic formula:

$$h = \frac{-75 \pm \sqrt{75^2 - 4 \times 37 \times -2}}{2 \times 37}$$
$$= 0.0263 \quad \text{or} \quad -2.053$$

But h must be small, and so the solution is $x = 2.0263$

that is, $x = 2.03$ correct to two decimal places

If you ignore the term in h^2 (which is small), Equation (i) would become $75h - 2 = 0$.

So $h = \frac{2}{75} = 0.027$

We still obtain the solution:

$x = 2.03$ correct to 2 decimal places

The method is clearly a powerful one.

Exercise 12(c)

1. The equation $x^3 + 2x = 34$ has a solution close to $x = 3$. By substituting $x = 3 + h$ into the equation and ignoring h^3, find a closer approximation to this solution.
2. The equation $x^4 - 3x^2 = 3$ has a solution close to $x = 2$. Let $x = 2 - h$ and ignoring powers of h higher than h^2, find and solve a quadratic equation in h. Hence, solve the original equation.

12.5 Iterative methods

When solving an equation $f(x) = 0$, it is possible to rearrange this equation in a variety of ways to give

$$x = F(x) \quad \text{(i)}$$

For example, the equation $x^2 - 2x - 5 = 0$ could be rearranged:

$$x = \frac{x^2 - 5}{2}, \quad \text{so} \quad F(x) = \frac{x^2 - 5}{2}$$

or if you divide the original equation by x, you get:

$$x - 2 - \frac{5}{x} = 0$$

Hence $\quad x = 2 + \frac{5}{x}, \quad$ so $\quad F(x) = 2 + \frac{5}{x}$

Equation (i) can be altered to an **iterative** (repeating) method, by writing it as:

$$x_{n+1} = F(x_n)$$

A value x_0 is found which is close to the solution of $f(x) = 0$, and hence, x_1 can be found. $F(x_1)$ will give x_2 and so on. If $F(x)$ satisfies the condition $|F'(\alpha)| < 1$ whenever α is close to the exact solution of $f(x) = 0$, then the iterative formula $x_{n+1} = F(x_n)$ will converge to the required solution.

Look at the quadratic equation $x^2 - 2x - 5 = 0$.

$$\begin{aligned}
\text{If} \quad f(x) &= x^2 - 2x - 5 \\
f(0) &= -5 \\
f(1) &= -6 \\
f(2) &= -5 \\
f(3) &= -2 \\
f(4) &= 3
\end{aligned}$$

So there is a solution for $3 \leq x \leq 4$.

Using the rearrangement:

$$x = \frac{x^2 - 5}{2}$$

if you try $x_0 = 3.5$,

$$x_1 = \frac{(3.5)^2 - 5}{2} = 3.625$$

$$x_2 = \frac{(3.625)^2 - 5}{2} = 4.07$$

$$x_3 = \frac{(4.07)^2 - 5}{2} = 5.78$$

Clearly, this method is not converging (getting closer) to the solution, it is **diverging** (getting further away).

Now $\quad F(x) = \dfrac{x^2 - 5}{2}$

so $\quad F'(x) = x$

Hence $\quad |F'(x)|$ is not less than 1 if $3 \leq x \leq 4$

The other formula $x = 2 + \frac{5}{x}$, gives $F(x) = 2 + \frac{5}{x}$

Hence, $F(x) = -\frac{5}{x^2}$

$\therefore \quad |F'(x)| < 1$ if $x^2 > 5$

so $\quad x > 2.24$

Hence we can take $x_0 = 3.5$,

$$\text{so} \quad x_1 = 2 + \frac{5}{3.5} = 3.429$$

$$x_2 = 2 + \frac{5}{3.429} = 3.458$$

$$x_3 = 2 + \frac{5}{3.458} = 3.446$$

$$x_4 = 2 + \frac{5}{3.446} = 3.451$$

$$x_5 = 2 + \frac{5}{3.451} = 3.449$$

The method converges quite slowly to the solution:

$x = 3.45$ (2 decimal places)

Another way of locating the solution of an equation is to use a graphical approach.

To solve $f(x) = 0$, rearrange the equation into the form $g(x) = h(x)$.

If you sketch $y = g(x)$ and $y = h(x)$, the points of intersection of the graph will help you locate the roots.

Example 12.4

Show that the cubic equation $2x^3 + x - 4 = 0$ has only one solution, $x = \alpha$. Find an integer n such that $n \leq \alpha \leq n + 1$.

By rearranging the equation to give a converging iteration formula, find the root correct to 2 decimal places.

Solution

Rearrange $2x^3 + x - 4 = 0$ to $2x^3 = 4 - x$.

The graphs of $y = 2x^3$ and $y = 4 - x$ clearly intersect at one point and certainly $0 < \alpha < 4$ (see Figure 12.4).

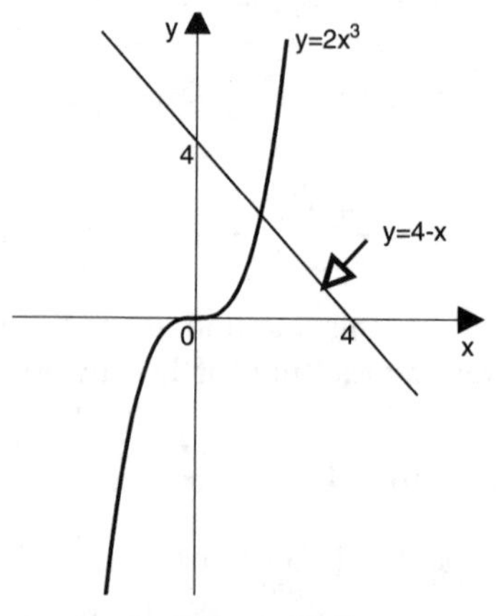

Figure 12.4

$$\text{If} \quad f(x) = 2x^3 + x - 4$$
$$f(0) = -4$$
$$f(1) = -1$$
$$f(2) = 14$$
$$\therefore \quad 1 \leq \alpha \leq 2$$

$$\text{Now} \quad x = \sqrt[3]{\frac{4-x}{2}},$$

$$\text{that is,} \quad x_{n+1} = \sqrt[3]{\frac{4-x_n}{2}}$$

Let $x_0 = 1$, $x_1 = 1.145$, $x_2 = 1.126$, $x_3 = 1.128$

that is, $x = 1.13$. (2 dec. places)

Example 12.5

The iteration formula:

$$x_{n+1} = \frac{2x_n{}^3 + 2}{3x_n{}^2 + 2}$$

converges to a value α. Find α correct to 3 significant figures, starting with $x_0 = 1$, and find the cubic equation which has α as a solution. By sketching two suitable graphs, show that α is the only solution of this cubic equation.

Solution

$$\text{If } x_0 = 1, \quad x_1 = \frac{2 \times 1^3 + 2}{3 \times 1^2 + 2} = 0.800$$
$$x_2 = \frac{2 \times 0.8^3 + 2}{3 \times 0.8^2 + 2} = 0.771$$
$$x_3 = \frac{2 \times 0.771^3 + 2}{3 \times 0.771^2 + 2} = 0.771$$

Therefore $\alpha = 0.771$ (3 dec. places)

Since the iteration has now converged to this value, x_n and x_{n+1} will always equal α.

$$\text{So} \quad \alpha = \frac{2\alpha^3 + 2}{3\alpha^2 + 2}$$
$$\therefore \quad \alpha(3\alpha^2 + 2) = 2\alpha^3 + 2$$
$$3\alpha^3 + 2\alpha = 2\alpha^3 + 2$$
$$\alpha^3 + 2\alpha - 2 = 0$$

Therefore α is a solution of the equation:

$$x^3 + 2x - 2 = 0$$

The equation can be rearranged to $x^3 = 2 - 2x$.

A sketch of the two graphs $y = x^3$ and $y = 2 - 2x$ shows that they can only cross at one point (see Figure 12.5). Hence the cubic equation only has one solution.

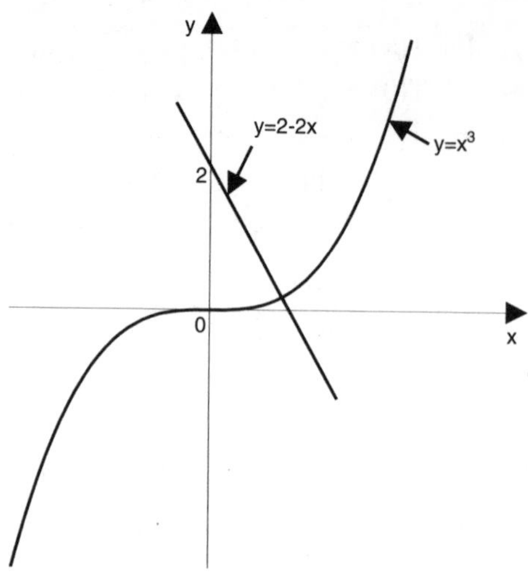

Figure 12.5

12.6 Newton–Raphson method

This is one of the most well-known iterative methods for solving an equation $f(x) = 0$. Suppose that we have located an approximation $x = \alpha$ to the solution of the equation.

The point $P(\alpha, f(\alpha))$ on the curve will be close to the point where the graph crosses the x-axis. Construct the tangent to the curve at P. This tangent will cut the x-axis at a point β, which will be even closer to the required solution (see Figure 12.6).

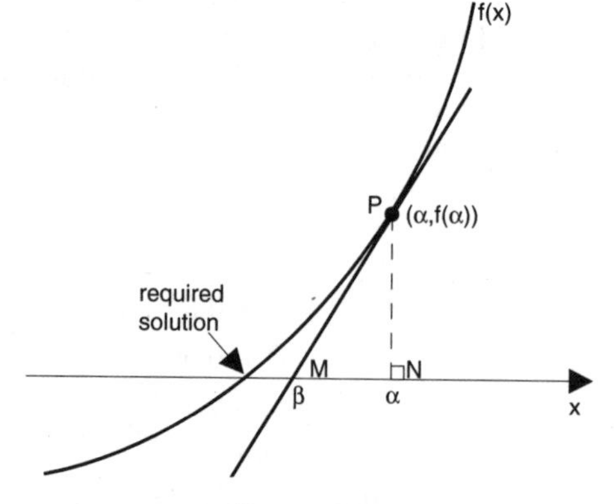

Figure 12.6

The gradient of the tangent at P is given by $f'(\alpha)$

But the gradient of the line $= \dfrac{PN}{MN} = \dfrac{f(\alpha)}{\alpha - \beta}$

$$\therefore \quad f'(\alpha) = \frac{f(\alpha)}{\alpha - \beta}$$

$$\therefore \quad \alpha - \beta = \frac{f(\alpha)}{f'(\alpha)}$$

So $$\beta = \alpha - \frac{f(\alpha)}{f'(\alpha)}$$

Clearly, this process could be repeated by drawing a tangent at $(\beta, f(\beta))$ and finding a value even closer to the correct value.

This process can be summarised by an iterative formula:

$$x_{n+1} = x_n - \frac{f(x_n)}{f'(x_n)} \tag{N2}$$

x_0 is taken as the initial approximation to the solution. The following example shows how the method works, and how to locate x_0.

Example 12.6

Show that the equation $x^4 - x^3 + 11 = 0$ has a solution between $x = 2$ and $x = 3$. Use the Newton–Raphson method to find this solution correct to two decimal places.

Solution

If $\quad f(x) = x^4 - x^3 - 11$

then $\quad f(2) = 2^4 - 2^3 - 11 = -3$

and $\quad f(3) = 3^4 - 3^3 - 11 = 43$

The function changes sign from -3 to $+43$, and so there is a solution between $x = 2$ and $x = 3$.

Now $\quad f'(x) = 4x^3 - 3x^2$

Hence the Newton–Raphson iteration will be:

$$x_{n+1} = x_n - \frac{(x_n{}^4 - x_n{}^3 - 11)}{(4x_n{}^3 - 3x_n{}^2)}$$

The solution is much nearer to $x = 2$, because $f(2) = -3$ is numerically much smaller than $f(3) = 43$.

Let $\quad x_0 = 2.1$

$$\therefore \quad x_1 = 2.1 - \frac{(2.1^4 - 2.1^3 - 11)}{(4 \times 2.1^3 - 3 \times 2.1^2)} = 2.134$$

$$x_2 = 2.134 - \frac{(2.134^4 - 2.134^3 - 11)}{(4 \times 2.134^3 - 3 \times 2.134^2)} = 2.133$$

The solution is 2.13 to 2 decimal places.

ACTIVITY 7

Square roots

To find $\sqrt{N}$ is the same as solving the equation $x^2 = N$. This could be solved by the Newton–Raphson method, that is, use $f(x) = x^2 - N$.

(i) Set up the iteration.

(ii) Try and write a computer program that will find $\sqrt{N}$ to 10 decimal places.

(ii) Using a spreadsheet approach, see if you can tabulate the $\sqrt{N}$ for a range of values of N. Investigate other problems of this type.

Example 12.7

On a single diagram, sketch the graphs of $y = 2\ln(5x)$ and $y = \dfrac{5}{x^2}$ for $x > 0$.

Show that the equation $2\ln(5x) = \dfrac{5}{x^2}$ has one root. If this root is denoted by α, and $n < \alpha < n + 1$, find n.

Use the iteration $x_{n+1} = \sqrt{\dfrac{5}{2\ln(5x_n)}}$ to find α correct to 3 significant figures.

Solution

The two graphs are shown in Figure 12.7. You can see that they only cross at one point, hence the equation has only one solution.

To locate the solution, let $f(x) = 2\ln(5x) - \dfrac{5}{x^2}$.

You cannot put $x = 0$ into this function.

So find
$$f(0.1) = -501$$
$$f(1) = -1.78$$
$$f(2) = 3.36$$

Therefore, $1 < \alpha < 2$, that is, $n = 1$ (see Figure 12.8).

The values of $f(1)$ and $f(2)$ suggest α is nearer to 1.

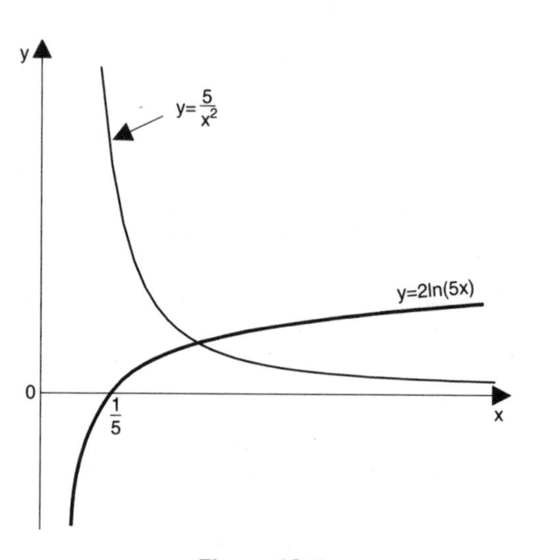

Figure 12.7

Let $x_0 = 1.4$

$$x_1 = \sqrt{\frac{5}{2\ln(5 \times 1.4)}} = 1.13$$

$$x_2 = \sqrt{\frac{5}{2\ln(5 \times 1.13)}}$$

$$x_3 = \sqrt{\frac{5}{2\ln(5 \times 1.2)}}$$

$$x_4 = \sqrt{\frac{5}{2\ln(5 \times 1.18)}}$$

$$= 1.186$$

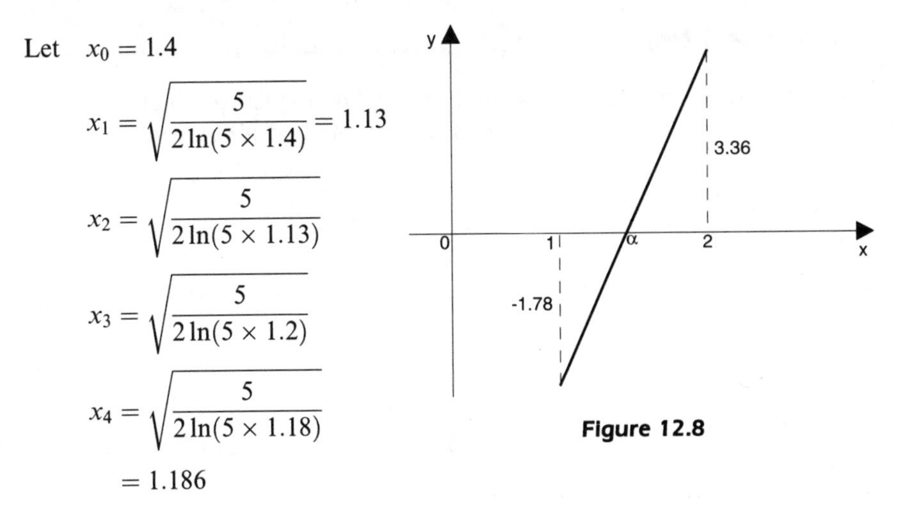

Figure 12.8

Therefore $\alpha = 1.19$ (3 significant figures).

The method converges quite slowly, and so an alternative solution could be obtained using the Newton–Raphson process. It may also be worth using linear interpolation to find x_0.

Here $\dfrac{3.36}{2-\alpha} = \dfrac{1.78}{\alpha - 1}$

Hence $3.36\alpha - 3.36 = 3.56 - 1.78\alpha$

$$\therefore \quad 5.14\alpha = 6.92$$

$$\alpha = 1.35$$

This result is not much better than that guessed in the previous example.

To set up the Newton–Raphson iteration, let

$$f(x) = 2\ln(5x) - \frac{5}{x^2}$$

then $$f'(x) = \frac{2}{5x} \times 5 + \frac{10}{x^3} = \frac{2}{x} + \frac{10}{x^3}$$

Hence the Newton–Raphson iteration will be:

$$x_{n+1} = x_n - \frac{2\ln(5x_n) - \dfrac{5}{{x_n}^2}}{\dfrac{2}{x_n} + \dfrac{10}{{x_n}^3}}$$

If $x_0 = 1.35$

$$x_1 = 1.35 - \frac{f(1.35)}{f'(1.35)} = 1.35 - \frac{1.076}{5.546} = 1.160$$

$$x_2 = 1.16 - \frac{f(1.16)}{f'(1.16)} = 1.16 - \frac{-0.2}{8.131}$$

As before, the root $= 1.19$ (2 decimal places)

Exercise 12(d)

For the following equations, find the smallest positive solution. Use an iterative method and the Newton–Raphson method if you can.

1 $x^3 + x - 3 = 0$

2 $2(x + \sin x) = 1$

3 $x^4 + x^2 - 4 = 0$

4 $x^3 = 4$

5 $\tan x = 4 \sin 2x$

Miscellaneous Examples 12

1 If $H = \dfrac{4x}{3y + 2x}$, and x and y can be measured to an accuracy of 1%, find the percentage error possible in calculating H.

2 If $f(4) = 0.4$ and $f(5) = -0.16$, find by linear interpolation an approximation to the solution of the equation $f(x) = 0$.

3 Find an approximation to 1 decimal place to the solution of the equation $3 \sin x = x^2 - 1$ for which $-\pi \leq x \leq 0$.

4 Find an iterative formula which will solve the equation $x^3 + 2x - 2 = 0$. Justify that the iteration converges to the solution, and find this solution correct to 3 decimal places.

Revision Problems 12

1 A jeweller has some gold in the shape of a solid sphere of radius 2 cm. The gold needs to be reshaped into a solid cylinder of length $\frac{4}{3}$ cm with a solid hemisphere exactly covering each end, as in Figure 12.9.

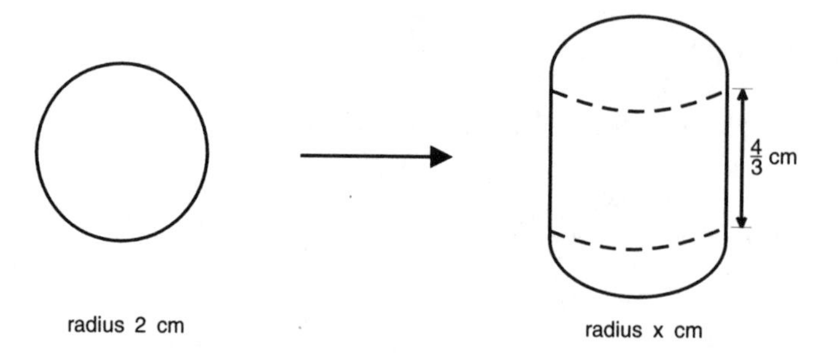

Figure 12.9

(a) Show that, if x cm is the radius of the new cylinder and hemispheres, then x satisfies the equation

$$x^3 + x^2 - 8 = 0$$

(the volume of a sphere of radius r is $\frac{4}{3}\pi r^3$).

(b) Show that the equation $x^3 + x^2 - 8 = 0$ has a root between $x = 1.7$ and $x = 1.8$.

(c) Show that the equation $x^3 + x^2 - 8 = 0$ can be rearranged as:

$$x = \sqrt{8 - x^3}$$

Using an iterative technique with:

$$x_{n+1} = \sqrt{8 - x_n{}^3} \quad \text{and} \quad x_1 = 1.7$$

calculate the values of x_2, x_3, x_4 and x_5. Comment on the appropriateness of this method for finding a root of the equation:

$$x^3 + x^2 - 8 = 0$$

and use the Newton–Raphson method once to find a second approximation. Give your answer correct to two decimal places.

2 (a) Show that there is a root of the equation $8 \sin x - x = 0$ lying between 2.7 and 2.8.

(b) Taking 2.8 as a first approximation to this root, apply the Newton–Raphson procedure once to $f(x) \equiv 8 \sin x - x$ to obtain a second approximation, giving your answer to 2 decimal places.

(c) Explain, with justification, whether or not this second approximation is correct to 2 decimal places.

(d) Evaluate $f\left(\frac{5\pi}{2}\right)$, and hence, by sketching suitable graphs, determine the number of roots of the equation $8 \sin x - x = 0$ in the range $x > 0$.

(LEAG)

13 Probability

For centuries, people have been interested in the theory of games. When mathematicians decided to apply the laws of probability to them, a whole new field of mathematics came into being. This chapter looks at the basic ideas of arrangement, selection and Venn diagrams which can be used to help analyse those situations in life which involve uncertainty, from the simple rolling of dice ultimately to the rules that govern the motion of atomic particles.

DO YOU KNOW

$n!$ means $n(n-1) \times \ldots 2 \times 1$

13.1 Arrangements (often called permutations)

Here is a simple problem. If you have three letters A, B and C, how many ways can you arrange these? If we list the answers, they are:

ABC, ACB, BCA, CAB, CBA, BAC.

In other words, there are six possible **arrangements** or **permutations.** Let us try to look systematically at how we arrived at this.

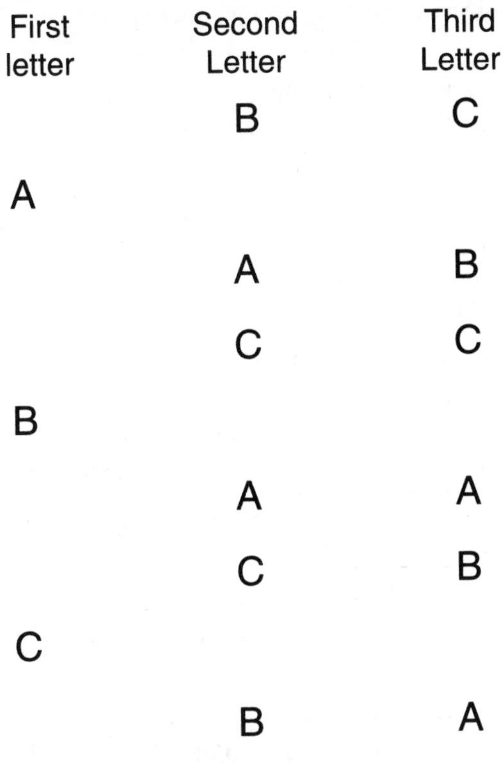

Figure 13.1

There are three ways of choosing the first letter. For each of these three ways, there are two ways of choosing the second letter, giving 3×2 ways of choosing the first two letters, and for each of these six ways, there is only one way of choosing the last. Hence the total number of arrangements is:

$3 \times 2 \times 1 = 3!$ (see Chapter 3 about factorials)

If we have n letters (or objects) to arrange, we have the result:

the total number of ways of arranging n different objects $= n!$ (S1)

What happens if you have, say, five letters, and only make three letter arrangements (no repetition of letters is allowed)?

The number of ways of choosing the first letter $= 5$
The number of ways of choosing the second letter $= 4$
The number of ways of choosing the third letter $= 3$
Hence the total number of ways $= 5 \times 4 \times 3 = 60$

It is not immediately obvious how they can be written using factorials, but we can employ a clever trick:

$$5 \times 4 \times 3 = \frac{5 \times 4 \times 3 \times \not{2} \times \not{1}}{\not{2} \times \not{1}} = \frac{5!}{2!}$$

If we have n objects, and only arrange them in groups of r ($r < n$), we have the result:

the number of ways of arranging r different objects selected from

n different objects is: $\dfrac{n!}{(n-r)!}$ (S2)

This is written ${}^nP_r$ or ${}_nP_r$

It is important to realise that these results assume that the objects are *all different*. Suppose, in the previous example, you could write down each letter as many times as you liked.

The number of ways of choosing the first letter = 5
" " " " " " " second letter = 5
" " " " " " " third letter = 5

Hence the total number of arrangements is $5 \times 5 \times 5 = 125$.

The result follows that if you arrange r different objects, chosen from n but repetition of the letters *is* allowed:

the total number of arrangements $= n^r$ (S3)

The situations that arise if objects are repeated lead to more difficult calculations.

MEMORY JOGGER

It is important with all problems to do with arrangements that you do not just rely on formulae, but break the problem down into individual parts

Suppose you were given the letters $ABBBCD$. How many six-letter arrangements can we make from these letters?

If the letters are all different, there are 6! ways. Here there are 3 Bs. giving only 4 different letters. Suppose, however, you hang labels on the Bs to distinguish them, then:

$AB_1CB_2B_3D$ would be the same as $AB_2CB_1B_3D$, etc.

In fact, there are 3! (because there are 3 Bs) ways of this particular arrangement. This is, of course, true for all the different arrangements, and so the total number of arrangements $= \dfrac{6!}{3!} = 120$.

This result can be generalised as follows.

The number of arrangements of n objects of which there are r_1 of one type the same, r_2 of another the same and so on:

$$= \frac{n!}{r_1!r_2!\ldots} \qquad \text{(S4)}$$

The use of this formula is illustrated in the following problem.

Example 13.1

How many arrangements can be made using all the letter of the word MISSISSIPPI?

Solution

Using equation (S4), $n = 11$,
the number of Ss $r_1 = 4$,
the number of Is $r_2 = 4$,
the number of Ps $r_2 = 2$.

Hence the number of arrangements:

$$= \frac{11!}{4!4!2!} = 34\,650$$

Example 13.2

How many four-letter arrangements can be made from the letters of the word BANANA.

Solution

The first important thing to realise about this type of problem is that it cannot be solved using just one formula. The four-letter arrangement will be governed by the number of As in them.

If there is only one A, the letters are $BANN$,

giving $\frac{4!}{2!} = 12$ arrangements,
↑
because there are 2 Ns

If there are two As, we could have the letters $BAAN$,

giving $\frac{4!}{2!} = 12$ arrangements;

or the letters AANN,

giving $\frac{4!}{2!2!} = 6$ arrangements.

If, however, there are three As, we could have the letters $BAAA$,

giving $\frac{4!}{3!} = 4$ arrangements;

or the letters $NAAA$,

giving 4 arrangements.

The total number of four-letter arrangements $= 12 + 12 + 6 + 4 + 4 = 38$.

> **MEMORY JOGGER**
>
> If you are trying to find the number of arrangements subject to a certain restraint, it may be easier to find the number of arrangements without that restraint, and subtract this from the total number of arrangements.

Example 13.3

A red, a blue, a green and two yellow discs are arranged in a straight line. In how many ways can they be arranged so that the yellow discs are *not* next to each other.

Solution

The number of ways of arranging the five discs

$$= \frac{5!}{2!} = 60$$

If the two yellow discs *are* next to each other, then *four different* units are being arranged.

Hence the number of arrangements $= 4! = 24$.

Hence the number of ways in which the discs are not next to each other

$$= 60 - 24 = 36$$

When an arrangement is in the form of a circle (or closed curve), a slightly different technique is needed. The following examples show you what you need to do.

Example 13.4

A group of five people sit down at a round table. How many different arrangements are possible if it doesn't matter which chairs they are occupying?

Solution

Figure 13.2 shows two different seating positions, but they are the same arrangement, relative to each other.

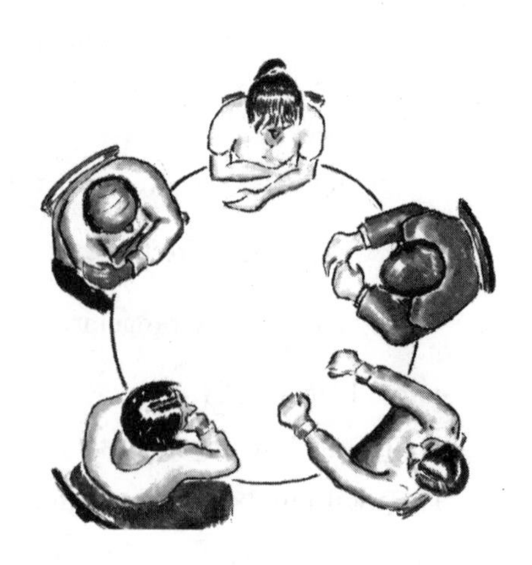

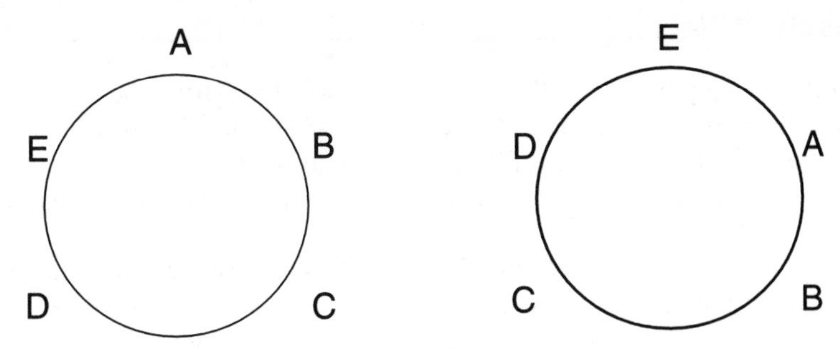

Figure 13.2

To solve this situation, you just fix one of the people and arrange the other four.

Hence the number of arrangements $= (5 - 1)! = 4! = 24$

In general, we have the result that:

the number of ways of arranging n objects in a circle is $(n - 1)!$ (S5)

Example 13.5

Six differently coloured beads are arranged on a circle wire. How many arrangements of the beads are possible?

Solution

It is easy to rush at this and say the answer is $(6 - 1)! = 120$. However, the two arrangements shown in Figure 13.3 would be different at a round table, but because the ring could be turned about the axis shown, the arrangements are the same.

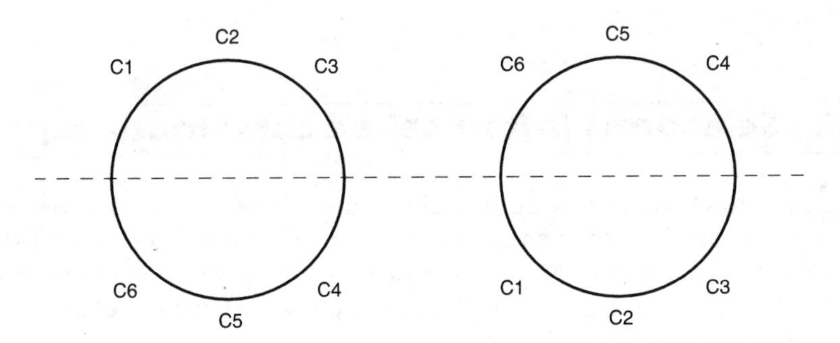

Figure 13.3

Hence the number of arrangements $= 120 \div 2 = 60$.

Exercise 13(a)

1. In how many ways can the letters of the word PEOPLE be arranged?
2. Two hundred and fifty people each buy a raffle ticket. There are six prizes to be won. In how many different ways can the prizes be given?
3. How many four-letter arrangements can be made using the letters from the word AVERAGE?
4. Four red, three blue and two black counters are arranged on a table in a straight line. How many different arrangements are possible? If the counters are arranged in a circle, how many arrangements are possible?
5. A group of five friends sit in adjacent seats at the cinema. Ben and Josh do not want to sit next to each other. In how many ways can they occupy the five seats?

6. A four-digit number is written down using only the digits 1, 2, 3, 5, 6 and 8 without repetition. Find:
   (i) How many different numbers are possible.
   (ii) How many odd numbers are possible.
   (iii) How many numbers divisible by 5 are possible.
7. In how many ways can three differently coloured beads be threaded on to a circular wire? Justify your answer carefully.
8. Eight people sit at a round table. Two of the people must not sit next to each other. In how many ways is this possible?

13.2 Selections (often called combinations)

When you are choosing a team, initially it doesn't matter in what order you choose the people. It is only when, say, a batting *order* is being constructed that you would be making an arrangement. Situations where order doesn't matter are referred to as *selections* (or *combinations*). If you were arranging r objects chosen from n, the number of arrangements

$$= {}^nP_r = \frac{n!}{(n-r)!}$$

However, a given selection of r objects can be arranged in $r!$ different arrangements of r. Hence the number of selections:

$$= {}^nP_r \div r!$$

$$= \frac{n!}{(n-r)!} \div r!$$

$$= \frac{n!}{(n-r)!r!}$$

This is denoted by

$${}^nC_r = \frac{n!}{(n-r)!r!} \qquad \text{(S6)}$$

(Also written ${}_nC_r$ or $\binom{n}{r}$.)

As with the formulae for arrangements, this formula must be used with care, particularly where some of the objects are repeated.

Example 13.6

Jamie has twelve different vehicles in his toy box. He chooses three. How many ways can he choose them?

If there are seven lorries and five cars, in how many ways can he choose two lorries and a car?

Solution

Here $n = 12$ and $r = 3$.

The total $= {}^{12}C_3 = \frac{12!}{9!3!} = 220$

In the second part he is certainly choosing 3 from 12, but ${}^{12}C_3$ is not the answer, because this would not ensure that he had two lorries and a car.

The lorries can be chosen in ${}^7C_2 = 21$ ways.

The car can be chosen in ${}^5C_1 = 5$ ways.

The total number of ways $= 21 \times 5 = 105$.

ACTIVITY 8

In Chapter 3, you were introduced to the idea of the binomial distribution. You should now be thinking why ${}^nC_r$ appears there and also in the work on selections. Try to think about what $(a+b)^n$ means.

Look at, say, $(a+b)^4$ as $(a+b)(a+b)(a+b)(a+b)$. Look at the number of ways you can select one letter from each bracket.

Exercise 13(b)

1 There are 10 books on a shelf. In how many ways can you select 4 of them?

2 A garden centre offers a total of 30 different shrubs, and 25 different perennials. Pam and Mike visit the centre to select 4 shrubs and 5 perennials. In how many different ways could they make the choice?

3 On a football pools coupon, you are required to select 12 draws from 40 results. How many selections are possible?

4 In how many ways can a quiz team of 2 men and 2 women be selected from a group of 5 men and 4 women? If the group of 9 people contains 1 married couple, and it is decided that only one or neither should be in the team, how many can be selected?

5 In a tennis club of 44 members, 25 are women. Find the number of ways of forming a committee of 4 members

(i) with no restrictions;
(ii) with at least 1 woman on the committee.

6 Prove ${}_nC_r = {}_nC_{n-r}$.

7 Four British candidates, 6 German candidates, 5 French candidates and 3 Italian candidates are available for selection on a European committee consisting of 5 people. Each nationality must have at least 1 representative. In how many ways can the committee be selected?

13.3 Probability

DO YOU KNOW

(i) The probability of an event happening

$$= \frac{\text{no. of ways event happens}}{\text{total number of outcomes}}$$

(ii) Probability is measured by a fraction or decimal which lies between 0 (impossible) and 1 (certain).

(iii) A probability tree is useful to show combined events.

13.4 Theoretical definition

If an experiment has a set of **equally likely** outcomes S (the **possibility space** or **sample space**). For a given event, E, the probability that E happens (written P(E)) is given by:

$$P(E) = \frac{n(E)}{n(S)} \qquad \text{(PR1)}$$

where $n(E)$ stands for the number of ways E can happen, and $n(S)$ the number of ways in which S can happen,

So $0 \leq P(E) \leq 1$

13.5 Venn diagrams

It is often possible to represent a probability situation in a diagram referred to as a **Venn diagram**. Each event is represented by a closed curve, and the total set of outcomes, S, by a rectangle.

(i) The region where two events overlap is written $A \cap B$ (A intersection B). $\cap$ represents AND.

If A and B have no points in common, that is, $A \cap B = \emptyset$ (the empty set) then events A and B are said to be **mutually exclusive** (they have nothing in common).

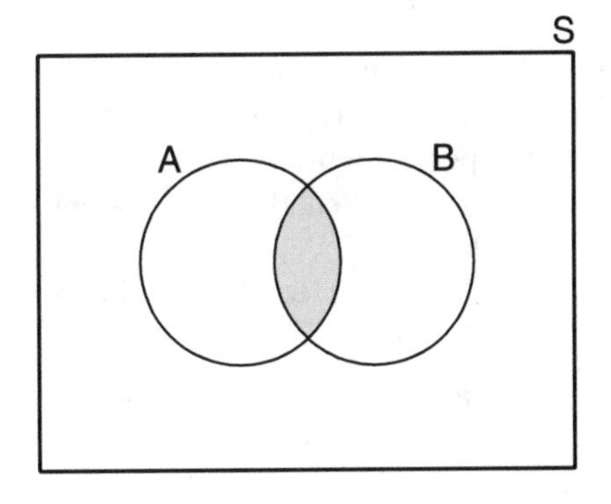

Figure 13.4

(ii) The combined region covered by two events, X and Y is written $X \cup Y$ (X union Y) $\cup$ represents OR (or BOTH)

$$P(X \cup Y) = P(X) + P(Y) - P(X \cap Y) \qquad \text{(PR2)}$$

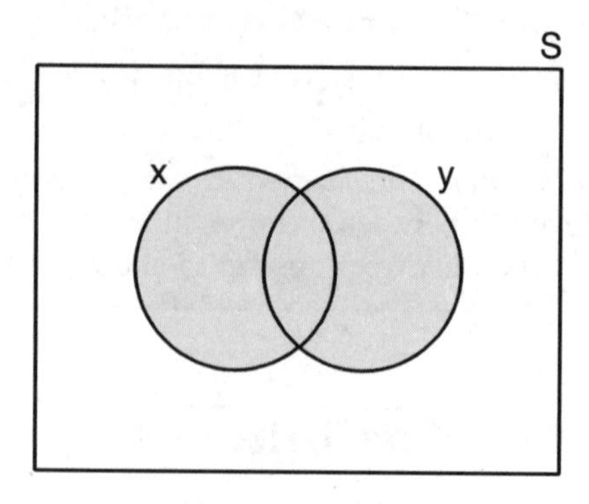

Figure 13.5

The probability that X or Y happens is the probability of X happening + the probability of Y happening – the probability that they both happen together.

If X and Y are mutually exclusive, $X \cap Y = \emptyset$ and $P(X \cap Y) = 0$

So $P(X \cup Y) = P(X) + P(Y)$ (PR3)

MEMORY JOGGER

You can only add probabilities for OR if the events are *mutually exclusive*

(iii) The region outside a set A represents the event A not happening written $\bar{A}$ (or A')

It is called the **complement** of A.

Since the total probability inside S is 1:

$$P(\bar{A}) = 1 - P(A) \qquad \text{(PR4)}$$

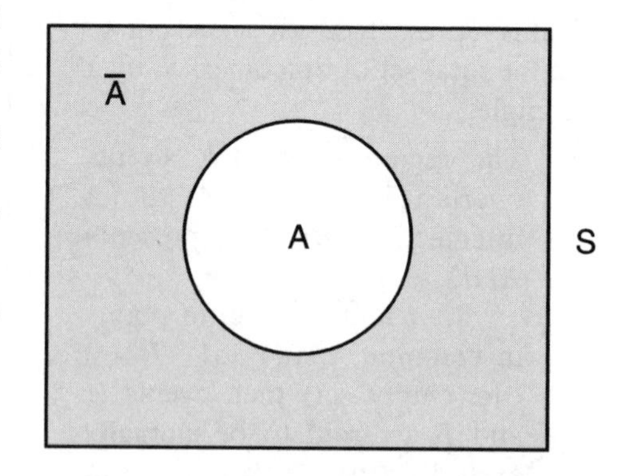

Figure 13.6

Example 13.7

Two events, A and B, are such that $P(A) = 0.4$, $P(B) = 0.5$, $P(A \text{ and } B) = 0.1$. Find:

(i) $P(A \text{ or } B)$ (ii) $P(A \text{ and not } B)$

Solution

Now because $P(A \cap B) = 0.1$, you can find the probabilities of all regions in the Venn diagram (see Figure 13.7)

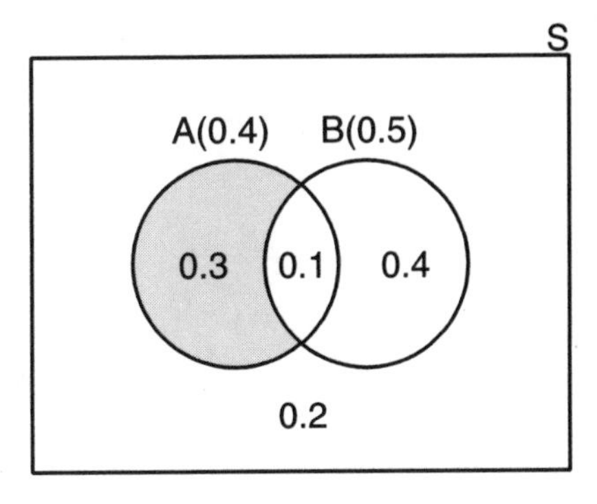

Figure 13.7

(i) $P(A \text{ or } B) = 0.3 + 0.1 + 0.4 = 0.8$
(ii) $P(A \text{ and not } B) = P(A \cap B') = 0.3$ (shown shaded)

13.6 Combined events

The probability of combined events can be shown on a **tree diagram**. **Combined events** occur when one event, B, follows another event, A. If the outcome of the second event, B, is not affected by the outcome of event A, then A and B are said to be independent. The probability of A and B can be found from the formula:

$$P(A \cap B) = P(A)P(B)$$

MEMORY JOGGER

You can only multiply probabilities when using the word AND if the events are independent

These combined events can most easily be represented on a **tree diagram**. Study the next example carefully.

Example 13.8

A bag contains 6 red discs and 4 blue discs. If 3 discs are taken from the bag, and replaced each time, what is the probability of 2 of the discs being red?

Solution

Since the discs are replaced each time, you do not affect the probabilities of red and blue on each draw. The different possibilities are represented on the tree shown in Figure 13.8.

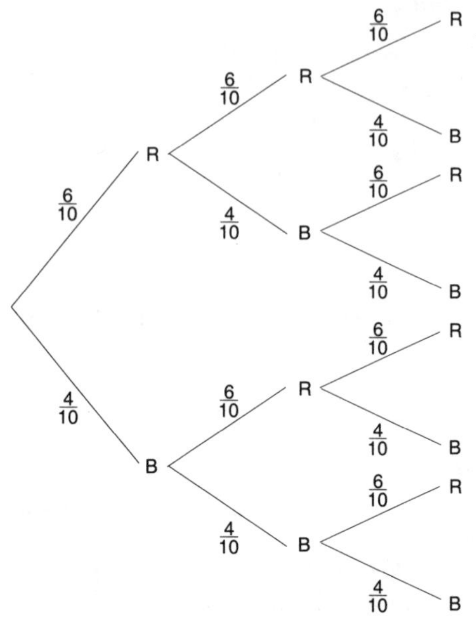

Figure 13.8

Hence, two reds are obtained for the outcomes: (red and red and blue) or (red and blue and red) or (blue and red and red).

Using the addition and multiplication rules, we have the probability:

$$= \tfrac{6}{10} \times \tfrac{6}{10} \times \tfrac{4}{10} + \tfrac{6}{10} \times \tfrac{4}{10} \times \tfrac{6}{10} + \tfrac{4}{10} \times \tfrac{6}{10} \times \tfrac{6}{10}$$
$$= 0.432$$

If the problem is now altered to the extent that the discs are not replaced, then clearly the probabilities on the second draw will be different. (There will be only 9 discs in the second draw). This idea is considered in the next section.

13.7 Conditional probability

The ideas behind **conditional probability** centre around the statement: 'What is the probability of event X happening given that event Y has already happened?' This will be written $P(X|Y)$. The order in which X and Y happen is important. If you consider the previous problem of taking discs from a bag that contained 6 red and 4 blue discs, without replacing them, then to find the probability that the first

2 discs are red, we must use the modified tree shown here. The branch for R followed by R has the probability of $\frac{5}{9}$ marked on it (see Figure 13.9). This is the probability that the second disc is red, given that the first disc is red, that is,

$$\text{P(2nd red|first red)} = \tfrac{5}{9}$$

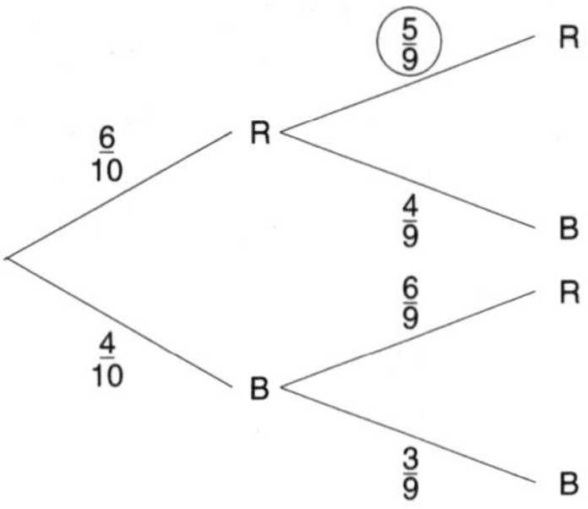

Figure 13.9

Hence the probability that the first two discs are red is $\frac{6}{10} \times \frac{5}{9}$.

This can be written:

$$\text{P(1st red} \cap \text{2nd red)} = \text{P(1st red).P(2nd red|1st red).}$$

If we denote 1st red by event Y, and 2nd red by event X,

then $\quad P(Y \cap X) = P(Y)P(X|Y) \qquad$ (PR5)

This rearranges to the conditional probability formula:

$$P(X|Y) = \frac{P(X \cap Y)}{P(Y)} \qquad \text{(PR6)}$$

If X and Y are independent, then $P(X|Y) = P(X)$.

Hence $\quad P(X \cap Y) = P(X)P(Y)$ if X and Y are **independent** events $\qquad$ (PR7)

Example 13.9

Two events, A and B, are not independent, and $P(A) = 0.5$, $P(B) = 0.7$. It is also given that $P(A \cup B) = 0.85$. Find:

(i) $P(A \cap B)$ $\qquad$ (ii) $P(B|A)$ $\qquad$ (iii) $P(A \cup B')$

Solution

(i) If $x = P(A \cap B)$, then the Venn diagram in Figure 13.10 shows us that since $P(A \cup B) = 0.85$,

$$0.85 = (0.5 - x) + x + (0.7 - x)$$

Hence: $\quad 0.85 = 1.2 - x, \quad$ so $x = 0.35$

Hence: $\quad P(A \cap B) = 0.35$

You could also have used the formula:

$$P(A \cup B) = P(A) + P(B) - P(A \cap B)$$

However, the Venn diagram in Figure 13.10 gives a clear picture of what is happening.

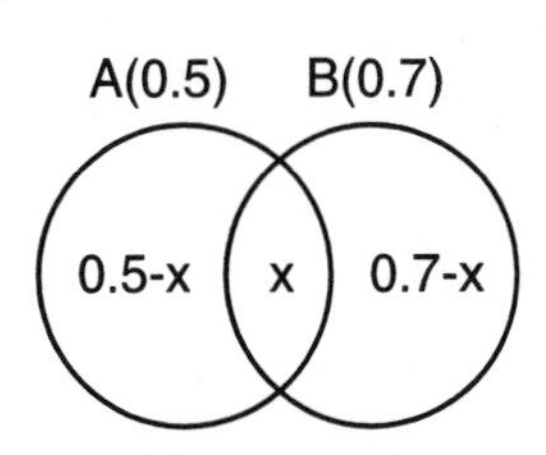

Figure 13.10

(ii) $P(B|A) = \dfrac{P(B \cap A)}{P(A)} = \dfrac{0.35}{0.5} = 0.7$

(iii) The Venn diagram in Figure 13.11 shows the set $A \cup B'$ shaded. The probability of the unshaded part is $0.7 - x = 0.35$

Hence $P(A \cup B') = 1 - 0.35$
$= 0.65$

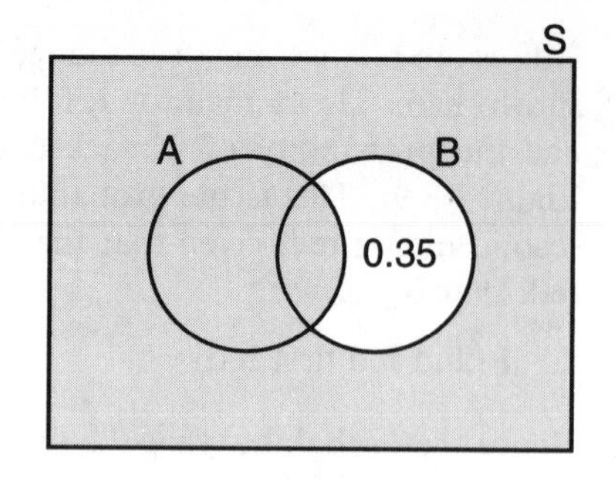

Figure 13.11

The following example is typical of how we might use the idea of conditional probability in practice.

Example 13.10

On average, Tim travels to work by bus 10% of the time, train 60% of the time, and car the rest of the time. If he travels by bus there is a probability of 0.1 that he will be late. If he travels by train, there is a probability 0.2 that he will be late, and if he travels by car, there is a probability of 0.3 that he is late. On a given day, Tim arrived late for work.

(i) Which form of transport is he most likely to have used?
(ii) What is the probability that Tim came by bus?

Solution

(i) It is sensible to draw a tree diagram for this problem; it is shown in Figure 13.12.

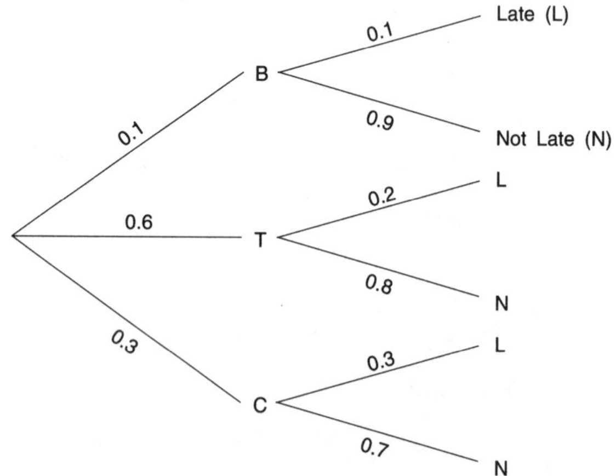

Figure 13.12

$$\begin{aligned} P(\text{bus} \cap \text{late}) &= 0.1 \times 0.1 = 0.01 \\ P(\text{train} \cap \text{late}) &= 0.6 \times 0.2 = 0.12 \\ P(\text{car} \cap \text{late}) &= 0.3 \times 0.3 = 0.09 \end{aligned}$$

He is most likely to have travelled by train, because this is the largest probability.

(ii) The second situation can be described using conditional probability.

$$P(B|L) = \frac{P(B \cap L)}{P(L)} = \frac{0.01}{0.01 + 0.09} = \tfrac{1}{22}$$

There is an alternative way of looking at this question, which often provides a nice simple method. Suppose you analysed the outcome of 100 trips to work. Then the tree diagram showing the number of outcomes is shown in Figure 13.13.

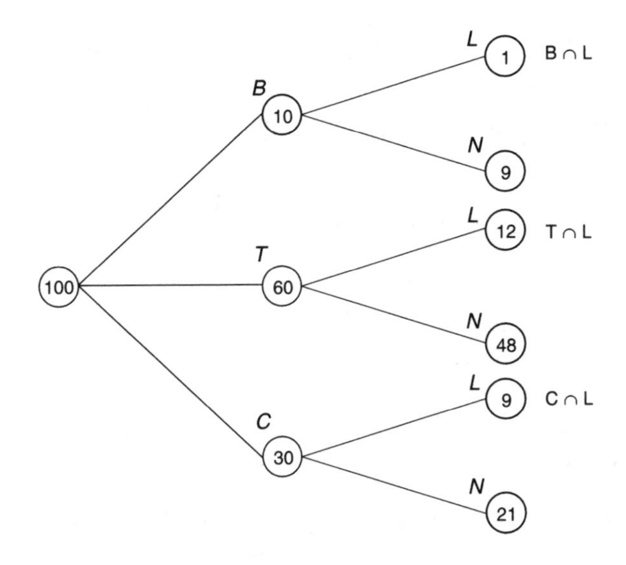

Figure 13.13

The number of times late $= 1 + 12 + 9 = 22$.

Of these, 1 was by bus.

Hence: $P(B|L) = \frac{1}{22}$

Exercise 13(c)

1 A and B are two events such that $P(A) = \frac{1}{4}$, $P(A|B) = \frac{1}{3}$ and $P(B|A) = \frac{1}{5}$. Find:

(i) $P(A \cap B)$ (ii) $P(B)$ (iii) $P(A \cup B)$

2 A die is rolled twice. Find the probability that

(i) a score of three or less is obtained each time;
(ii) a total score of three or less is obtained.

3 H and K are independent events. If $P(H) = 0.4$ and $P(K) = 0.6$, find:

(i) $P(H \cap K)$ (ii) $P(H \cup K)$

4 The probability that I catch my train each morning is 0.8. What is the probability that on three consecutive days I have missed the train at least once.

5 Two events, X and Y, are mutually exclusive. If $P(X) = \frac{2}{3}$ and $P(Y) = \frac{1}{6}$, find:

(i) $P(X \cup Y)$ (ii) $P(X \cap \bar{Y})$

6 In a group of 80 people, 37 own a cat, 24 own a dog, and 16 own a cat and a dog. Find:

(i) the probability that a person chosen at random owns a cat given they own a dog;
(ii) the probability that a person chosen at random owns a dog given that they own a cat.

7 Events A and B are such that $P(A) = \frac{1}{4}$, $P(A|B) = \frac{1}{5}$ and $P(\bar{B}|\bar{A}) = \frac{2}{3}$. By drawing a suitable tree diagram, find:

(i) $P(\bar{A}|B)$ (ii) $P(B \cap A)$ (iii) $P(A \cup \bar{B})$

Miscellaneous Examples 13

1 In how many ways can the letters of the word GENERAL be arranged? What is the probability that the arrangement begins and ends in E?

2 Harry and Karen have been invited to a dinner party. There are 8 guests altogether, and they are going to sit at a round table. What is the probability that Harry and Karen end up sitting next to each other?

3 From a group of 7 boys and 8 girls, a quiz team of 4 is selected. In how many ways can it be selected if the team is to consist of 2 boys and 2 girls?

4 A tennis knockout is entered by n people. How many matches are needed to produce a winner?

5 How many four-digit numbers can be made from the numbers 2, 3, 5, 8 and 9 without repetition. What is the probability that the number will be greater than 5000?

6 The events A and B are such that $P(A) = \frac{3}{5}$, $P(B) = \frac{3}{10}$ and $P(\bar{A}|\bar{B}) = \frac{2}{5}$. Find:

(i) $P(A \cup B)$ (ii) $P(A \cap B)$ (iii) $P(A|B)$

Investigate whether or not A and B are independent or mutually exclusive.

Revision Problems 13

1 A player has two dice, which are indistinguishable in appearance. One die is fair, so that the probability of getting a six on any throw is $\frac{1}{6}$, and one is biased in such a way that the probability of getting a six on any throw is $\frac{1}{3}$.

(i) The player chooses one of the dice at random and throws it once.

(a) Find the probability that a six is thrown.

(b) Show that the conditional probability that the die is the biased one, given that a six is thrown, is $\frac{2}{3}$.

(ii) The player chooses one of the dice at random and throws it twice.

(a) Show that the probability that two sixes are thrown is $\frac{5}{72}$.

(b) Find the conditional probability that the die is the biased one, given that two sixes are thrown.

(iii) The player chooses one of the dice at random and throws it n times. Show that the conditional probability that the die is the biased one, given that n sixes are thrown, is:

$$\frac{2^n}{2^n + 1}$$

(UCLES)

2 (i) Two fair dice are thrown, and events A, B and C are defined as follows:

A: the sum of the two scores is odd;
B: at least one of the two scores is greater than 4;
C: the two scores are equal.

Find, showing your reasons clearly in each case, which two of these three events are: (a) mutually exclusive; (b) independent.

Find also $P(C|B)$, making your method clear.

(ii) Two players A and B regularly play each other at chess. When A has the first move in a game, the probability of A winning that game is 0.4, and the probability of B winning that game is 0.2. When B has the first move in a game, the probability of B winning that game is 0.3 and the probability of A winning that game is 0.2. Any game of chess that is not won by either player ends in a draw.

(a) Given that A and B toss a fair coin to decide who has first move in a game, find the probability of the game ending in a draw.

(b) To make their games more enjoyable, A and B agree to change the prodedure for deciding who has the first move in a game. As a result of their new procedure, the probability of A having first move in any game is p. Find the value of p which gives A and B equal changes of winning each game.

(UCLES)

14 Further differentiation

All the work on differentiation that we have looked at so far has involved the expression that is to be differentiated being written in the form $y = f(x)$. We say that y is given *explicitly* in terms of x. However, there are other ways in which functions can be written, and we are now in a position once again to increase the range of functions that we can differentiate.

14.1 Implicit functions

When we write down a function in the form $y = 4x^2 + \sin x$, then y is given **explicitly**. This means substitution of a value of x immediately gives the value of y. However, if you look at the equation:

$$x^2 + 6xy + y^3 = 8$$

then the situation is more complicated. For example, if you put $x = 1$,

$$1 + 6y + y^3 = 8$$

that is,

$$y^3 + 6y - 7 = 0$$

This means there may be three values of y for one value of x. In fact, this equation factorises:

$$(y - 1)(y^2 + y + 7) = 0$$

In this case, $y = 1$ and $y^2 + y + 7 = 0$ has no solution.

An equation in the form $x^2 + 6xy + y^3 = 8$ is said to be an **implicit** expression, because the value of y is not given immediately. Curves of implicit functions are often very complicated, with loops or several arcs. However, because the equation still represents a curve, it must be possible to differentiate it to find its gradient.

The simplest type of implicit function would be something like:

$$y^2 = 8x^3 + 4x$$

To differentiate the L.H.S., use the chain rule:

$$2(y)^1 \times \frac{dy}{dx} = 24x^2 + 4$$

↓

(this is the derivative of what is inside the squared function)

that is, $$2y\frac{dy}{dx} = 24x^2 + 4$$

$$\frac{dy}{dx} = \frac{12x^2 + 2}{y}$$

The answer contains x and y, which is nearly always the case with an implicit function.

If you differentiate y^n with respect to x, you get:

$$\frac{d}{dx}y^n = ny^{n-1}\frac{dy}{dx} \qquad \text{(D13)}$$

Let us return to the example at the beginning of this section:

$$x^2 + 6xy + y^3 = 8$$

The $6xy$ term is a product.

Hence: $$2x \quad + \quad 6x \quad \times \quad \frac{dy}{dx} \quad + \quad y \quad \times \quad 6 + 3y^2\frac{dy}{dx} = 0$$

	↑	↑	↑	↑
		diff	2nd	diff
	1st	2nd		1st

You can now factorise this, hence

$$\frac{dy}{dx}(6x + 3y^2) = -2x - 6y$$

$$\therefore \qquad \frac{dy}{dx} = \frac{-2(x + 3y)}{3(2x + y^2)}$$

We can now apply this idea to the usual range of applications of differentiation.

Example 14.1

Find the equation of the tangent to the curve $x^2 + 4xy + 2y^2 = 7$ at the point $(1, 1)$.

Solution

$$x^2 + 4xy + 2y^2 = 7$$

Differentiate:

$$2x + 4x\frac{dy}{dx} + 4y + 4y\frac{dy}{dx} = 0$$

$x = 1, y = 1$

$$\therefore \quad 2 + 4\frac{dy}{dx} + 4 + 4\frac{dy}{dx} = 0$$

$$\therefore \quad 8\frac{dy}{dx} = -6, \; \frac{dy}{dx} = -\frac{3}{4}$$

Using the straight-line equation $y - y_1 = m(x - x_1)$ we get:

$$(y - 1) = -\frac{3}{4}(x - 1)$$

$$4y - 4 = -3x + 3$$

$$\therefore \quad 4y + 3x = 7$$

Example 14.2

Find the equation of the normal to the curve $y + x \sin y = 2$ which passes through the point where $y = \frac{\pi}{4}$. If this line cuts the x-axis at P and the y-axis at Q, find the area of triangle OPQ where O is the origin of coordinates. (Work to 3 significant figures throughout.)

Solution

$y + x \sin y = 2$

Differentiate

$$\frac{dy}{dx} + x \cos y \frac{dy}{dx} + \sin y = 0$$

$$\therefore \quad \frac{dy}{dx}(1 + x \cos y) = -\sin y$$

$$\therefore \quad \frac{dy}{dx} = \frac{-\sin y}{1 + x \cos y}$$

We need to find what x is equal to when $y = \frac{\pi}{4}$,

$$\therefore \quad \frac{\pi}{4} + x \sin\frac{\pi}{4} = 2$$

$$\therefore \quad 0.785 + 0.707x = 2$$

$$x = 1.72$$

$$\therefore \quad \frac{dy}{dx} = \frac{-\sin\frac{\pi}{4}}{1 + 1.72\cos\frac{\pi}{4}} = -0.319$$

The gradient of the normal $= \dfrac{-1}{\text{gradient of tangent}}$ (using $m_1 m_2 = -1$)

$$= \frac{-1}{-0.319} = 3.13$$

$\therefore$ The equation of the normal is:

$$\left(y - \frac{\pi}{4}\right) = 3.13(x - 1.72)$$

$$y - 0.785 = 3.13x - 5.38$$

$$y = 3.13x - 4.60$$

If $x = 0, \quad y = -4.60$

$y = 0, \quad x = 1.47$

The area of the triangle $= \frac{1}{2} \times 1.47 \times 4.6$

$= 3.38 \text{ units}^2$

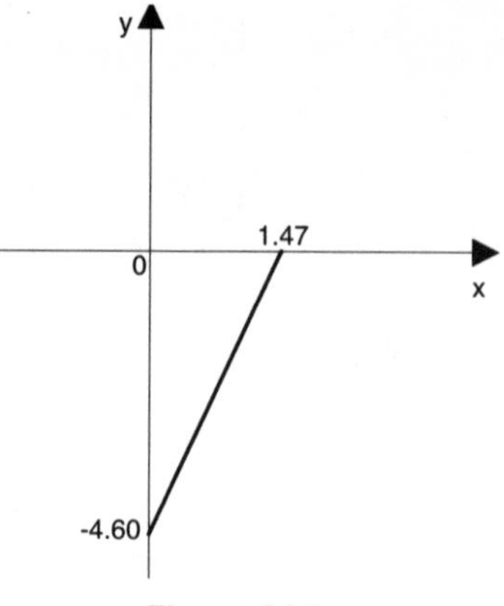

Figure 14.1

Exercise 14(a)

1 Find $\frac{dy}{dx}$ in the following cases:

(i) $2y^2 + xy = 4$ (ii) $x^2 + 6y^2 + 2xy = 0$
(iii) $x \sin 2y + x^2 = 0$ (iv) $x \tan y = 1$
(v) $2(x - y)^2 + x = 1$ (vi) $x^3 + y^3 = xy$
(vii) $x \sin^2 y = 1$ (viii) $(2x + 3y)^3 = 1$
(ix) $x \tan^2 y = y$ (x) $3x^2 \cos y = y^2$

2 Evaluate $\frac{dy}{dx}$ at the given points for the given functions:

(i) $x^2 + xy = 2y^2$ at $(1, 1)$ (ii) $x \sin y + y^2 = 1 + \frac{\pi^2}{4}$ at $\left(1, \frac{\pi}{2}\right)$
(iii) $x^2 + y^2 = 5$ at $(2, 1)$ (iv) $(x - y)^2 = 16$ at $(5, 1)$
(v) $x(1 + \sin y)^2 = 8$ at $\left(2, \frac{\pi}{2}\right)$ (vi) $y^3 + \sin^2 x = xy$ at $(\pi, \sqrt{\pi})$

3 Find the equation of the tangent and normals to the following curves at the given point:

(i) $x^2 + 6xy + y^2 = 8 : (1, 1)$

(ii) $(x - 2y)^2 + y^3 = 1 : (2, 1)$

(iii) $\frac{3x + 3y}{x - y} = 2 + y : (2, -1)$

14.2 Parametric differentiation

In Chapter 6, we looked at how to represent a curve using parametric coordinates. We can now look at the problem of differentiating functions expressed in this way.

Suppose $y = f(t)$ and $x = g(t)$. If you change t by an amount δt, then x will change by an amount δx, and y will change by an amount δy, so:

$$y + \delta y = f(t + \delta t) \text{ and } x + \delta x = g(t + \delta t)$$

$$\therefore \quad \delta y = f(t + \delta t) - f(t) \text{ and } \delta x = g(t + \delta t) - g(t)$$

$$\therefore \quad \frac{\delta y}{\delta x} = \frac{f(t + \delta t) - f(t)}{g(t + \delta t) - g(t)} = \frac{f(t + \delta t) - f(t)}{\delta t} \div \frac{g(t + \delta t) - g(t)}{\delta t}$$

$$\therefore \quad \text{as } \delta t = 0 \quad \frac{dy}{dx} = \frac{df}{dt} \div \frac{dg}{dt}$$

In other words:

$$\frac{dy}{dx} = \frac{\frac{dy}{dt}}{\frac{dx}{dt}} \qquad \text{(D14)}$$

or

$$\frac{dy}{dx} = \frac{dy}{dt} \times \frac{dt}{dx}$$

Again, we can explore the usual applications of the calculus to this type of problem.

Example 14.3

A curve has parametric equation $x = 2t^3 + 1, y = 3t^2 - 1$. Find:

(i) $\dfrac{dy}{dx}$ in terms of t;

(ii) the equation of the tangent to the curve at the point $(17, 11)$.

Solution

(i) $\dfrac{dy}{dt} = 6t \qquad \dfrac{dx}{dt} = 6t^2$

Hence: $$\frac{dy}{dx} = \frac{\frac{dy}{dt}}{\frac{dx}{dt}} = \frac{6t}{6t^2} = \frac{1}{t}$$

(ii) In order to find the gradient, you need to find t.

Using the x value, $2t^3 + 1 = 17$ (you could also put $3t^2 - 1 = 11$)

$$2t^3 = 16$$

$$t^3 = 8$$

therefore: $t = 2$

$$\therefore \quad \frac{dy}{dx} = \tfrac{1}{2}$$

We have $y - 11 = \tfrac{1}{2}(x - 17)$

$$\therefore \quad 2y - 22 = x - 17$$

$$2y = x + 5$$

Example 14.4

The parametric equations of a curve are:

$$x = 6t, \qquad y = \frac{2}{t^2}$$

where t takes all values, and $t \neq 0$. The points P and Q have parameters p and q.

(i) Find the coordinates of the mid point of the line PQ.
(ii) If the gradient of PQ is 1, find a relationship between p and q, in its simplest form.
(iii) Find the equation of the tangent which has a gradient of $\frac{2}{3}$.

Solution

(i) Using the mid point formula $\left(\frac{x_1 + x_2}{2}, \frac{y_1 + y_2}{2}\right)$ we have:

the mid point is

$$\left(\frac{6p + 6q}{2}, \frac{\frac{2}{p^2} + \frac{2}{q^2}}{2}\right) = \left(3(p + q), \left(\frac{1}{p^2} + \frac{1}{q^2}\right)\right)$$

(ii) P is $\left(6p, \frac{2}{p^2}\right)$ and Q is $\left(6q, \frac{2}{q^2}\right)$

The gradient of PQ is $\dfrac{\frac{2}{p^2} - \frac{2}{q^2}}{6p - 6q} = \dfrac{2\left[\frac{1}{p^2} - \frac{1}{q^2}\right]}{6(p - q)}$

This can be simplified as follows:

$$= \frac{{}^{1}\cancel{2}\left[\frac{q^2 - p^2}{p^2q^2}\right]}{{}_{3}\cancel{6}\,(p - q)} = \frac{q^2 - p^2}{3p^2q^2(p - q)} = \frac{{}^{-1}\cancel{(q - p)}\,(q + p)}{3p^2q^2\cancel{(p - q)}} = -\frac{(q + p)}{3p^2q^2}$$

If the gradient is 1,

then $-\dfrac{(q+p)}{3p^2q^2} = 1$

so: $-(q+p) = 3p^2q^2$

or: $3p^2q^2 + q + p = 0$

(iii) To find the gradient, we need to use the fact that:

$$\frac{dy}{dx} = \frac{\frac{dy}{dt}}{\frac{dx}{dt}} = -\frac{4}{t^3} \div 6 = -\frac{2}{3t^3}$$

If the gradient is $\frac{2}{3}$, then:

$\frac{2}{3} = \dfrac{-2}{3t^3}$ $\quad \therefore \quad t^3 = -1$

and so $\quad t = -1$

Substituting into x and y, the point on the curve will be at $x = -6$, $y = 2$.

The equation of the tangent can be found using $(y - y_1) = m(x - x_1)$,

that is, $(y - 2) = \frac{2}{3}(x + 6)$

This simplifies to $3y = 2x + 18$.

Example 14.5

Find the equation of the tangent to the curve given by $y = t^3 + 1$ and $x = t + 2$ at the point $P(1, 0)$. Does this tangent cut the curve at any other points?

Solution

$x = t + 2 \quad \therefore \quad \dfrac{dx}{dt} = 1$

$y = t^3 + 1 \quad \therefore \quad \dfrac{dy}{dt} = 3t^2$

so: $\dfrac{dy}{dx} = 3t^2$

At P, $x = 1$, so $t + 2 = 1$, $\therefore t = -1$

$\therefore \quad \dfrac{dy}{dx} = 3$

Equation of the tangent is $y - 0 = 3(x - 1)$

that is, $\quad y = 3x - 3$

If this line cuts the curve again, then the parametric coordinates must satisfy the equation of the tangent,

so: $\quad t^3 + 1 = 3(t + 2) - 3$

that is, $\quad t^3 - 3t - 2 = 0$

At first sight, we are faced with a cubic equation in t, but we know that $t = -1$ is one solution, hence $t + 1$ is a factor. Use the techniques of Chapter 2:

$$(t+1)(t^2 - t - 2) = 0$$

that is, $(t+1)(t+1)(t-2) = 0$

Hence another solution is $t = 2$, which means that the tangent also cuts the curve again at the point $(4, 9)$.

Exercise 14(b)

1 Find $\frac{dy}{dx}$ in the following cases, simplifying your answer if possible.

(i) $x = 4t, y = 2t^2 + 1$

(ii) $x = \sqrt{t},\ y = \frac{1}{\sqrt{t}}$

(iii) $x = 4\cos 3\theta, y = 2\sin 3\theta$

(iv) $x = ct, y = \frac{c}{t}$ where c is a constant

(v) $x = 4\sec\alpha,\ y = 3\tan\alpha$

(vi) $x = \frac{1+t}{1-t},\ y = \frac{2t}{1-t}$

(vii) $x = 2t + 3t^2,\ y = 2t - 3t^2$

(viii) $x = \frac{\sin\theta}{1+\cos\theta},\ y = \frac{\cos\theta}{1+\sin\theta}$

2 Find the equation of the tangent and normal to the curve given parametrically by $x = 3t + 1, y = 2t^2 + 5$ at the point $(7, 13)$.

3 Find the equation of the tangent to the curve $y = at^2, x = 2$ at at the point with parameter t. This line cuts the x-axis at P and the y-axis at Q. Find the area of the triangle OPQ.

4 Find the equation of the tangent to the curve $x = 4t - 1, y = t^3 + 1$ at the point $(-5, 0)$. Find the point where this tangent meets the curve again.

14.3 To differentiate $y = a^x$ and $y = e^x$

Before looking at this in detail, it is suggested that you work through the following activity.

ACTIVITY 9

This investigation looks at graphs of the form $y = a^x$. Find a piece of graph paper as large as possible. Set up Cartesian axes, with x from -2 to 2 in steps of 0.5, and y from 0 to 10.

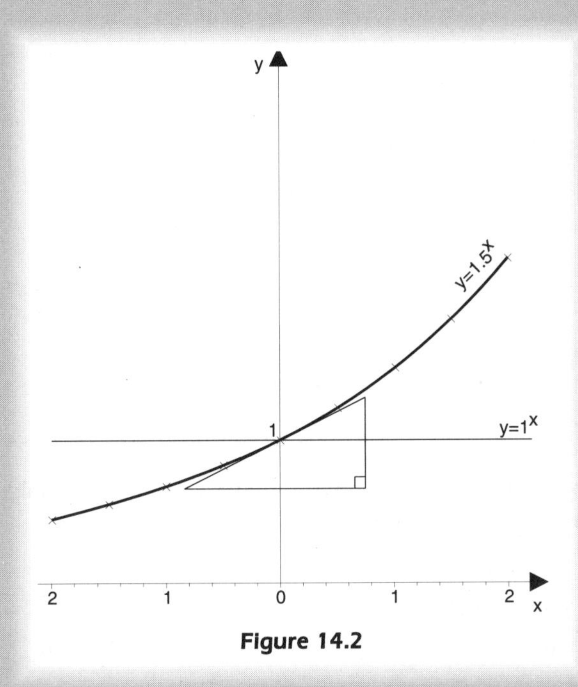

Figure 14.2

Choose a value of a between 1 and 3, say 1.5, and complete the table of values as shown below.

x	−2	−1.5	−1	− 0.5	0	0.5	1	1.5	2
$y = 1.5^x$	0.44	0.54	0.67	0.82	1	1.22	1.5	1.84	2.25

Plot the graph as shown in Figure 14.2. Now draw a tangent at $x = 0$ on the graph and find the gradient (it is in fact 0.41 (2 dec. places)) at x = 0. Hence, if a = 1.5, the gradient of $y = a^x$ at x = 0 is 0.41. We will denote this value by G.

As accurately as you can, repeat this process for several other values of a given to complete the following table.

a	1	1.5	2	2.5	3	3.5
gradient (G) at $x = 0$	0	0.41				

Having completed the table, you will be able to complete the graph of G against a (started in Figure 14.3).

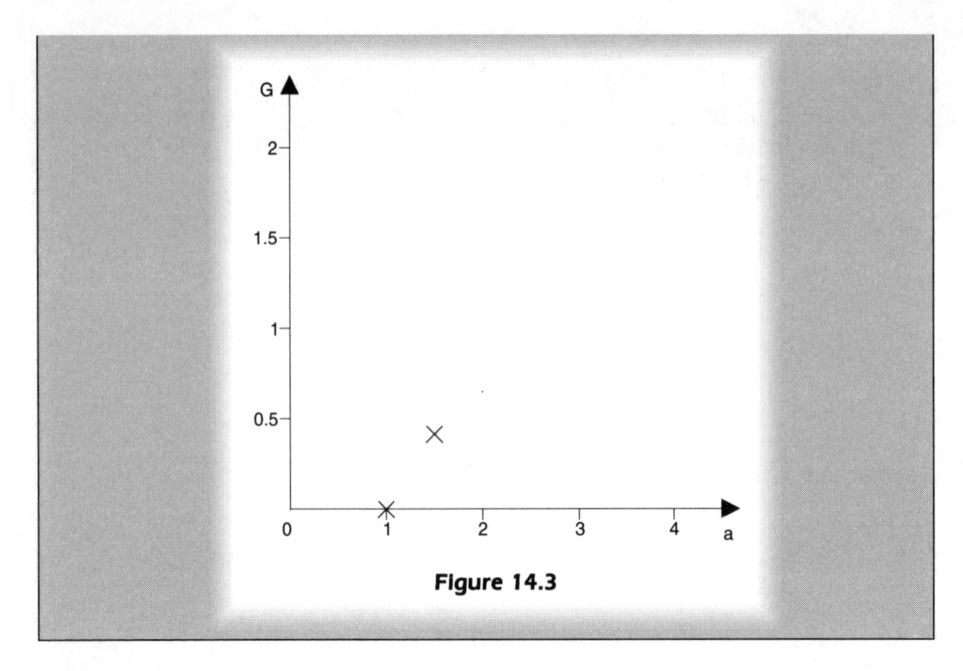

Figure 14.3

Having worked through the above activity, we can now analyse the situation using the notation of the calculus.

If $\qquad y = a^x \qquad$ (i)

Increase x by δx, then

$$y + \delta y = a^{x+\delta x} \qquad \text{(ii)}$$

$$\text{(i)} - \text{(ii)} \qquad \delta y = a^{x+\delta x} - a^x$$

$$= a^x(a^{\delta x} - 1)$$

$$\therefore \qquad \frac{\delta y}{\delta x} = a^x \frac{(a^{\delta x} - 1)}{\delta x}$$

Since $a^0 = 1$, the expression $\dfrac{a^{\delta x} - 1}{\delta x}$ can be written $\dfrac{a^{\delta x} - a^0}{\delta x - 0}$.

This is simply the gradient of the chord joining the points $(0, a^0)$ and $(\delta x, a^{\delta x})$. (See Figure 14.4)

As $\delta x \to 0$, the chord becomes the tangent at $x = 0$.

In Activity 9 you were able to see that, for example, if $a = 1.5$, this gradient $= 0.41$.

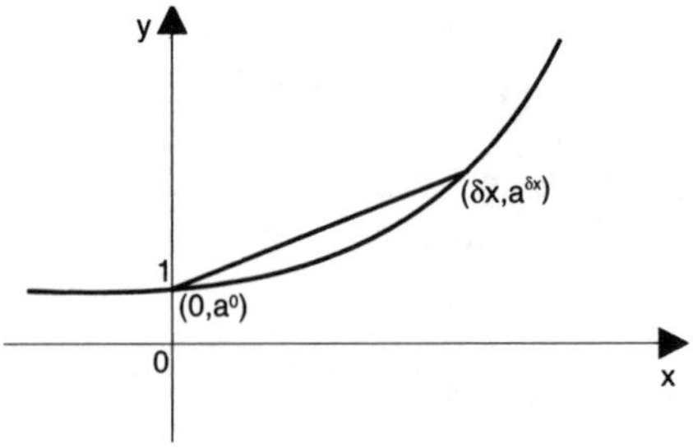

Figure 14.4

So if $y = 1.5^x$,

$$\frac{dy}{dx} = 0.41 \times 1.5^x$$

For the moment, we can summarise this by saying that if $y = a^x$, and G is the gradient of the curve at $x = 0$, then:

$$\frac{dy}{dx} = Ga^x$$

Now, looking at the second Table in Activity 9 and the graph in Figure 14.3, you can see that there will be a value of a which we denote by e, for which $G = 1$.

Hence: if $y = e^x$,

$$\frac{dy}{dx} = e^x. \qquad \text{(D15)}$$

In particular, if $y = e^{ax}$,

$$\frac{dy}{dx} = ae^{ax}. \qquad \text{(D 15a)}$$

Let us see if we can calculate e.

Since $G = \lim_{\delta x \to 0} \dfrac{e^{\delta x} - 1}{\delta x} = 1$ in this case,

then $\dfrac{e^{\delta x} - 1}{\delta x} = 1 + k$, where k is small and δx has not become exactly zero.

$\therefore \quad e^{\delta x} - 1 = \delta x + k\delta x.$

Now k and δx are small, so $k\,\delta x$ is very small and can be ignored.

Hence: $e^{\delta x} = 1 + \delta x$

Let $\delta x = \dfrac{1}{n}$, so $e^{\frac{1}{n}} = 1 + \dfrac{1}{n}$,

that is, $e = \left(1 + \dfrac{1}{n}\right)^n$

Now δx is very small, and so $n = \dfrac{1}{\delta x}$ is very large.

$n = 1$ $\qquad e = \left(1 + \dfrac{1}{1}\right)^1 = 2$

$n = 10$ $\qquad e = \left(1 + \dfrac{1}{10}\right)^{10} = 2.5937$

$n = 100$ $\qquad e = \left(1 + \dfrac{1}{100}\right)^{100} = 2.7048$

$n = 1000$ $\qquad e = (1.001)^{1000} = 2.7169$

$n = 10\,000$ $\qquad e = (1.0001)^{10000} = 2.7181$

$n = 100\,000$ $\qquad e = (1.00001)^{100000} = 2.7183$

Notice how the value converges. By increasing n, you can find e to any degree of accuracy. It is one of the most important numbers in mathematics. In fact, $e = 2.718$ (3 decimal places).

Example 14.6

Differentiate with respect to x:

(i) e^{4x} (ii) $6e^{-x^2}$ (iii) xe^{2x}

Solution

(i) If $y = e^{4x}$

$u = 4x, \quad y = e^u$

$$\therefore \quad \frac{du}{dx} = 4, \quad \frac{dy}{du} = e^u \qquad \therefore \quad \frac{dy}{dx} = 4 \times e^u = 4e^{4x}.$$

(ii) $y = 6e^{-x^2}$.

Use the chain rule:

$$\frac{dy}{dx} = 6 \times e^{-x^2} \times -2x$$
$$= -12xe^{-x^2}$$

(iii) $y = xe^{2x}$

$$\frac{dy}{dx} = \underset{\substack{\uparrow \\ \text{1st}}}{x} \times \underset{\substack{\uparrow \\ \text{diff 2nd}}}{2e^{2x}} + \underset{\substack{\uparrow \\ \text{2nd}}}{e^{2x}} \times \underset{\substack{\uparrow \\ \text{diff 1st}}}{1}$$

$$= e^{2x}(2x + 1)$$

Example 14.7

The parametric equation of a curve is given by:

$$x = 5t + e^{-2t}, \quad y = 4t - e^{3t}.$$

Find the coordinates of a turning point on the curve.

Solution

$$\frac{dy}{dt} = 4 - 3e^{3t} \qquad \frac{dx}{dt} = 5 - 2e^{-2t}$$

So: $\dfrac{dy}{dx} = \dfrac{4 - 3e^{3t}}{5 - 2e^{-2t}}$

At a turning point, $\dfrac{dy}{dx} = 0.$

Therefore: $\quad 4 - 3e^{3t} = 0$

that is, $\quad 4 = 3e^{3t}$

$$e^{3t} = \tfrac{4}{3},$$

$\therefore \quad 3t = \ln\tfrac{4}{3}$

$$t = \tfrac{1}{3}\ln\tfrac{4}{3} = 0.09589.$$

Substitute this value into the parametric equation to find the coordinates:

so: $\quad x = 5 \times 0.09589 + e^{-2\times 0.09589} = 1.30$

similarly, $\quad y = -0.95$

The turning point is $(1.3, -0.95)$

Example 14.8

A factory is looking at mathematical models to predict the number of workers it needs in relationship to its wage rate. A suggested model is:

$$N = 180e^{-0.006W},$$

where N is the number of workers and £W the average weekly pay.

The unions want to maximise the total weekly wage bill. Find the number of workers employed, and the value of W.

Solution

If £T is the total weekly wage bill, then:

$$T = N \times W,$$

that is, $\quad T = 180We^{-0.006W}$

$$\frac{dT}{dW} = 180e^{-0.006W} + 180W \times -0.06e^{-0.006W}$$

$$= 180e^{-0.006W}[1 - 0.006W]$$

$$= 0 \text{ if } 1 = 0.006W, \quad W = 166.67.$$

If $\quad W > 166.67 \quad \dfrac{dT}{dW} < 0,$

$\quad W < 166.67 \quad \dfrac{dT}{dW} > 0.$

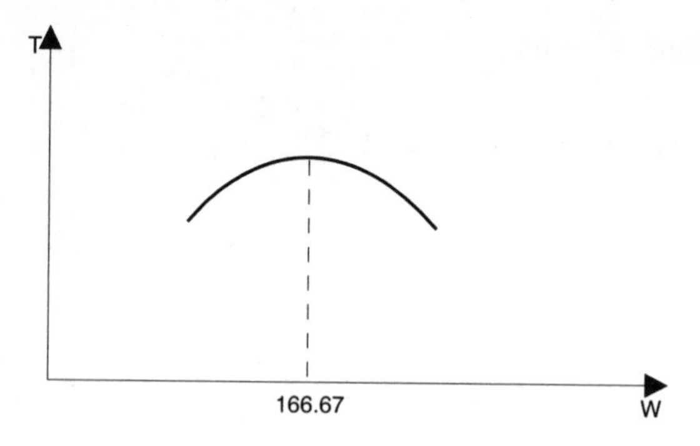

Figure 14.5

A graph of T against W shows that this is a maximum point.
To find N, substitute into the formula for N:

$$N = 180e^{-0.006 \times 166.67}$$
$$= 66 \qquad \text{(to the nearest whole number)}$$

Hence 66 workers are employed, earning on average £166.67 per week.

Exercise 14(c)

1 Differentiate with respect to x and simplify your answers where possible:

(i) e^{3x} (ii) xe^{-2x} (iii) $2x \sin x$

(iv) $e^x \cos^2 x$ (v) $\dfrac{e^x}{1+x^2}$ (vi) $e^x \tan 2x$

(vii) $\dfrac{x^3 e^{2x}}{(1+x^2)}$ (viii) $e^{-\sin x}$ (ix) $(1+e^{2x})^3$

(x) $e^{-x} \sin 2x \cos 2x$

2 Find $\dfrac{dy}{dx}$ for the following parametric equations:

(i) $x = e^{2t}$, $y = 1 + e^{-2t}$

(ii) $x = e^t \sin t$, $y = e^t \cos t$

(iii) $x = e^{3\theta}$, $y = e^{-2\theta}$

(iv) $x = 1 + e^{-t} \sin t, y = 1 - e^{-t} \cos t$

3 Find the equation of the tangent and normal to the curve $y = xe^x$ at the point where $x = 1$.

4 Find the coordinates of the turning point on the graph of $y = x + e^{-x}$. Sketch the graph.

5 The volume V of gas in a container is given by the formula $V = 10 + te^{-2t}$, where t is the time. Find the rate of increase of V when $t = 0.1$.

14.4 Differentiation of ln x

If $y = \ln x$, then it follows that $e^y = x$.

We can treat this as an implicit function.

So: $$e^y \frac{dy}{dx} = 1$$

that is, $$\frac{dy}{dx} = \frac{1}{e^y} = \frac{1}{x}$$

Hence: if $y = \ln x$, $\dfrac{dy}{dx} = \dfrac{1}{x}$.

This is a very important result, because it also enables us to complete the integration rule:

so: $$\int x^{-1} dx = \ln x + c. \qquad \text{(I14)}$$

Example 14.9

Differentiate the following expressions with respect to x, simplifying your answers if possible:

(i) $\ln 4x$ (ii) $\ln \cos x$ (ii) $\ln \sec x$ (iv) $\ln \sqrt{\dfrac{1+x}{1-x}}$

Solution

(i) $y = \ln 4x$, let $u = 4x$, then $y = \ln u$, $\dfrac{du}{dx} = 4$ and $\dfrac{dy}{du} = \dfrac{1}{u}$,

$$\therefore \quad \frac{dy}{dx} = \frac{1}{u} \times 4 = \frac{1}{4x} \times 4 = \frac{1}{x}.$$

(ii) $y = \ln \cos x$ let $u = \cos x$, then $\dfrac{du}{dx} = -\sin x$, $y = \ln u$ and $\dfrac{dy}{du} = \dfrac{1}{u}$,

$$\therefore \quad \frac{dy}{dx} = \frac{1}{u} \times -\sin x = \frac{1}{\cos x} \times -\sin x$$

$$= \frac{-\sin x}{\cos x} = -\tan x.$$

(iii) $y = \ln \sec x$, let $u = \sec x$, $\dfrac{du}{dx} = \sec x \tan x$, $y = \ln u$, $\dfrac{dy}{du} = \dfrac{1}{u}$,

$$\therefore \quad \frac{dy}{dx} = \frac{1}{u} \times \sec x \tan x = \frac{1}{\sec x} \times \sec x \tan x \quad = \tan x.$$

(iv) This would be a difficult question to do as a function of a function. Using the logarithm rules, it is better to expand the logarithm before you differentiate.

So: $$y = \ln \sqrt{\frac{1+x}{1-x}} = \frac{1}{2} \ln \frac{1+x}{1-x} = = \frac{1}{2}[\ln(1+x) - \ln(1-x)]$$

$$\therefore \quad \frac{dy}{dx} = \frac{1}{2}\left[\frac{1}{1+x} \times 1 - \frac{1}{1-x} \times -1\right]$$

$$= \frac{1}{2}\left[\frac{(1-x)+(1+x)}{(1+x)(1-x)}\right]$$

$$= \frac{1}{(1+x)(1-x)} = 1 - x^2$$

Exercise 14(d)

Differentiate the following functions with respect to x, simplifying your answers where possible:

1 $\ln 3x$ **2** $x \ln x$ **3** $\ln \cos x$

4 $\ln \sec x$ **5** $\ln(1+x^2)$ **6** $\ln \dfrac{(1-x)}{(1+x)}$

7 $\ln(1+e^x)$ **8** $e^x \ln x$ **9** $\ln \sin^2 x$

10 $(\ln x)^2$

11 Find $\dfrac{dy}{dx}$ if $x = 1 + \ln t$, $y = t^2 + 2 \ln t$

12 Find the equation of the tangent at the point $t = 2$ on the curve given by $x = 1 + e^t$, $y = \ln t$. Give your answers to 3 significant figures.

13 Find the coordinates of the turning point on the graph $y = x^2 \ln x$, $x \neq 0$. State whether it is a maximum or a minimum.

14.5 Differentiation of inverse trigonometric functions

If $y = \sin^{-1} x$, then $\sin y = x$. This is now an implicit function:

$\therefore$ using the chain rule,

$$\cos y \frac{dy}{dx} = 1$$

$$\therefore \quad \frac{dy}{dx} = \frac{1}{\cos y} = \frac{1}{\sqrt{1 - \sin^2 y}} = \frac{1}{\sqrt{1 - x^2}}$$

The other inverse functions follow in a similar fashion.

SUMMARY

$$\frac{d}{dx}\sin^{-1} x = \frac{1}{\sqrt{1-x^2}} \qquad \text{(D16)}$$

$$\frac{d}{dx}\cos^{-1} x = \frac{-1}{\sqrt{1-x^2}} \qquad \text{(D17)}$$

$$\frac{d}{dx}\tan^{-1} x = \frac{1}{1-x^2} \qquad \text{(D18)}$$

Example 14.10

Differentiate the following functions with respect to x. Simplify your answers if possible.

(i) $\sin^{-1} 4x$ (ii) $\cos^{-1}(2x+1)$ (iii) $\tan^{-1}(1-3x)$

Solution

(i) $y = \sin^{-1} 4x$, let $u = 4x$, then $y = \sin^{-1} u$,

$$\therefore \quad \frac{dy}{du} = \frac{1}{\sqrt{1-u^2}} \text{ and } \frac{du}{dx} = 4.$$

Using the chain rule, $\dfrac{dy}{dx} = \dfrac{1}{\sqrt{1-u^2}} \times 4 = \dfrac{4}{\sqrt{1-16x^2}}$.

(ii) $y = \cos^{-1}(2x+1)$, let $u = 2x+1$, then $y = \cos^{-1} u$

$$\therefore \quad \frac{dy}{du} = -\frac{1}{\sqrt{1-u^2}} \text{ and } \frac{du}{dx} = 2,$$

$$\text{so} \quad \frac{dy}{dx} = -\frac{1}{\sqrt{1-u^2}} \times 2$$

$$= -\frac{2}{\sqrt{1-(2x+1)^2}} = \frac{2}{\sqrt{1-4x^2-4x-1}}$$

$$= -\frac{-2}{\sqrt{-4x-4x^2}}$$

It is worth looking closely at this answer, because it would appear that the answer contains the square root of a negative quantity. However, the original expression was $\cos^{-1}(2x+1)$, and in order for this to have an answer, then clearly, $-1 \leq 2x+1 \leq 1$.

Hence: $-2 \leq 2x \leq 0$

that is, $-1 \leq x \leq 0$

For these values of x, $-4x - 4x^2 \geq 0$, and so the square root can be evaluated.

(iii) $y = \tan^{-1}(1 - 3x)$, let $u = 1 - 3x$

$$\therefore \quad y = \tan^{-1} u$$

$$\frac{dy}{du} = \frac{1}{1+u^2} \times -3 = \frac{-3}{1+(1-3x)^2} = \frac{-3}{9x^2 - 6x + 2}$$

Exercise 14(e)

Differentiate the following functions, simplifying your answers if possible:

1 $\tan^{-1} 2x$ **2** $\cos^{-1} 3x$ **3** $\sin^{-1} 4x$

4 $\frac{1}{2}\tan^{-1}(x+1)$ **5** $\cos^{-1}(1-x)$ **6** $\sin^{-1}(3x+1)$

7 $x\sin^{-1} x$ **8** $e^x \cos^{-1} x$ **9** $\cos^{-1}(1-x^2)$

10 $\tan^{-1}(1+x^3)$

Miscellaneous Examples 14

1 Find the equation of the normal to the curve $(x-2y)^3 = x^2 + 2$ at the point $(5, 1)$.

2 The tangent to the curve $y = e^{x^2} + 1$ at the point where $x = 1$ meets the axes of coordinates at P and Q. Find the area of the triangle OPQ where O is the origin.

3 Find the coordinates of the points on the curve $y = \frac{1+e^x}{1-e^x}$ where the tangent is parallel to the line $y = x$.

4 Find the coordinates of the stationary point on the curve $y = 4x^2 - \ln \frac{1}{2}x$ and determine the nature of the point.

5 The function f is defined for all values of x by $f(x) = 2 - e^{-x}$. Show that $\frac{dy}{dx} > 0$ for all values of x, and sketch the graph.

6 Show that $\frac{d}{dx}\left\{2x\sqrt{45 - x^2} + 8\sin^{-1}\frac{x}{2}\right\} = 2\sqrt{4 - x^2}$.

Hence: evaluate $\int_0^2 \frac{1}{2}\sqrt{4 - x^2}dx$.

Revision Problems 14

1 The equation of a curve is $(x-y)^2 + 3(x+y)^2 = 16$. Show that the gradient of the curve is given by the equation $\frac{dy}{dx} = -\frac{(2x+y)}{(2y+x)}$

(i) Find the coordinates of the points on the curve at which the tangents are parallel to either the x or y axes.

(ii) Find where the curve cuts the axes and hence find the equation of the tangents at these points.

2 In a medical treatment, 500 milligrammes of a drug are administered to a patient. At time t hours after the drug is administered, X milligrammes of the drug remain in the patient. The doctor has a mathematical model which states that:

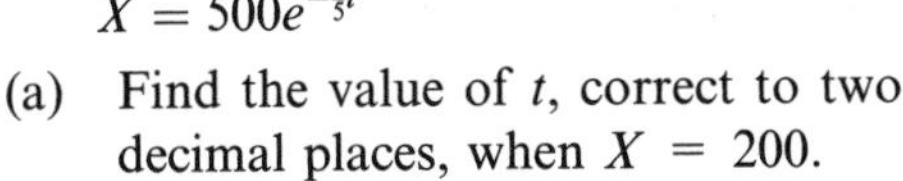

$$X = 500e^{-\frac{1}{5}t}$$

(a) Find the value of t, correct to two decimal places, when $X = 200$.

(b) (i) Express $\frac{dX}{dt}$ in terms of t.

(ii) Hence show that when $X = 200$, the rate of decrease of the amount of the drug remaining in the patient is 40 milligrammes per hour. (NEAB)

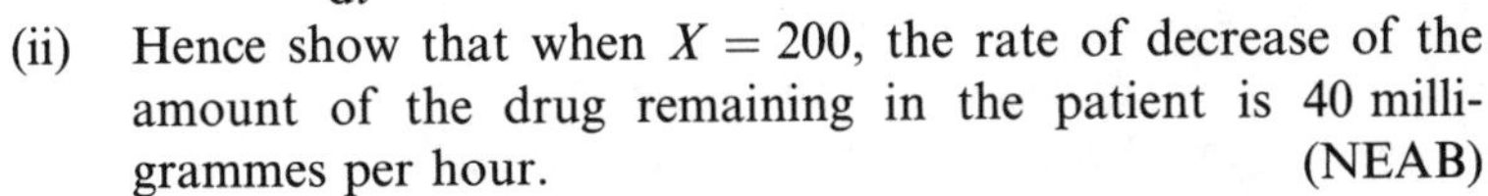

3 The curve C has parametric equations given by:

$$x = \frac{4}{t}, \quad y = 2t, \quad t \neq 0$$

(i) Sketch the graph of C.

(ii) Find an expression for $\frac{dy}{dx}$ in terms of t.

(iii) Find the equation of the normal to the curve at the point $P(-2, -4)$.

(iv) The normal in part (iii) meets the curve again at Q. Find the coordinates of Q.

4 The diagram in Figure 14.6 shows the curve with equation:

$$y = \frac{\ln x}{x^2} \quad \text{for } x > 0$$

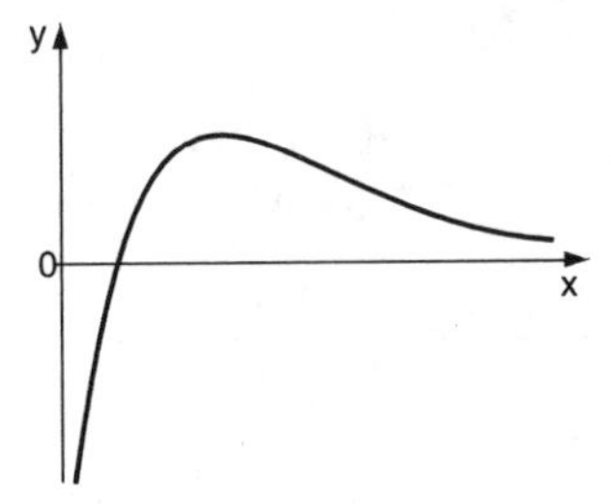

Figure 14.6

(a) State the x coordinate of the point where the curve crosses the x-axis.

(b) Show that:

$$\frac{dy}{dx} = \frac{1 - 2\ln x}{x^3}$$

(c) Find the coordinates of the maximum point of the curve and calculate the value of $\frac{d^2y}{dx^2}$ there.

(d) The finite region bounded by the curve, the x-axis and the line $x = 2$ has area equal to:

$$\int_b^2 \frac{\ln x}{x^2}\,dx.$$

Evaluate this integral, leaving your answer in terms of natural logarithms.

(e) Use the trapezium rule with three ordinates to obtain an estimate of:

$$\int_b^2 \frac{\ln x}{x^2}\,dx,$$

giving your answer to three significant figures.
With the aid of a diagram, explain briefly why the trapezium rule will lead to an underestimate of this integral.

(AEB 94)

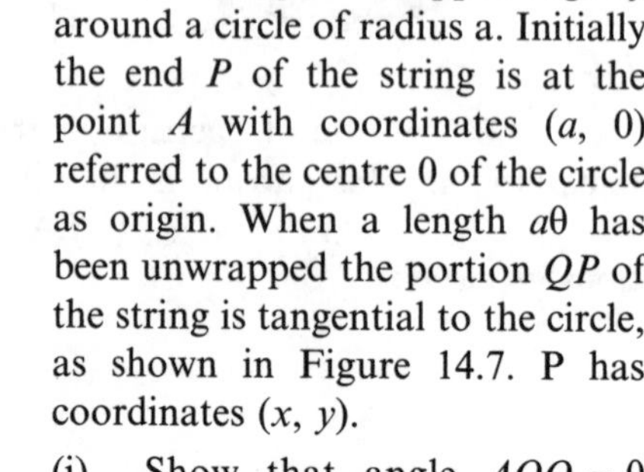

5 A thin string is wrapped tightly around a circle of radius a. Initially the end P of the string is at the point A with coordinates $(a, 0)$ referred to the centre 0 of the circle as origin. When a length $a\theta$ has been unwrapped the portion QP of the string is tangential to the circle, as shown in Figure 14.7. P has coordinates (x, y).

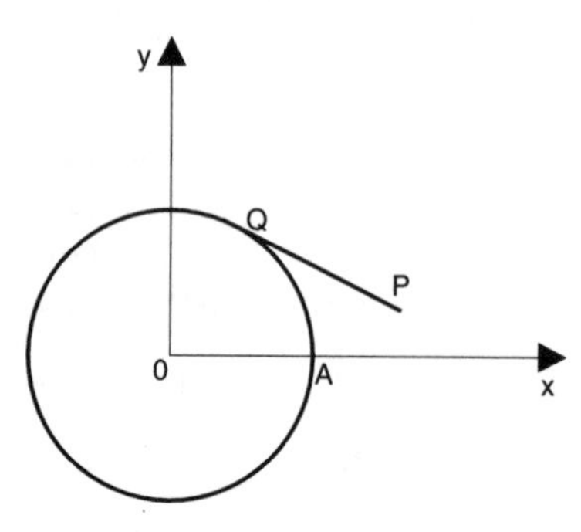

Figure 14.7

(i) Show that angle $AOQ = \theta$ and deduce that

$$x = a\cos\theta + a\theta\sin\theta.$$

(ii) Find the corresponding expression for y in terms of a and θ.

(iii) Show that $x^2 + y^2 = a^2(1 + \theta^2)$

(iv) Find $\dfrac{dy}{dx}$ in terms of θ.

(v) Given that θ is sufficiently small for θ^3 and higher powers of θ to be neglected, express x as a polynomial in θ.

(UCLES)

15 Further integration

The final chapter in this book looks at further applications of the integral calculus.

It is probably true to say that most integration nowadays is done using numerical techniques by calculator or computer. However, it is important that you fully understand the theory behind the techniques. In the final section, simple differential equations are studied. These can be used for anything from predicting the weather to designing a mathematical model to describe the economy.

15.1 Integrating e^x, a^x and $\frac{1}{x}$

We can now complete our work on the calculus for the functions considered in Chapter 14.

Since: $\frac{d}{dx}e^{ax} = ae^{ax}$, it follows that:

$$\int e^{ax}\,dx = \frac{1}{a}e^{ax} + c$$

Also, since $\frac{d}{dx}a^x = a^x \ln a$, then

$$\int a^x\,dx = \frac{1}{\ln a}a^x + c$$

Finally in this section, because $\frac{d}{dx}(\ln x) = \frac{1}{x}$, we can now complete the integration formula for x^n, because:

$$\int \frac{1}{x} \, dx = \ln x + c$$

and in general:

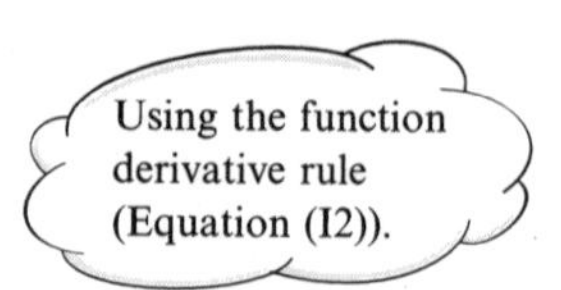

$$\int \frac{1}{ax+b} \, dx = \frac{1}{a} \ln(ax+b) + c$$

You need to be careful with this integral in the definite situation. If $\frac{1}{ax+b} < 0$ between the limits, then use the formula:

$$\int \frac{1}{ax+b} \, dx = \frac{1}{a} \ln |ax+b| + c \tag{I15}$$

See Example 15.3.

The following examples show these functions in use.

Example 15.1

Find the area enclosed between the curves $y = e^{2x}$, $y = e^{4x} + 1$ and the lines $x = -\frac{1}{2}$ and $x = 1$.

Solution

The area is shown in Figure 15.1:

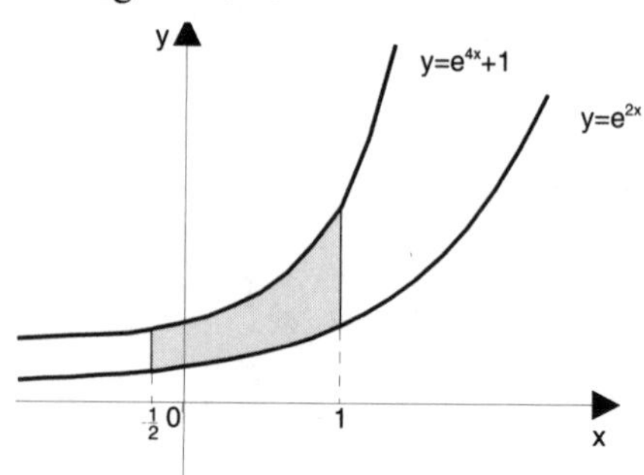

Figure 15.1

$$\begin{aligned}
\text{that is, area} &= \int_{-\frac{1}{2}}^{1} (e^{4x} + 1) - e^{2x} \, dx \\
&= \left[\tfrac{1}{4}e^{4x} + x - \tfrac{1}{2}e^{2x}\right]_{-\frac{1}{2}}^{1} \\
&= \left[\tfrac{1}{4}e^{4} + 1 - \tfrac{1}{2}e^{2}\right] - \left[\tfrac{1}{4}e^{-2} - \tfrac{1}{2} - \tfrac{1}{2}e^{-1}\right] \\
&= 10.96 - -0.65 \\
&= 11.6 \text{ (3 sig. figs)}
\end{aligned}$$

Example 15.2

Find:

(i) $\displaystyle\int \frac{x+2}{x^2}\,dx$ (ii) $\displaystyle\int_{0.1}^{0.4} \frac{1}{2x+1} - \frac{1}{3x+1}\,dx$

Solution

(i) $$\int \frac{x+2}{x^2}\,dx = \int \frac{1}{x} + \frac{2}{x^2}\,dx$$
$$= \int \frac{1}{x} + 2x^{-2}\,dx = \ln x + \frac{2x^{-1}}{-1} + c$$
$$= \ln x - \frac{2}{x} + c$$

(ii) $$\int_{0.1}^{0.4} \frac{1}{2x+1} - \frac{1}{3x+2}\,dx = \left[\tfrac{1}{2}\ln(2x+1) - \tfrac{1}{3}\ln(3x+1)\right]_{0.1}^{0.4}$$
$$= \left[\tfrac{1}{2}\ln 1.8 - \tfrac{1}{3}\ln 2.2\right] - \left[\tfrac{1}{2}\ln 1.2 - \tfrac{1}{3}\ln 1.3\right]$$
$$= 0.03107 - 0.003706$$
$$= 0.0274 \text{ (3 sig. figs)}$$

Example 15.3

Find the area enclosed between the curve $y = \dfrac{1}{2x-1}$ and the lines $x = -2$ and $x = 0$. Show clearly how you need to take care in carrying out the integration.

Solution

This is the situation in which $\dfrac{1}{2x-1}$ is negative between the limits (see Figure 15.2).

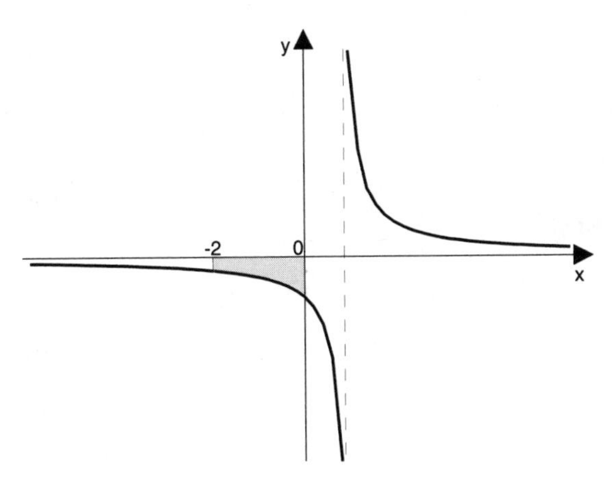

Figure 15.2

Hence: $\displaystyle\int_{-2}^{0} \frac{1}{2x-1}\,\mathrm{d}x = \left[\tfrac{1}{2}\ln|2x-1|\right]_{-2}^{0}$

$$= \tfrac{1}{2}\ln|-1| - \tfrac{1}{2}\ln|-5|$$
$$= -0.805$$

The answer is negative because the area is *below* the x-axis.

Hence the area $= 0.805$ units^2.

Example 15.4

Evaluate the following integrals by means of the function derivative rule.

(i) $\displaystyle\int \frac{x^2}{1+x^3}\,\mathrm{d}x$ (ii) $\int \cot x\,\mathrm{d}x$ (iii) $\int xe^{-x^2}\,\mathrm{d}x$

Solution

(i) Since $\dfrac{\mathrm{d}}{\mathrm{d}x}(1+x^3) = 3x^2$, then $\dfrac{\mathrm{d}}{\mathrm{d}x}(\ln(1+x^3)) = \dfrac{1}{1+x^3} \times 3x^2$

$$= \frac{3x^2}{1+x^3} \qquad \text{(3 times too great)}$$

Hence: $\displaystyle\int \frac{x^2}{1+x^3}\,\mathrm{d}x = \tfrac{1}{3}\ln(1+x^3) + c$

(ii) $\cot x = \dfrac{\cos x}{\sin x}$.

Now $\dfrac{\mathrm{d}}{\mathrm{d}x}(\sin x) = \cos x$, so $\dfrac{\mathrm{d}}{\mathrm{d}x}(\ln \sin x) = \dfrac{1}{\sin x} \times \cos x = \cot x$

Hence: $\displaystyle\int \cot x\,\mathrm{d}x = \ln \sin x + c$

(iii) $\dfrac{\mathrm{d}}{\mathrm{d}x}(x^2) = 2x \quad \therefore \quad \dfrac{\mathrm{d}}{\mathrm{d}x}\left(e^{-x^2}\right) = e^{-x^2} \times (-2x) = -2xe^{-x^2}$

hence: $\displaystyle\int xe^{-x^2} = -\tfrac{1}{2}e^{-x^2} + c$

Exercise 15(a)

1 Integrate the following functions with respect to x:

(i) e^{2x} (ii) $3e^{-2x}$ (iii) 2^x (iv) $\dfrac{1}{2x}$

(v) $\dfrac{1}{3x-1}$ (vi) $\dfrac{4}{x+6}$ (vii) $\dfrac{1}{(x+2)^3}$ (viii) $\dfrac{1}{e^{4x}}$

(ix) $(e^x)^3$ (x) xe^{x^2}

2 Find the following definite integrals:

(i) $\int_0^1 e^{2x}\,dx$ (ii) $\int_1^2 4e^{\frac{1}{2}x}\,dx$ (iii) $\int_1^2 \frac{1}{3x}\,dx$

(iv) $\int_4^5 \frac{2}{1-x}\,dx$ (v) $\int_0^1 \frac{4}{x+3}\,dx$ (vi) $\int_{0.5}^{0.8} \frac{1}{x^2}-\frac{1}{x}\,dx$

(vii) $\int_1^2 \frac{1}{e^{2t}}\,dt$ (viii) $\int_0^1 ye^{-y^2}\,dy$ (ix) $\int_0^{\frac{\pi}{2}} 2\cos x e^{\sin x}\,dx$

(x) $\int_{0.1}^{0.2} \tan x\,dx$ (xi) $\int_0^1 \frac{4x}{1+3x^2}\,dx$ (xii) $\int_0^1 4^x\,dx$

3 Find the area enclosed between the curves $y = e^x$, $y = e^{-x}$, the lines $x = -3$ and $x = -2$ and the x-axis.

4 Find the volume generated when the area under the curve $y = e^{2x} + 1$ between the lines $x = 0$ and $x = 2$ is rotated by 360° about the x-axis.

5 Find the volume generated when the area bounded by the lines $y = e^x$, $y = 4$, and the y-axis is rotated by 360° about the x-axis.

6 The motion of a particle moving in a straight line is given by the equation $s = 8 + 4e^{2t}$, where s (metres) is the distance t seconds after passing a fixed point 0.

(i) Find the velocity of the particle after 3 seconds.
(ii) How far does the particle move in the fourth second?

7 Find the area enclosed by the curve $y = \frac{2}{3x-1}$, the x-axis, and the lines $x = 2$ and $x = 3$.

15.2 Integration by substitution

We have already come across a simple form of the substitution rule, in Chapter 8. (the function derivative rule). We state this rule again for convenience.

$$\text{If} \quad \int f(x)\,dx = g(x)$$

$$\text{then} \quad \int f(ax+b)\,dx = \frac{1}{a}g(ax+b)$$

Now consider $\mathbf{I} = \int 3x^2(x^3 - 1)^4\,dx$.

Let $u = x^3 - 1$ (known as the substitution)

$$\frac{du}{dx} = 3x^2$$

This will be written $du = 3x^2\,dx$.

(Strictly speaking, this is not allowed, but it can be proved that in the context of integration this is acceptable.)

Replace the function of x and the dx in the integral (in two stages here to make it clear).

$$\mathbf{I} = \int 3x^2u^4\,dx = \int u^4\,du$$

The integral is now extremely easy:

$$\mathbf{I} = \tfrac{1}{5}u^5 + c$$

The final answer must be a function of x because the integral is with respect to x, and so u is replaced:

$$\text{hence:}\quad \mathbf{I} = \tfrac{1}{5}(x^3 - 1)^5 + c$$

Choosing the substitution is a skill most students take a long time to master. You should not worry if you find it difficult. Most people do. Work carefully through the following examples and try to see what decides the substitution.

(i) $\displaystyle\int \frac{3x}{(1+x^2)^3}\,dx$ Let $u = 1 + x^2$, $\dfrac{du}{dx} = 2x$

$$\therefore\quad du = 2x\,dx$$

$$\text{hence:}\quad \mathbf{I} = \int \frac{\frac{3}{2}du}{u^3}$$

$$= \int \tfrac{3}{2}u^{-3}\,du$$

$$= \frac{3u^{-2}}{-4} + c$$

$$= \frac{-3}{4(1+x^2)^2} + c$$

(ii) $\int (1+2x)^9 x^2\,dx$ Let $u = 1 + 2x$, $\dfrac{du}{dx} = 2$

$$\therefore \quad du = 2dx$$

hence: $\mathbf{I} = \int u^9 x^2\,dx$

$$= \int u^9 \left(\frac{u-1}{2}\right)^2 \cdot \frac{du}{2}$$

$\left(\text{Here we have rearranged } u \text{ to get } x = \dfrac{u-1}{2}\right)$

$$= \tfrac{1}{8}\int u^9(u^2 - 2u + 1)\,du$$

$$= \tfrac{1}{8}\int u^{11} - 2u^{10} + u^9\,du$$

$$= \tfrac{1}{8}\left[\frac{u^{12}}{12} - \frac{2u^{11}}{11} + \frac{u^{10}}{10}\right] + c$$

$$= \tfrac{1}{8}\left[\frac{(1+2x)^{12}}{12} - \frac{2(1+2x)^{11}}{11} + \frac{(1+2x)^{10}}{10}\right] + c$$

This could be simplified if needed, but it involves quite a lot of work. You might like to try and prove that it equals:

$$\frac{(1+2x)^{10}}{5280}(220x^2 - 20x + 1)$$

(iii) $\int \sec x \tan x (1 + \sec^3 x)^2\,dx$ Let $u = \sec x$, $\dfrac{du}{dx} = \sec x \tan x$

$$\therefore \quad du = \sec x \tan x\,dx$$

hence: $\mathbf{I} = \int (1 + u^3)^2\,du$

$$= \int 1 + u^6 + 2u^3\,du$$

$$= u + \frac{u^7}{7} + \frac{u^4}{2} + c$$

$$= \sec x + \frac{\sec^7 x}{7} + \frac{\sec^4 x}{2} + c$$

When evaluating a definite integral by means of a substitution, alter the limits as you go along, and don't rewrite your answer using the original variable. The following example shows you how this works.

Example 15.5

Find:

(i) $\displaystyle\int_1^2 \frac{3x^2}{1+x^3}\,dx$ (ii) $\displaystyle\int_3^4 \frac{1}{x\ln x}\,dx$

Solution

(i) Let $u = 1 + x^3$, so $\dfrac{du}{dx} = 3x^2$, hence: $du = 3x^2\,dx$

Now if $x = 1$, $u = 1 + 1^3 = 2$

$x = 2$, $u = 1 + 2^3 = 9$

Hence the integral becomes:

$$\int_2^9 \frac{1}{u}\,du = \Big[\ln u\Big]_2^9$$

$$= \ln 9 - \ln 2 = 1.50$$

(ii) Let $u = \ln x$ then $\dfrac{du}{dx} = \dfrac{1}{x}$ $\therefore du = \dfrac{1}{x}dx$

If $x = 3$, $u = \ln 3$

$x = 4$, $u = \ln 4$

$$\therefore \text{ the integral } = \int_{\ln 3}^{\ln 4} \frac{1}{u}\,du = \Big[\ln u\Big]_{\ln 3}^{\ln 4}$$

$$= \ln(\ln 4) - \ln(\ln 3) = 0.233$$

Exercise 15(b)

1 Integrate the following functions with respect to x, using the given substitution:

(i) $x(x^2+1)^9 : u = x^2 + 1$ (ii) $\dfrac{3x}{(1+x^2)^3} : u = 1 + x^2$

(iii) $(2x-1)(x+1)^{10} : u = x + 10$ (iv) $\sin^3 x : u = \cos x$

(v) $6x^2 e^{3x^3} : t = 3x^3$ (vi) $\cot x : u = \sin x$

(vii) $\dfrac{4\ln x}{3x} : y = \ln x$ (viii) $\dfrac{e^x}{(1+e^x)^2} : t = e^x$

(ix) $\dfrac{2x}{(1-x^2)^4} : u = 1 - x^2$

(x) $\sec^2 x(1+\tan x)^3 : u = 1 + \tan x$

2 By means of a suitable substitution, evaluate the following:

(i) $\displaystyle\int \frac{x^2}{1+x^3}\,\mathrm{d}x$ (ii) $\displaystyle\int_0^1 xe^{x^2+1}\,\mathrm{d}x$

(iii) $\displaystyle\int_0^1 (x-3)(2x+1)^9\,\mathrm{d}x$ (iv) $\displaystyle\int_0^{\frac{\pi}{2}} \cos^3 x\,\mathrm{d}x$

(v) $\displaystyle\int_0^1 \frac{x^2}{(1+x^3)^4}\,\mathrm{d}x$ (vi) $\displaystyle\int_0^1 \sec^2 x\sqrt{1+\tan x}\,\mathrm{d}x$

15.3 Integrals involving inverse trigonometric functions

(i) $\displaystyle \mathbf{I} = \int \frac{\mathrm{d}x}{a^2+x^2}$ Let $x = a\tan\theta$ $\therefore$ $\dfrac{\mathrm{d}u}{\mathrm{d}x} = a\sec^2\theta$

that is, $\mathrm{d}x = a\sec^2\theta\,\mathrm{d}\theta$

$$\therefore \quad \mathbf{I} = \int \frac{a\sec^2\theta\,\mathrm{d}\theta}{a^2+a^2\tan^2\theta} = \int \frac{a\sec^2\theta}{a^2\sec^2\theta}\,\mathrm{d}\theta$$

$$= \int \frac{1}{a}\,\mathrm{d}\theta$$

$$= \frac{\theta}{a} + c = \frac{1}{a}\tan^{-1}\frac{x}{a} + c$$

(ii) $\displaystyle \mathbf{I} = \int \frac{\mathrm{d}x}{\sqrt{a^2-x^2}}$ Let $x = a\sin\theta$, $\dfrac{\mathrm{d}x}{\mathrm{d}\theta} = a\cos\theta$

$\therefore$ $\mathrm{d}x = a\cos\theta\,\mathrm{d}\theta$

Hence: $\displaystyle \mathbf{I} = \int \frac{a\cos\theta\,\mathrm{d}\theta}{\sqrt{a^2-a^2\sin^2\theta}}$

$$= \int \frac{a\cos\theta}{a\cos\theta}\,\mathrm{d}\theta$$

$$= \int 1\,\mathrm{d}\theta$$

$$= \theta = \sin^{-1}\frac{x}{a} + c$$

SUMMARY

$$\int \frac{dx}{a^2+x^2} = \frac{1}{a}\tan^{-1}\frac{x}{a} + c \qquad \text{(I16)}$$

$$\int \frac{dx}{\sqrt{a^2-x^2}} = \sin^{-1}\frac{x}{a} + c \qquad \text{(I17)}$$

We can use these standard results together with the function derivative rule and completing the square to tackle slightly more difficult situations, as the following example shows.

Example 15.6

Evaluate:

(i) $\displaystyle\int \frac{dx}{x^2+2x+8}$ (ii) $\displaystyle\int \frac{dx}{\sqrt{15+4x-4x^2}}$

Solution

(i) Completing the square, $x^2+2x+8 = (x+1)^2+7 = (x+1)^2+(\sqrt{7})^2$

$$\therefore \int \frac{dx}{(x+1)^2+(\sqrt{7})^2} = \frac{1}{\sqrt{7}}\tan^{-1}\frac{(x+1)}{\sqrt{7}} \qquad \text{(Using Equation (I16))}$$

(ii)

$$\begin{aligned} 15+4x-4x^2 &= -4(x^2-x-\tfrac{15}{4}) \\ &= -4\left((x-\tfrac{1}{2})^2-4\right) \\ &= 16-4(x-\tfrac{1}{2})^2 \\ &= 16-(2x-1)^2 = 4^2-(2x-1)^2 \end{aligned}$$

$$\therefore \int_{\frac{1}{2}}^{1} \frac{dx}{\sqrt{4^2-(2x-1)^2}} = \left[\tfrac{1}{2}\sin^{-1}\frac{(2x-1)}{4}\right]_{\frac{1}{2}}^{1} \qquad \text{(Using Equation (I17))}$$

$$= \left[\tfrac{1}{2}\sin^{-1}\tfrac{1}{4}\right] - \left[0\right] = 0.126$$

Remember to use radians

Exercise 15(c)

Evaluate:

1 $\displaystyle\int \frac{1}{x^2+9}\,dx$

2 $\displaystyle\int \frac{1}{\sqrt{2-x^2}}\,dx$

3 $\displaystyle\int \frac{1}{\sqrt{4-9x^2}\,dx}$

4 $\displaystyle\int \frac{1}{x^2+6x+13}\,dx$

5 $\displaystyle\int_0^1 \frac{1}{\sqrt{16-x^2}}\,dx$

6 $\displaystyle\int_0^2 \frac{1}{2x^2+8x+25}\,dx$

15.4 Integration by parts

Look again at the formula for differentiating a product:

$$\frac{\mathrm{d}}{\mathrm{d}x}u(x)v(x) = u(x)\frac{\mathrm{d}v}{\mathrm{d}x} + v(x)\frac{\mathrm{d}u}{\mathrm{d}x} \qquad \text{(i)}$$

$$\text{Let } \frac{\mathrm{d}v}{\mathrm{d}x} = \mathrm{f}(x), \text{ then } v = \int \mathrm{f}(x)\,\mathrm{d}x$$

So Equation (i) can be written:

$$\frac{\mathrm{d}}{\mathrm{d}x}\left[u(x)\int \mathrm{f}(x)\,\mathrm{d}x\right] = u(x)\mathrm{f}(x) + \frac{\mathrm{d}u}{\mathrm{d}x}\int \mathrm{f}(x)\,\mathrm{d}x$$

or rearranging:

$$u(x)\mathrm{f}(x) = \frac{\mathrm{d}}{\mathrm{d}x}\left[u(x)\int \mathrm{f}(x)\,\mathrm{d}x\right] - \frac{\mathrm{d}u}{\mathrm{d}x}\int \mathrm{f}(x)\,\mathrm{d}x$$

Now integrate each of the three terms:

$$\int u(x)\mathrm{f}(x)\,\mathrm{d}x = u(x)\int \mathrm{f}(x)\,\mathrm{d}x - \int\left(\frac{\mathrm{d}u}{\mathrm{d}x}\int \mathrm{f}(x)\,\mathrm{d}x\right)\mathrm{d}x \qquad \text{(I18)}$$

This formula enables us to integrate the product of the two functions $u(x)\mathrm{f}(x)$. It is referred to as **integrating by parts**.

It looks quite a difficult formula to remember. Written in words, it is often more easy to use:

$\int$(first × second) dx = first × integral of the second

−integral of (derivative of the first × integral of the second)

Example 15.7

Integrate the following functions with respect to x:

(i) $x \sin x$ (ii) $x^2 \cos x$ (iii) $e^x \sin x$ (iv) $\ln x$

Solution

(i) Here x is the first function, and $\sin x$ the second function. The integral of the second function is therefore $-\cos x$.

Hence: $\int x \sin x\,\mathrm{d}x = x \times -\cos x - \int 1 \times -\cos x\mathrm{d}x$

(x ↑ first; $-\cos x$ ↑ integral of 2nd; 1 ↑ derivative of 1st; $-\cos x$ ↑ integral of 2nd)

$$= -x\cos x + \int \cos x\,\mathrm{d}x$$

$$= -x\cos x + \sin x + c$$

(ii) $\int x^2 \cos x \, dx = x^2 \sin x - \int 2x \times \sin x \, dx$

$$= x^2 \sin x - \int 2x \sin x \, dx$$

The remaining integral also requires integration by parts again. Take care with various negative signs.

$$\therefore \quad \int x^2 \cos x \, dx = x^2 \sin x - 2\left[x \times -\cos x - \int 1 \times -\cos x \, dx\right]$$

$$= x^2 \sin x - 2\left[-x \cos x + \int \cos x \, dx\right]$$

$$= x^2 \sin x + 2x \cos x - 2\int \cos x \, dx$$

$$= x^2 \sin x + 2x \cos x - 2 \sin x + c$$

(iii) $\int e^x \sin x \, dx = e^x \times -\cos x - \int e^x \times -\cos x \, dx = -e^x \cos x + \int e^x \cos x \, dx$

The formula for parts needs to be used again.

If $\mathbf{I} = \int e^x \sin x \, dx$, then we have:

$$\mathbf{I} = -e^x \cos x + \left[e^x \times \sin x - \int e^x \times \sin x \, dx\right]$$

$$= -e^x \cos x + e^x \sin x - \int e^x \sin x \, dx$$

It would appear that we are not getting anywhere, but you can see that in fact we have returned to the original integral **I**. Hence:

$$\mathbf{I} = -e^x \cos x + e^x \sin x - \mathbf{I}$$

$$\therefore \quad 2\mathbf{I} = -e^x \cos x + e^x \sin x$$

$$\therefore \quad \mathbf{I} = \tfrac{1}{2}e^x[-\cos x + \sin x] + c \qquad \text{(Remember to add } +c\text{)}$$

It is worth looking at how this answer can be simplified:

$$\mathbf{I} = \tfrac{1}{2}e^x[-\cos x + \sin x]$$

This contains an expression of the form $A \cos x + B \sin x$, see Section 9.11.

In this example, $A = -1$, $B = 1 \quad \therefore R = \sqrt{(-1)^2 + 1^2} = \sqrt{2}$.

$$\text{So} \quad \mathbf{I} = \frac{\sqrt{2}e^x}{2}\left[-\frac{1}{\sqrt{2}}\cos x + \frac{1}{\sqrt{2}}\sin x\right]$$

$$\therefore \quad \mathbf{I} = \frac{\sqrt{2}}{2}e^x \sin\left(x - \frac{\pi}{4}\right) + c$$

(iv)

MEMORY JOGGER

Look out for a disguised product written $f(x) \times 1$.

At first sight, this does not look like a product. However, $\ln x$ can be written $\ln x \times 1$.

$$\therefore \quad \int \ln x \, dx = \int \ln x \times 1 \, dx$$

$$= \ln x \times x - \int \frac{1}{x} \times x \, dx$$

$$= x \ln x - \int 1 \, dx$$

$$= x \ln x - x + c$$

Exercise 15(d)

1 Integrate the following functions with respect to x:

(i) xe^{2x} (ii) $x^2 \cos x$ (iii) $x^2 \ln x$ (iv) $x^2 e^{-x}$

(v) $e^{-x} \sin 2x$ (vi) $x \tan^{-1} x$ (vii) $\tan^{-1} x$ (viii) $\ln(x+2)$

2 Find:

(i) $\int_0^1 x^3 e^{2x} \, dx$ (ii) $\int_0^{\frac{\pi}{2}} e^x \sin 2x \, dx$ (iii) $\int_1^2 x^3 \ln x \, dx$

15.5 Partial fractions

The method of partial fractions is an algebraic technique that allows a rational function to be split up into simpler component parts. With ordinary fractions, for example, $\frac{8}{15}$ can be written $\frac{1}{3} + \frac{1}{5}$, or $\frac{31}{12}$ can be written $2 + \frac{1}{3} + \frac{1}{4}$.

(i) $$\frac{4x}{(x-2)(x+2)}$$

Assume this has come from $\dfrac{A}{(x-2)}$ and $\dfrac{B}{(x+1)}$.

$$\text{Let} \quad \frac{4x}{(x-2)(x+1)} \equiv \frac{A}{(x-2)} + \frac{B}{(x+1)}$$

Multiply through by the common denominator $(x-2)(x+1)$,

$$\therefore \quad 4x \equiv A(x+1) + B(x-2)$$

The $\equiv$ sign means it is true for all values of x.

The values of x where $(x-2)$ and $(x+1)$ are zero are used as follows:

Let $x = 2 \qquad 8 = 3A + 0 \quad \therefore \quad A = \frac{8}{3}$

Let $x = -1 \quad -4 = 0 - 3B \quad \therefore \quad B = \frac{4}{3}$

$$\text{Hence} \quad \frac{4x}{(x-2)(x+1)} \equiv \frac{8}{3(x-2)} + \frac{4}{3(x+1)}$$

(ii) $\dfrac{3x^2+1}{x(x-1)^2}$

If a bracket is raised to a power (in this case 2), you must assume it has come from $\dfrac{A}{(x-1)^2}$ and $\dfrac{B}{(x-1)}$

$$\text{Hence let} \quad \frac{3x^2+1}{x(x-1)^2} \equiv \frac{A}{(x-1)^2} + \frac{B}{(x-1)} + \frac{c}{x}$$

Multiply through by the common denominator $x(x-1)^2$:

$$3x^2 + 1 \equiv Ax + B(x-1)x + C(x-1)^2$$

Let $x = 1 \quad 4 = A + 0 + 0 \quad \therefore \quad A = 4$

Let $x = 0 \quad 1 = 0 + 0 + C \quad \therefore \quad C = 1$

There is no obvious choice for x which gives B, so just use a convenient value.

Let $x = 2, \quad 13 = 2A + 2B + C$

$\therefore \quad 13 = 8 + 2B + 1$

$B = 2$

$$\text{Hence} \quad \frac{3x^2+1}{x(x-1)^2} = \frac{4}{(x-1)^2} + \frac{2}{(x-1)} + \frac{1}{x}$$

(iii) $\dfrac{4x-2}{(x-3)(x^2+4)}$

If the bracket does not factorise, as here x^2+4 does not factorise, you can only assume it has come from $\dfrac{Ax+B}{x^2+4}$

$$\text{So let} \quad \frac{4x-2}{(x-3)(x^2+4)} \equiv \frac{Ax+B}{x^2+4} + \frac{C}{x-3}$$

Multiply through by the common denominator $(x-3)(x^2+4)$:

$$4x - 2 \equiv (Ax + B)(x - 3) + C(x^2 + 4)$$

Let $x = 3$ $\qquad 10 = 0 + 13C \qquad \therefore \quad C = \frac{10}{13}$

Let $x = 0$ $\qquad -2 = -3B + 4C$

$\therefore \quad -2 = -3B + \frac{40}{13} \qquad \therefore \quad B = \frac{22}{13}$

Let $x = 1$ $\qquad 2 = -2A - 2B + 5C$

$\therefore \quad -2 = -2A - \frac{44}{13} + \frac{50}{13} \qquad \therefore \quad A = -\frac{10}{13}$

Hence $$\frac{4x - 2}{(x-3)(x^2+4)} \equiv \frac{\frac{-10}{13} \times + \frac{22}{13}}{x^2 + 4} + \frac{\frac{10}{13}}{x - 3}$$

$$= \frac{-10x + 22}{13(x^2 + 4)} + \frac{10}{13(x - 3)}$$

(iv) $\dfrac{x^3}{x^2 - 4}$

MEMORY JOGGER

If the degree of the top line $\geq$ degree of the bottom line, you must divide first, and work on the remainder

The **degree** of x^3 is 3. The **degree** of $x^2 - 4$ is 2:

$$\begin{array}{r|l} & x \\ x^2 - 4 & x^3 \\ & \underline{x^3 - 4x} \\ & \quad 4x \end{array}$$

The remainder is $4x$, hence:

$$\frac{x^3}{x^2 - 4} = x + \frac{4x}{x^2 - 4}$$

$$= x + \frac{4x}{(x-2)(x+2)}$$

Now apply the partial fractions technique to $\dfrac{4x}{(x-2)(x+2)}$ *only*.

Hence $$\frac{x^3}{x^2 - 4} = x + \frac{2}{(x-2)} + \frac{2}{(x+2)}$$

Example 15.8

Find $\displaystyle\int \frac{2x-1}{(x-2)(x+3)}\,\mathrm{d}x$

Solution

First express the function to be integrated in partial fractions.

$$\text{Let} \quad \frac{2x-1}{(x-2)(x+1)} \equiv \frac{A}{(x-2)} + \frac{B}{(x+1)}$$

$$\therefore \quad 2x - 1 \equiv A(x+1) + B(x-2)$$

$$\text{Let } x = -1 \quad -3 = 0 - 3B \quad \therefore \quad B = 1$$

$$\text{Let } x = 2 \qquad 3 = 3A + 0 \quad \therefore \quad A = 1$$

$$\text{Hence} \quad \int \frac{2x-1}{(x-2)(x+3)}\,\mathrm{d}x = \int \frac{1}{x-2} + \frac{1}{x+3}\,\mathrm{d}x$$

$$= \ln(x-2) + \ln(x+3) + \ln k$$

Note that we write $C = \ln k$; this enables all of the R.H.S. to be combined easily to give:

$$= \ln k(x-2)(x+3)$$

A slightly more subtle use of this technique is given in the following example:

Example 15.9

Evaluate: $\displaystyle\int \frac{4x}{(x-2)(x+1)}\,\mathrm{d}x$

Solution

At the beginning of this section, we showed that:

$$\frac{4x}{(x-2)(x+1)} \equiv \frac{8}{3(x-2)} + \frac{4}{3(x+1)}$$

$$\text{Hence} \quad \int \frac{4x}{(x-2)(x+1)}\,\mathrm{d}x = \int \frac{8}{3(x-2)} + \frac{4}{3(x+1)}\,\mathrm{d}x$$

$$= \tfrac{8}{3}\ln(x-2) + \tfrac{4}{3}\ln(x+1) + \tfrac{4}{3}\ln k$$

Note how $+C$ has been written $\frac{4}{3}\ln k$. This enables us to simplify as follows:

$$= \tfrac{4}{3}[2\ln(x-2) + \ln(x+1) + \ln k] = \tfrac{4}{3}\ln k(x+1)(x-2)^2$$

Example 15.10

Evaluate: $\displaystyle\int_2^3 \frac{x}{(x-3)(x^2+1)}\,\mathrm{d}x$ correct to 3 significant figures.

Solution

Expand $\dfrac{x}{(x-3)(x^2+1)}$ into partial fractions.

Let $\dfrac{x}{(x-3)(x^2+1)} \equiv \dfrac{A}{(x-3)} + \dfrac{Bx+C}{x^2+1}$

$\therefore \quad x \equiv A(x^2+1) + (Bx+C)(x-3)$

Let $x = 3$ $\quad 3 = 10A \quad \therefore \; A = \frac{3}{10}$

Let $x = 0$ $\quad 0 = A - 3C \quad \therefore \; C = \frac{A}{3} = \frac{1}{10}$

Compare coefficients of x^2 $\quad 0 = A + B \quad \therefore \; B = -\frac{3}{10}$

The integral can now be written:

$$\int_4^5 \frac{3}{10(x-3)} + \frac{-3x+1}{10(x^2+1)}\,\mathrm{d}x$$

$$= \left[\tfrac{3}{10}\ln(x-3)\right]_4^5 + \int_4^5 \frac{-3x}{10(x^2+1)} + \frac{1}{10(x^2+1)}\,\mathrm{d}x$$

$$= \left[\tfrac{3}{10}\ln(x-3) - \tfrac{3}{20}\ln(x^2+1) + \tfrac{1}{10}\tan^{-1}x\right]_4^5$$

$$= \left[\tfrac{3}{10}\ln 2 - \tfrac{3}{20}\ln 26 + \tfrac{1}{10}\tan^{-1}5\right] - \left[\tfrac{3}{10}\ln 1 - \tfrac{3}{20}\ln 17 + \tfrac{1}{10}\tan^{-1}4\right]$$

$$= -0.1434 - -0.7174 = 0.574$$

Example 15.11

Expand the function $\dfrac{2x}{(x-3)(2x+1)}$ as a power series as far as the term in x^2. State the range of values of x for which the expansion is valid.

Solution

$$\frac{2x}{(x-3)(2x+1)} = \frac{6}{7(x-3)} + \frac{2}{7(2x+1)} \text{ using partial fractions}$$

$$= -\tfrac{6}{7}(3-x)^{-1} + \tfrac{2}{7}(1+2x)^{-1}$$

$$= -\tfrac{2}{7}\left(1-\frac{x}{3}\right)^{-1} + \tfrac{2}{7}(1+2x)^{-1}$$

$$= -\tfrac{2}{7}\left[1 + (-1)\left(-\frac{x}{3}\right) + \frac{(-1)(-2)}{2}\left(-\frac{x}{3}\right)^2\right]$$
$$+\tfrac{2}{7}\left[1 + (-1)(2x) + \tfrac{(-1)(-2)}{2}(2x)^2\right]$$
$$= \tfrac{2}{7}\left[(1 - 2x + 4x^2) - (1 + \tfrac{1}{3}x + \tfrac{1}{9}x^2)\right]$$
$$= -\tfrac{2}{3}x + \tfrac{10}{9}x^2$$

It is valid if $-1 < 2x < 1$, or $-\frac{1}{2} < x < \frac{1}{2}$.

Exercise 15(e)

1 Express in partial fractions:

(i) $\dfrac{x}{(1+x)(1-2x)}$ (ii) $\dfrac{2x-1}{x(1+2x)}$ (iii) $\dfrac{x}{(1+x)^2(1-x)}$

(iv) $\dfrac{x+1}{(x-3)(x^2+1)}$ (v) $\dfrac{x^3}{(x-1)(x+2)}$ (vi) $\dfrac{x}{(x-1)^2(x^2+2)}$

2 Evaluate:

(i) $\displaystyle\int_3^4 \frac{x}{(x-2)(x-1)}\,dx$ (ii) $\displaystyle\int_0^1 \frac{x}{(1+x)^2(1+2x)}\,dx$

(iii) $\displaystyle\int_{0.1}^{0.2} \frac{x^3}{(1+x^2)(1-x)}\,dx$

15.6 Differential equations

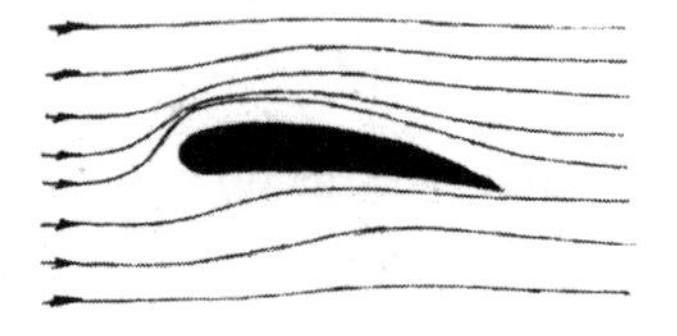

Many problems in science are solved through differential equations.

In this section, we shall look at how to form simple differential equations as well as how to solve them.

Type I

$$\frac{dy}{dx} = f(x)$$

Here we can solve this differential equation by straightforward integration.

So $y = \displaystyle\int f(x)\,dx$

Hence if $\frac{dy}{dx} = x^2 + 1$, then the solution of this differential equation is given by:

$$y = \int x^2 + 1\,dx$$

So $y = \frac{x^3}{3} + x + c$

Notice that the solution must contain the constant of integration c. This is referred to as the **general solution**.

If you draw all the **solution curves** for different values of c (a few are shown in Figure 15.3) you get the **family** of solution curves. Very often, this set of curves, which in fact represents the differential equation, actually describes a physical situation. For example, fluid flow, or magnetic field lines.

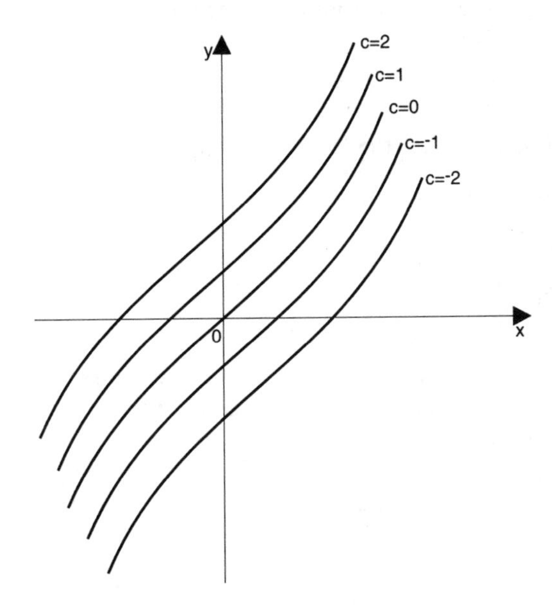

Figure 15.3

Type II

$$\frac{dy}{dx} = g(y)$$

In order to solve this type of equation, it must be written in a way that at first sight is not allowed. Remember that $\frac{dy}{dx}$ does not mean $dy \div dx$, and so, strictly speaking, it is not possible to write this differential equation as

$$\frac{dy}{g(y)} = dx \quad \text{or} \quad \frac{1}{g(y)}\,dy = dx$$

However, it can be proved that if integral signs are included, then

$$\int \frac{1}{g(y)}\,dy = \int dx \quad \text{is allowed.}$$

Consider the example:

$$\frac{dy}{dx} = 2y$$

Write this $\int \frac{1}{y}\,dy = \int 2\,dx$

$$\therefore \quad \ln y = 2x + C$$

If a logarithm appears in the solution, it is more convenient to write C as $\ln A$, where A is also a constant.

Hence: $\ln y = 2x + \ln A$

The reason for this is so that the following simplification can be carried out:

$$\ln y - \ln A = 2x$$

$$\ln \frac{y}{A} = 2x$$

$$\therefore \quad \frac{y}{A} = e^{2x}, \text{ or } y = Ae^{2x}$$

The family of solution curves is illustrated in Figure 15.4. Note that the x-axis, where $A = 0$, is a solution of the original differential equation.

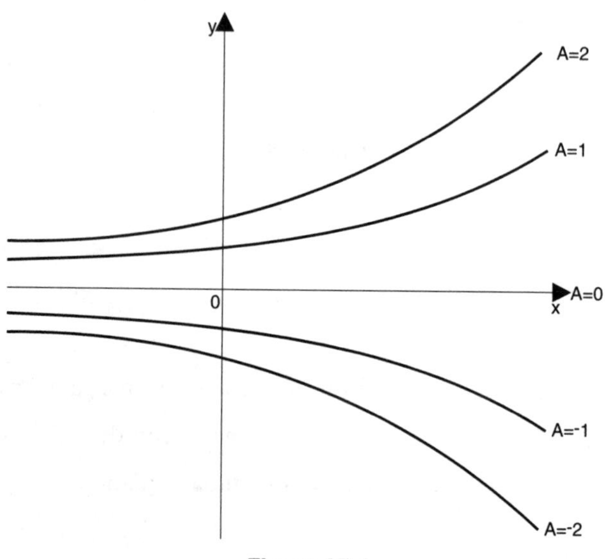

Figure 15.4

Type III

Types I and II can be combined together whenever $\frac{dy}{dx}$ can be written:

$$\frac{dy}{dx} = f(x)g(y)$$

then $$\int \frac{1}{g(y)}\,dy = \int f(x)\,dx$$

This technique is called **separating the variables**.

Consider the example:

$$\frac{dy}{dx} = (1 + y^2)e^x$$

$$\therefore \quad \int \frac{1}{1+y^2}\,dy = \int e^x\,dx$$

hence $$\tan^{-1} y = e^x + c$$

or $$y = \tan(e^x + c)$$

Example 15.12

The gradient of a curve at the point (x, y) is given by the expression $y(x + 1)$. Find the equation of the curve given that it passes through the point $(0, 4)$.

Solution

Since the gradient is $\frac{dy}{dx}$, we have:

$$\frac{dy}{dx} = y(x + 1)$$

that is, $$\int \frac{dy}{y} = \int (x + 1)\,dx$$

$$\therefore \quad \ln y = \frac{x^2}{2} + x + c$$

Now $y = 4$ when $x = 0$,

$$\therefore \quad \ln 4 = 0 + c$$

The equation is

$$\ln y = \frac{x^2}{2} + x + \ln 4$$

$$\text{or} \quad \ln y - \ln 4 = \frac{x^2}{2} + x$$

$$\ln \frac{y}{4} = \frac{x^2}{2} + x$$

$$\therefore \quad \frac{y}{4} = e^{\frac{x^2}{2}+x}$$

$$\text{or} \quad y = 4e^{\frac{x^2}{2}+x}$$

MEMORY JOGGER

Always combine logarithms in the solution of a differential equation

Example 15.13

The relationship between the exposed area, A, of an engineering part and the force, F, on that part is given by the equation:

$$\frac{dF}{dA} = kA + 10 \quad \text{where } k \text{ is a constant}$$

Solve this equation. Using the fact that when $A = 2$, $F = 40$, and when $A = 10$, $F = 200$, find F when $A = 12$.

Solution

$$\int dF = \int kA + 10\, dA$$

$$\therefore \quad F = \frac{kA^2}{2} + 10A + C$$

$A = 2, \quad F = 40 \qquad \therefore \quad 40 = 2k + 20 + C, \quad 20 = 2k + C \qquad$ (i)

$A = 10, F = 200 \qquad \therefore \quad 200 = 50k + 100 + C, \; 100 = 50k + C \qquad$ (ii)

(ii) – (i) $\quad 80 = 48k, \quad k = \frac{5}{3}$

Substitute into (i) $\quad 20 = \frac{10}{3} + C, \quad C = \frac{50}{3}$

Hence $\quad F = \dfrac{5A^2}{6} + 10A + \frac{50}{3}$

If $A = 12$, $F = \frac{5}{6} \times 12^2 + 120 + \frac{50}{3}$

$= 256\frac{2}{3}$

Example 15.14

A garden water tank is in the shape of a cuboid with a square base of side 1.4 m, and height 2.8 m. There is a small hole in the bottom of the tank from which water is leaking. After t minutes, when the depth of the water is h metres, the rate of escape of the water is $40\sqrt{h}$ m^3/minute. Show that:

$$\frac{dh}{dt} = -\frac{1000}{49}\sqrt{h}$$

If the tank is full initially, how long will it take to empty?

Solution

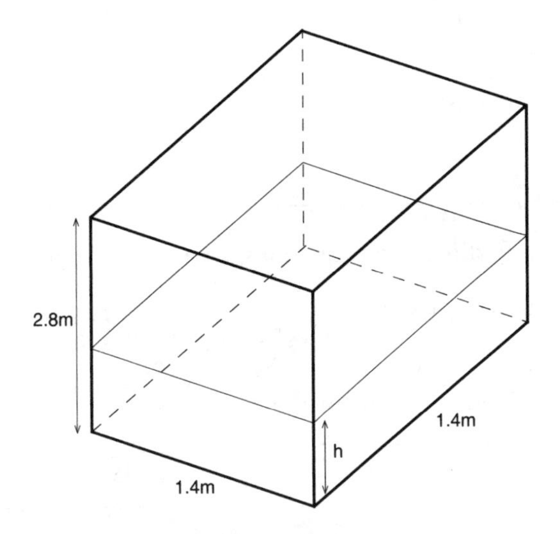

Figure 15.5

At time t, the volume $V = 1.4 \times 1.4h$ m^3

$$= 1.96h \text{ m}^3$$

Now $\frac{dV}{dt} = 1.96 \frac{dh}{dt}$. m^3/minute.

But this is caused by the *leakage* $40\sqrt{h}$

$$\therefore \quad \frac{dV}{dt} = 40\sqrt{h} = -1.96\frac{dh}{dt}, \quad \text{(note the negative sign)}$$

that is, $\frac{dh}{dt} = \frac{-1000}{49}\sqrt{h}.$

MEMORY JOGGER

If a quantity y is *decreasing* with time, $\frac{dy}{dt} < 0$.

So $$\int \frac{dh}{\sqrt{h}} = \frac{-1000}{49}\int dt,$$

that is, $$2\sqrt{h} = \frac{-1000}{49}t + C.$$

$t = 0,\ h = 280 \quad \therefore \quad 2\sqrt{280} = C.$

The tank is empty when $h = 0$,

$$\therefore \qquad 0 = \frac{-1000}{49}t + 2\sqrt{280}$$

$$\therefore \qquad t = \frac{2\sqrt{280} \times 49}{1000} \text{ minutes}$$

$$= 1.6 \text{ minutes}$$

Exercise 15(f)

Solve the following differential equations, subject to the given conditions.

1. $\dfrac{dy}{dx} = x^2;\ x = 2,\ y = 1$
2. $x\dfrac{dy}{dx} = y;\ x = 3,\ y = 2$
3. $\dfrac{dy}{dx} = y\sin x;\ x = \dfrac{\pi}{4},\ y = 1$
4. $x^2\dfrac{dy}{dx} = y^2 + 1;\ y = 0,\ x = 4$
5. $\dfrac{dy}{dx} = y(x - 1);\ x = 3,\ y = 7$
6. $\cos x\dfrac{dy}{dx} = \sin x;\ x = 0,\ y = \dfrac{\pi}{6}$

Miscellaneous Examples 15

Evaluate the following integrals:

1 $\displaystyle\int 3^{-x}\,dx$

2 $\displaystyle\int x^2 \sin 3x\,dx$

3 $\displaystyle\int \frac{2x - 1}{x^2 + 1}\,dx$

4 $\displaystyle\int xe^{x^2}\,dx$

5 $\displaystyle\int \frac{dx}{\sqrt{9 - 4x^2}}$

6 $\displaystyle\int x^3 \ln 2x\,dx$

7 $\displaystyle\int_0^{\frac{\pi}{4}} \sin^3 4x\,dx$

8 $\displaystyle\int_2^3 \frac{x}{(x - 1)^2(x + 1)}\,dx$

9 $\displaystyle\int_0^1 (3x - 1)(x + 1)^4\,dx$

10 $\displaystyle\int_0^{\frac{\pi}{4}} \sin t \cos 2t\,dt$

11 $\int_0^1 \frac{e^{-2x}}{1+e^{-2x}}\mathrm{d}x$ **12** $\int_0^{\frac{\pi}{4}} e^x \sin 2x \mathrm{d}x$

13 $\int_{0.1}^{0.2} \frac{x^2}{\sqrt{1-x^3}}\mathrm{d}x$ **14** $\int_0^{\frac{\pi}{4}} \sin^3 4\theta \, \mathrm{d}\theta$

15 $\int_1^2 \tan 4y \, \mathrm{d}y$

16 Expand $\mathrm{f}(x) = \frac{x}{(x-2)(1+x)}$ in partial fractions. Hence, using the binomial expansion, find a, b and c if $\mathrm{f}(x) = a + bx + cx^2$ where powers of x^3 and higher are neglected.

17 Expand the function $\frac{x-1}{(1+x^2)(1+2x)}$ as a power series as far as the term in x^3. State the range of values of x for which the expansion is valid.

Revision Problems 15

1 A much simplified model for the filling of a new reservoir is as follows.

Initially, the reservoir is empty. At time t years after starting to be filled, the volume of water it contains is V cubic metres and the depth of water is h metres, where V is proportional to h.

The rate at which water enters the reservoir is constant, but there is a loss of water (due to leakage and evaporation), the rate of which is proportional to h.

Show that this model leads to the differential equation:

$$\frac{\mathrm{d}h}{\mathrm{d}t} = \lambda - \mu h, \text{ where } \lambda \text{ and } \mu \text{ are constants.}$$

Given that $\lambda = 300$ and $\lambda = 1.5$, solve this equation for h as a function of t.

It is possible to start drawing water from the reservoir when the depth reaches 150 metres. Calculate the value of t when this occurs.

(Oxford & Cambridge)

2 A function f is defined by the equation:

$$f(x) = \frac{3(x-1)}{x(x+3)}$$

for all real positive values of x.

(i) Figure 15.6 shows a sketch (not drawn to scale) of the graph of $y = f(x)$. Write down the coordinates of the point marked A.

(ii) Express $f(x)$ in partial fractions in the form:

$$\frac{a}{x} + \frac{b}{x+3}$$

where a and b are numbers.

(iii) Use the partial fraction form of $f(x)$ to find an expression for $f'(x)$. Hence show that:

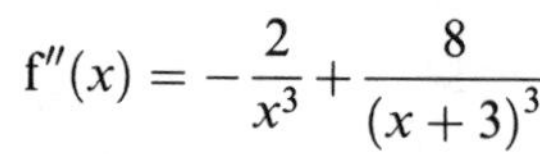

$$f''(x) = -\frac{2}{x^3} + \frac{8}{(x+3)^3}$$

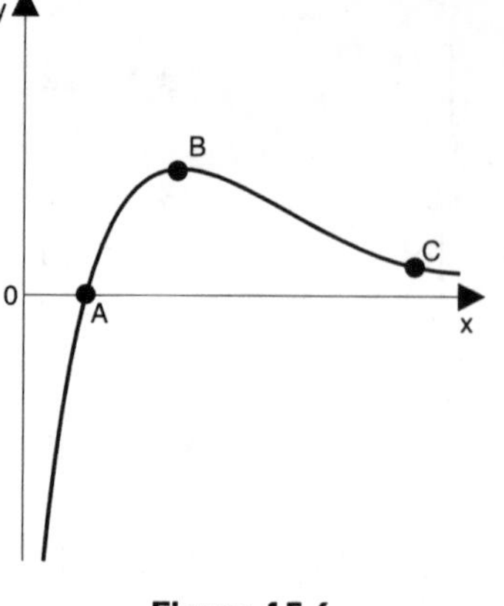

Figure 15.6

(iv) Find the x-coordinate of the maximum point marked B. (You are not required to prove that it is a maximum rather than a minimum.)

(v) Show that the x-coordinate of the point of inflexion marked C satisfies the equation $4x^3 - (x+3)^3 = 0$. Denoting the L.H.S. of this equation by $g(x)$, write an expression for $g'(x)$.

(vi) You are given that the equation in (v) has just one root, which is approximately equal to 5. Use the Newton–Raphson method once to find a better approximation.

(Oxford & Cambridge)

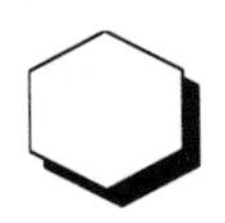

Summary of formulae

Powers

$x^m \times x^n = x^{m+n}$	P1
$x^m \div x^n = x^{m-n}$	P2
$(x^m)^n = x^{mn}$	P3
$x^0 = 1$	P4
$(xy)^m = x^m y^m$	P5
$a^{\frac{m}{n}} = \left(a^{\frac{1}{n}}\right)^n = (a^m)^{\frac{1}{n}}$	P6
$a^{\frac{1}{n}} = \sqrt[n]{a}$	P7
$a^{-n} = \dfrac{1}{a^n}$	P8

Logarithms

$a^x = y \Leftrightarrow x = \log_a y$	L1
$\log M + \log N = \log MN$	L2
$\log M - \log N = \log \dfrac{M}{N}$	L3
$\log M^n = n \log M$	L4
$\log_a 1 = 0$	L5
$\log_a a = 1$	L6
$\log \dfrac{1}{N} = -\log N$	L7
$N = a^{\log_a N}$	L8
$\log_a N = \dfrac{\log_b N}{\log_b a}$	L9
$\log_a b = \dfrac{1}{\log_a b}$	L10

Factors

$f(a)$ is the remainder when $f(x)$ is divided by $(x - a)$ F1

$f\left(\frac{b}{a}\right) = 0 \Rightarrow (ax - b)$ is a factor of $f(x)$ F2

Quadratic functions

$$x^2 + bx + c = \left(x + \frac{b}{2}\right)^2 + c - \frac{b^2}{4} \qquad \text{Q1}$$

If α, β are the roots of $ax^2 + bx + c = 0$,

then $\alpha + \beta = -\frac{b}{a}$ Q2

$\alpha\beta = \frac{c}{a}$ Q3

Arithmetic series

nth term $T_n = a + (n - 1)d$ AS1

sum of n terms $= \frac{n}{2}(2a + (n - 1)d)$ AS2

$= n \times$ (the average of first and last terms) (AS2a)

Geometric series

Sum of n terms $= \frac{a(r^n - 1)}{(r - 1)}$ $\quad r > 1$ GS1a

$= \frac{a(1 - r^n)}{(1 - r)}$ $\quad r < 1$ GS1b

nth term $T_n = ar^{n-1}$ GS2

Sum of infinite series $= \frac{a}{1 - r}$ $\quad$ if $-1 < r < 1$ GS3

Binomial expansion

$$(a+b)^n = \binom{n}{0}a^2 + \binom{n}{1}a^{n-1}b + \ldots \binom{n}{n}b^n \qquad \text{B1}$$

$$\text{where } \binom{n}{r} = \frac{n!}{r!(n-r)!} = {}^nC_r$$

$$r\text{th term} = \binom{n}{r-1}a^{n-r+1}b^r \qquad \text{B2}$$

$$\text{term in } b^r = \binom{n}{r}a^{n-r}b^r \qquad \text{B3}$$

n not a positive integer

$$(1+b)^n = 1 + nb + \frac{n(n-1)}{2!}b^2 + \frac{n(n-1)(n-2)}{3!}b^2 + \ldots \qquad |b| < 1 \qquad \text{B4}$$

Coordinate geometry

$$\text{Distance between points} = \sqrt{(x_1 - x_2)^2 + (y_1 - y_2)^2} \qquad \text{C1}$$

in two dimensions

$$= \sqrt{x_1 - x_2)^2 + (y_1 - y_2)^2 + (z_1 - z_2)^2} \qquad \text{C2}$$

in three dimensions

$$\text{Gradient of a line} = \frac{y_1 - y_2}{x_1 - x_2} \qquad \text{C3}$$

$$\text{Mid point} = \left(\frac{x_1 + x_2}{2}, \frac{y_1 + y_2}{2}\right) \qquad \text{C4a}$$

in two dimensions

$$= \left(\frac{x_1 + x_2}{2}, \frac{y_1 + y_2}{2}, \frac{z_1 + z_2}{2}\right) \qquad \text{C4b}$$

in three dimensions

A point which divides the points (x_1y_1) and (x_2y_2) in the ratio $m : n$ is:

$$\left(\frac{mx_2 + nx_1}{m+n}, \frac{my_2 + ny_1}{m+n}\right) \qquad \text{C5a}$$

in two dimensions

$$= \left(\frac{mx_2 + nx_1}{m+n}, \frac{my_2 + ny_1}{m+n}, \frac{mz_2 + nz_1}{m+n}\right) \qquad \text{C5b}$$

in three dimensions

The area of the triangle formed by the points (x_1, y_1), (x_2, y_2) and (x_3, y_3) is:

$$|\tfrac{1}{2}[x_1(y_2 - y_3) + x_2(y_3 - y_1) + x_3(y_1 - y_2)] \qquad \text{C6}$$

Perpendicular distance of (h, k) from $ax + by + c = 0$ is $\left|\dfrac{ah + bk + c}{\sqrt{a^2 + b^2}}\right|$ C7

Straight lines

Equation of a line through (x_1, y_1) with gradient m is $y - y_1 = m(x - x_1)$ SL1

The line joining $(x_1\, y_1)$ and $(x_2\, y_2)$ is:

$$\frac{y - y_1}{y_2 - y_1} = \frac{x - x_1}{x_2 - x_1} \qquad \text{SL2}$$

Perpendicular lines of gradient m_1 and m_2 are such that $m_1 m_2 = -1$ SL3

Differentiation I

$$\frac{\mathrm{d}}{\mathrm{d}x}(kx^n) = knx^{n-1} \qquad \text{D2}$$

$$\frac{\mathrm{d}}{\mathrm{d}x}(\mathrm{f} + \mathrm{g}) = \frac{\mathrm{df}}{\mathrm{d}x} + \frac{\mathrm{dg}}{\mathrm{d}x} \qquad \text{D3}$$

Chain rule: $$\frac{\mathrm{d}y}{\mathrm{d}x} = \frac{\mathrm{d}y}{\mathrm{d}u} \times \frac{\mathrm{d}u}{\mathrm{d}x} \qquad \text{D4}$$

Product rule: $$\frac{\mathrm{d}}{\mathrm{d}x}(uv) = u\frac{\mathrm{d}v}{\mathrm{d}x} + v\frac{\mathrm{d}u}{\mathrm{d}x} \qquad \text{D5}$$

Quotient rule: $$\frac{\mathrm{d}}{\mathrm{d}x}\left(\frac{u}{v}\right) = \frac{v\frac{\mathrm{d}u}{\mathrm{d}x} - u\frac{\mathrm{d}v}{\mathrm{d}x}}{v^2} \qquad \text{D6}$$

The circle

Equation centre 0: $x^2 + y^2 = r^2$ C8

centre (h, k): $(x - h)^2 + (y - k)^2 = r^2$ C9

Integration I

$$\int kx^n \,\mathrm{d}x = \frac{k}{n+1}x^{n+1} + c \qquad \text{I1a}$$

$\int kh'(x)\mathrm{f}(h(x))\,\mathrm{d}x = k\mathrm{g}(h(x))$

where $\int \mathrm{f}(x)\,\mathrm{d}x = \mathrm{g}(x)$ I2

Area between $y = \mathrm{f}(x)$ and x-axis $= \int_a^b y\,\mathrm{d}x$ I3

Area between y_1 and y_2 $= \int_a^b y_2 - y_1\,\mathrm{d}x$ I4

Area between $y = f(x)$ and y-axis $= \int_a^b x \, dy$ I5

Volume of revolution about x-axis $= \int_a^b \pi y^2 \, dx$ I6

between graphs $= \int_a^b \pi(y_2^2 - y_1^2) \, dx$ I8

Volume of revolution about y-axis $= \int_a^b \pi x^2 \, dy$ I7

between graphs $= \int_a^b \pi(x_2^2 - x_1^2) \, dy$ I9

Approximate integration

Trapezium rule $\int_a^b \pi f(x) \, dx = \frac{h}{2}[y_0 + 2y_1 + \ldots 2y_{n-1} + y_n]$ I10

Simpson's rule $= \frac{h}{3}[y_0 + y_n + 4(y_1 + y_3 + \ldots) + 2(y_2 + y_4 + \ldots)]$ I11

Sectors

In radians area $= \frac{1}{2}r^2\theta$ SR1

Arc length $= r\theta$ SR2

Trigonometric identities

$\sin^2 x + \cos^2 x \equiv 1$ TI1

$\operatorname{cosec}^2 x \equiv 1 + \cot^2 x$ TI2

$\sec^2 x \equiv 1 + \tan^2 x$ TI3

$\cos(A + B) \equiv \cos A \cos B - \sin A \sin B$ TI4

$\cos(A - B) \equiv \cos A \cos B + \sin A \sin B$ TI5

$\sin(A + B) \equiv \sin A \cos B + \cos A \sin B$ TI6

$\sin(A - B) \equiv \sin A \cos B - \cos A \sin B$ TI7

$\tan(A + B) \equiv \dfrac{\tan A + \tan B}{1 - \tan A \tan B}$ TI8

$\tan(A - B) \equiv \dfrac{\tan A - \tan B}{1 + \tan A \tan B}$ TI9

$$\cos 2A \equiv \cos^2 A - \sin^2 A \qquad \text{TI10}$$

$$\equiv 2\cos^2 A - 1 \qquad \text{TI10a}$$

$$\equiv 1 - 2\sin^2 A \qquad \text{TI10b}$$

$$\sin 2A \equiv 2 \sin A \cos A \qquad \text{TI11}$$

$$\tan 2A \equiv \frac{2 \tan A}{1 - \tan^2 A} \qquad \text{TI12}$$

$$\sin X + \sin Y \equiv 2 \sin\left(\frac{X+Y}{2}\right) \cos\left(\frac{X-Y}{2}\right) \qquad \text{TI13}$$

$$\sin X - \sin Y \equiv 2 \cos\left(\frac{X+Y}{2}\right) \sin\left(\frac{X-Y}{2}\right) \qquad \text{TI14}$$

$$\cos X + \cos Y \equiv 2 \cos\left(\frac{X+Y}{2}\right) \cos\left(\frac{X-Y}{2}\right) \qquad \text{TI15}$$

$$\cos X - \cos Y \equiv -2 \sin\left(\frac{X+Y}{2}\right) \sin\left(\frac{X-Y}{2}\right) \qquad \text{TI16}$$

$$\cos 2A \equiv \frac{1-t^2}{1+t^2}, \qquad t = \tan A \qquad \mathit{TI}17$$

$$\sin 2A \equiv \frac{2t}{1+t^2} \qquad \text{TI18}$$

$$\tan 2A \equiv \frac{2t}{1-t^2} \qquad \text{TI19}$$

Small angles

$$\sin\theta = \theta \qquad \text{SA1}$$

$$\cos\theta = 1 - \tfrac{1}{2}\theta^2 \qquad \text{SA2}$$

$$\tan\theta = \theta \qquad \text{SA3}$$

Differentiation II

$$\frac{\mathrm{d}}{\mathrm{d}x}(\sin x) = \cos x \qquad \text{D7}$$

$$\frac{\mathrm{d}}{\mathrm{d}x}(\cos x) = -\sin x \qquad \text{D8}$$

$$\frac{\mathrm{d}}{\mathrm{d}x}(\tan x) = \sec^2 x \qquad \text{D9}$$

$$\frac{\mathrm{d}}{\mathrm{d}x}(\sec x) = \sec x \tan x \qquad \text{D10}$$

$$\frac{d}{dx}(\operatorname{cosec} x) = -\operatorname{cosec} x \cot x \qquad \text{D11}$$

$$\frac{d}{dx}(\cot x) = -\operatorname{cosec}^2 x \qquad \text{D12}$$

$$\frac{d}{dx}(y^n) = ny^{n-1}\frac{dy}{dx} \qquad \text{D13}$$

If $y = f(t)$ and $x = g(t)$, $\dfrac{dy}{dx} = \dfrac{\frac{dy}{dt}}{\frac{dx}{dt}}$ D 14

$$\frac{d}{dx}e^{ax} = a\,e^{ex} \qquad \text{D15a}$$

$$\frac{d}{dx}\sin^{-1} x = \frac{1}{\sqrt{1-x^2}} \qquad \text{D16}$$

$$\frac{d}{dx}\cos^{-1} x = \frac{-1}{\sqrt{1-x^2}} \qquad \text{D17}$$

$$\frac{d}{dx} = \tan^{-1} x = \frac{1}{1+x^2} \qquad \text{D18}$$

Integration II

$$\int \cos nx\, dx = \frac{1}{n}\sin nx + c \qquad \text{I12}$$

$$\int \sin nx\, dx = \frac{-1}{n}\cos nx + c \qquad \text{I13}$$

$$\int \frac{1}{x}\, dx - \ln x + c \qquad \text{I14}$$

$$\int \frac{1}{ax+b}\, dx = \frac{1}{a}\ln ax + b + c \qquad \text{I15}$$

$$\int \frac{dx}{a^2+b^2} = \frac{1}{a}\tan^{-1}\frac{x}{a} \qquad \text{I16}$$

$$\int \frac{dx}{\sqrt{a^2-x^2}} = \sin^{-1}\frac{x}{a} \qquad \text{I17}$$

Integration by parts

$$\int uv\, dx = u\int v\, dx - \int \frac{du}{dx}\left(\int v\, dx\right) dx \qquad \text{I18}$$

Vectors

Length of $x\mathbf{i} + y\mathbf{j} + z\mathbf{k} = \sqrt{x^2 + y^2 + z^2}$ V1

Vector joining $\vec{AB} = \mathbf{b} - \mathbf{a}$ V2

Equation of a line through $\mathbf{a}$ in direction $\mathbf{d}$ is $\mathbf{r} = \mathbf{a} + \lambda\mathbf{d}$ V3

Scalar product $\mathbf{x}.\mathbf{y} = x_1y_1 + x_2y_2 + x_3y_3$ V4

Angle between vectors θ is given by:

$$\cos\theta = \frac{a_1b_1 + a_2b_2 + a_3b_3}{\sqrt{a_1^2 + a_2^2 + a_3^2}.\sqrt{b_1^2 + b_2^2 + b_3^2}} \qquad \text{V5}$$

$\mathbf{x}.\mathbf{y} = 0$ means $\mathbf{x}$ is perpendicular to $\mathbf{y}$ V6

Vector equation of a plane through $\mathbf{a}$, $\mathbf{b}$, $\mathbf{c}$ is

$$\mathbf{r} = \mathbf{a} + \lambda(\mathbf{b} - \mathbf{a}) + \mu(\mathbf{c} - \mathbf{a}) \qquad \text{V7}$$

or $\mathbf{r}.\hat{\mathbf{n}} = d$ V8

Numerical methods

Linear interpolation, closer solution to $f(x) = 0$ between a and b is

$$a + \frac{|f(a)|(b-a)}{|f(a)| + |f(b)|} \qquad \text{N1}$$

Newton–Raphson iteration to $f(x) = 0$:

$$x_{n+1} = x_n - \frac{f(x_n)}{f'(x_n)} \qquad \text{N2}$$

Arrangement and selection

$n! = n(n-1)(n-2)\ldots 2.1$ S1

$${}^nP_r = \frac{n!}{(n-r)!} \qquad \text{S2}$$

$${}^nC_r = \frac{n!}{r!(n-r)!} \qquad \text{S6}$$

Probability

$P(X \cup Y) = P(X) + P(Y) - P(X \cap Y)$ PR2

$P(\bar{A}) = 1 - P(A)$ PR4

$$P(X|Y) = \frac{P(X \cap Y)}{P(Y)} \qquad \text{PR6}$$

Answers

Exercise 1(a)

1 (i) 9 (ii) 16 (iii) $\frac{1}{2}$ (iv) $\frac{2}{3}$

(v) 1 (vi) $\frac{27}{125}$ (vii) -64 (viii) $\left(\frac{9}{4}\right)^{-\frac{1}{2}} = \frac{2}{3}$

(ix) 0.01 (x) $\left(\frac{9}{100}\right)^{\frac{3}{2}} = \frac{27}{1000}$ (xi) $\left(\frac{9}{4}\right)^{\frac{1}{2}} = \frac{3}{2}$ (xii) 5

2 (i) $8x^3 \div 9x^2 = \dfrac{8x}{9}$ (ii) $2x^{\frac{3}{2}} \times 27x^{\frac{3}{2}} = 54x^3$

(iii) $(16 \times 512) \div (\frac{1}{4} \times 2) = 16\,384$ (iv) $\dfrac{9^2 \times 5^3}{9^2 \times 5} = 25$

(v) $x^{\frac{1}{4}+\frac{2}{3}-\frac{1}{2}} = x^{\frac{5}{12}}$ (vi) $(8 \times 2)^{\frac{1}{2}} = 4$

(vii) $x^{\frac{1}{2}} \times x^{\frac{3}{2}} = x^2$ (viii) $2x^{\frac{1}{2}} \div 2x^2 = x^{-\frac{3}{2}}$

(ix) $x \div x^2 = x^{-1}$ (x) $32^{\frac{1}{2}} \times 16^{\frac{1}{2}} = 16\sqrt{2}$

Exercise 1(b)

1 (i) $6\sqrt{2}$ (ii) $4\sqrt{2} + 2\sqrt{6}$ (iii) 40

(iv) $\sqrt{2} + 2$ (v) $\sqrt{2 \times 196} = 14\sqrt{2}$ (vi) -1

(vii) $18 + 1 + 6\sqrt{2} = 19 + 6\sqrt{2}$

(viii) $\sqrt{6} + 2\sqrt{2} + 3\sqrt{3} + 6$

2 (i) $\dfrac{4\sqrt{2}}{2} = 2\sqrt{2}$ (ii) $\dfrac{\sqrt{8}}{8} = \dfrac{2\sqrt{2}}{8} = \dfrac{\sqrt{2}}{4}$

(iii) $\sqrt{2} - 1$ (iv) $\sqrt{3} + \sqrt{2}$

(v) $\sqrt{2} - 1 - 1 - \sqrt{2} = -2$ (vi) $\dfrac{4\sqrt{3} - 3\sqrt{2}}{48 - 18} = \dfrac{2\sqrt{3}}{15} - \dfrac{\sqrt{2}}{10}$

(vii) $1 - \frac{1}{2}\sqrt{3}$ (viii) $-(3 + 2\sqrt{2})$

Exercise 1(c)

1 (i) $\log_2 16 = 4$ (ii) $\log_5 25 = 2$ (iii) $\log_9 729 = 3$
(iv) $\log_8 4 = \frac{2}{3}$ (v) $\log_{10} 0.0001 = -4$ (vi) $\log_{16} 0.25 = -\frac{1}{2}$
(vii) $\log_7 1 = 0$ (viii) $\log_x t = 3$ (ix) $\log_t q = x$
(x) $\log_p y = \frac{1}{x}$

2 (i) $64 = 4^3$ (ii) $10\,000 = 10^4$ (iii) $y = x^t$ (iv) $q = 4^t$
(v) $4 = x^y$ (vi) $\frac{1}{16} = 4^{-2}$ (vii) $x^z = y$ (viii) $2 = e^t$
(ix) $4q = t^6$

3 (i) 3 (ii) 2 (iii) 3 (iv) -1
(v) 4 (vi) -2 (vii) -3 (viii) 9

4 (i) $\log 6$ (ii) $\log 3$
(iii) $\log 16$ (iv) $\log 9 + \log 8 = \log 72$
(v) $\log xyz$ (vi) $\log 4 + \log 3 = \log 12$
(vii) $\log x^2 + \log y = \log x^2 y$ (viii) $\log 10 + \log x = \log 10x$
(ix) $\log \dfrac{x+1}{x-1}$ (x) $\dfrac{\log 4^3}{\log 4} = \dfrac{3 \log 4}{\log 4} = 3$

5 (i) $\log x + \log y$ (ii) $\log p - \log 2 - \log q$
(iii) $\log x + \log y - \log t$ (iv) $2 \log x - \log y$
(v) $\frac{1}{2} \log p - \frac{1}{2} \log q$ (vi) $\log a^3 - \log 100 = 3 \log a - 2$
(vii) $\log 1 - \log 10 - \log x = -1 - \log x$
(viii) $\frac{1}{2} \log 4x^2 - \frac{1}{2} \log y^3 = \frac{1}{2} \log 4 + \log x - \frac{3}{2} \log y$

6 (i) $x \log 2 = \log 5$ $\therefore$ $x = 2.32$ (3 sig. figs)
(ii) $t \log 3 = \log 4$ $\therefore$ $t = 1.26$ (3 sig. figs)
(iii) $(x+2) \log 4 = \log 9$ $\therefore$ $x = \dfrac{\log 9}{\log 4} - 2 = -0.415$
(iv) Both are powers of 2, so $3t - 1 = 1 + t$ $\therefore$ $t = 1$
(v) $(2x - 1) \log 3 = (x + 1) \log 2$

$\therefore$ $x(2 \log 3 - \log 2) = \log 2 + \log 3$

giving $x = \dfrac{\log 6}{\log 4.5} = 1.19$

(vi) $(x + 1) \log 4 + (2x - 1) \log 3 = \log 8$

$\therefore$ $x(\log 4 + 2 \log 3) = \log 8 + \log 3 - \log 4$

$x \log 36 = \log 6$

$2x \log 6 = \log 6$

hence $x = \frac{1}{2}$

Exercise 1(d)

1 $x = e^3 = 20.1$ **2** $x^2 + x - 6 = 0$, only $x = 2$
3 $x = 2^6 = 64$ **4** $(x + 3)(x - 2) = 50$, $x = 7$

5 $\log_4 x + \dfrac{1}{\log_4 x} = 2$, $x = 4$

6 Change all to base 10:

$$\frac{\log_{10} x}{\log_{10} 2} = \frac{\log_{10} y}{\log_{10} 4} - 5$$

$$\frac{\log_{10} x}{\log_{10} 4} = \frac{\log_{10} y}{\log_{10} 2} + 3$$

$$\therefore \quad 3.32 \log_{10} x = 1.66 \log_{10} y - 5 \qquad \text{(i)}$$

$$1.66 \log_{10} x = 3.32 \log_{10} y + 3 \qquad \text{(ii)}$$

(i) × 1.66 − (ii) × 3.32 gives $0 = -8.2668 \log_{10} y - 18.26$

Hence $\log_{10} y = -2.2088$, $y = 6.18 \times 10^{-3}$

Substitute this into (i) to get $\log_{10} x = -2.6104$

Hence $x = 2.45 \times 10^{-3}$

Miscellaneous examples 1

1 (i) $\frac{1}{16}$ (ii) 4 (iii) 128 (iv) 1 000 000
(v) $\frac{1}{8}$ (vi) 4 (vii) $2\frac{1}{4}$ (viii) 25
(ix) 0.1 (x) $\frac{1}{16}$

2 (i) $2x^6$ (ii) $8x^3$ (iii) $32a^3$ (iv) x^2
(v) $32a$ (vi) xy^2 (vii) $4x^2y^2$ (viii) $64p^4$
(ix) $\dfrac{1}{x(x^2+1)}$ (x) $\dfrac{8}{x^{\frac{5}{2}}}$ or $8x^{-\frac{5}{2}}$

3 (i) $\log xy^2$ (ii) $\log \dfrac{t^3}{x^2}$ (iii) $\log \dfrac{\sqrt{A}}{y^2}$ (iv) $\log \sqrt{\dfrac{4x}{y^2}}$
(v) $\log \dfrac{x^4y^3}{t^2}$ (vi) $\log 100 + \log x^3 = \log 100x^3$

4 (i) $5 \log x + \log y$ (ii) $\frac{1}{2}\log x - \frac{1}{2}\log y$ (iii) $-\frac{1}{2}\log x$
(iv) $3 \log x - 3 \log y$ (v) $\frac{2}{3}\log x - \log y$

5 (i) $x = \dfrac{\log 3}{\log 4} = 0.792$
(ii) $(2x + 1) \log 2 = \log 4; x = 0.5$
(iii) $2^x = 2^{2x+6}$, $x = -6$
(iv) $(2x + 1) \log 3 + (x + 2) \log 2 = \log 8;\ x = -0.14$

6 (i) $\dfrac{\sqrt{2}}{2}$ (ii) $\sqrt{2} - 1$
(iii) $\dfrac{\sqrt{6}+\sqrt{3}}{3}$ (iv) $\dfrac{\sqrt{2}}{2}$

(v) $\frac{(1+\sqrt{5})^2}{-4}$ or $\frac{3+\sqrt{5}}{-2}$ (vi) $\sqrt{3}(\sqrt{2}-1) = \sqrt{6}-\sqrt{3}$

7 (i) $(x+1)(3x+1) = 65;\ x = 4$

(ii) $\frac{\log_2 x}{\log_2 4} + \log_2 x = 9 \quad \therefore \quad \log_2 x = 6,\ x = 64$

(iii) $\frac{3x+2}{x+1} = \frac{4x+1}{2};\ x = 1$

Revision problems 1

1 $t = 0;\ v = 7499, \quad$ so $7499 = k$

$t = 1; 6000 = 7499e^{-\lambda}$, hence $\lambda = +0.22$

$\therefore \quad t = 3;\ v = 7499e^{-3\times 0.22} \quad = £3876$

2 (a) 1881–1891: $\frac{33}{30} = 1.10$, 1891–1901: 1.12, 1901–1911: 1.11. Constant growth rate which suggests exponential growth.

(b) $41r^t$

(c) If $41r^T = 82$, then $r^T = 2$ so $T = \frac{\ln 2}{\ln r}$

(d) $r = 1.1$ gives $T = 7$

Exercise 2(a)

1 (i) $6x^2 - 5x - 4$ (ii) $t^3 + t^2 + t + 1$

(iii) $x^3 - 1$

(iv) $(3x-1)(2x^2-3x-2) = 6x^3 - 11x^2 - 3x + 2$

(v) $x^4 + x^3 - 2x^2 + x + 2$ (vi) $2t^4 + 2t^3 + 5t^2 + t + 2$

2 (i) $3x^2 + 7x + 14$ (ii) $4x^3 - 4x^2 + x + 6$

(iii) $7x^2 + 14x + 21$ (iv) $4x^2 + 4x + 2$

3 (i) 2 (ii) 13 (iii) 5

Exercise 2(b)

1 (i) $(x-2)(x+1)(x+4)$ (ii) $(x+3)(2x-1)(2x+1)$

(iii) $(x-1)(3x-2)(2x+5)$ (iv) $(2t+3)(t-4)(t-5)$

2 (i) $(x-1)(5x-7)$ (ii) $(2t+3)(3t-5)$

(iii) $(t-1)(t-2)(t-3)$ (iv) $(x-1)^2(x+1)(x^2+1)$

(v) $(2t+1)(t^2+1)$ (vi) $(y-2)(y-3)(y+4)$

3 (i) $x = 2$ (ii) $x = \pm\sqrt{3}$ (iii) $1, \frac{1}{3}, -\frac{1}{2}$ (iv) $-\frac{1}{4} \pm \sqrt{2}$

Exercise 2(c)

1 $\frac{17}{18}$

2 19

3 $15(x+1) - 40 = 12x + 24$, giving $x = 16\frac{1}{3}$

4 2.11, −0.356

5 Leads to quadratic $6x^2 - 5x - 2 = 0$; $x = 1.13, -0.295$

6 Squaring once leads to $9x - 9 = 2\sqrt{7x+1}\sqrt{2x-1}$; squaring again leads to $25x^2 - 142x + 85 = 0$. Hence $x = 5, -17$, but only $x = 5$ works.

7 $t^2 = \dfrac{-8 \pm \sqrt{64+24}}{2}$. Only $t^2 = \dfrac{-8+\sqrt{88}}{2}$ possible, hence $t = \pm 0.831$

8 $2x - 1 = 9(3x+5)$; $x = -1.84$

9 $4x^2 - x - 3 = 0$; $-\frac{3}{4}, 1$

10 $\frac{1}{6}$

Exercise 2(d)

1 (i) $(x+3)^2 + 3$; least at $(-3, 3)$
(ii) $\left(x - \frac{3}{2}\right)^2 - \frac{5}{4}$; least at $\left(\frac{3}{2}, -\frac{5}{4}\right)$
(iii) $\frac{5}{4} - \left(x + \frac{1}{2}\right)^2$; greatest at $\left(-\frac{1}{2}, \frac{5}{4}\right)$
(iv) $4(x+1)^2 + 11$; least at $(-1, 11)$
(v) $9\left(x - \frac{1}{3}\right)^2 - 12$; least at $\left(\frac{1}{3}, -12\right)$
(vi) $2\left(x - \frac{3}{4}\right)^2 - \frac{1}{8}$; least at $\left(\frac{3}{4}, -\frac{1}{8}\right)$
(vii) $-\frac{13}{8} - 6\left(x + \frac{1}{4}\right)^2$; greatest at $\left(-\frac{1}{4}, -\frac{13}{8}\right)$
(viii) $7\left(x - \frac{11}{4}\right)^2 - \frac{345}{28}$; least at $\left(\frac{11}{4}, -\frac{345}{28}\right)$

2 (i) 6 (ii) $-\frac{4}{5}$ (iii) $\frac{6}{5}$ (iv) $\frac{6}{25}$

3 (i) $X^2 + X + 4 = 0$ (ii) $4X^2 + 7X + 4 = 0$
(iii) $4X^2 + 7X + 4 = 0$ (iv) $8X^2 - 11X + 8 = 0$

Exercise 2(e)

1 $(1, 5), (5, 1)$

2 $(1, 1), \left(-\frac{4}{5}, -\frac{17}{10}\right)$

3 $(1, 2), \left(14\frac{11}{25}, -7\frac{3}{5}\right)$

4 $(\pm 4, \pm 1)$

5 $(3, 1), \left(\frac{6}{5}, -2\right)$

Miscellaneous examples 2

1 (i) $x^3 + 2x^2 + 3x + 2$ (ii) $6x^3 - 5x^2 - 8x + 3$
(iii) $3x^2 + 2xy - 8y^2$ (iv) $x^4 + 8x^2 + 16$
(v) $t^3 + t^2 - 14t - 8$ (vi) $t^4 - t^3 - 4t^2 - 11t - 3$

2 (i) $(x-2)(x+4)(x-1)$ (ii) $(x+4)(2x-3)(x+2)$
(iii) $(2x-1)(x+6)(x+7)$

3 (i) $(t+3)(t-5)(t+4)$ (ii) $(x-2)(x+7)(x-8)$

4 97

5 $f(-2) = -8 + 4k - 14 + 2 = 0$, $k = 5$

6 (i) $-3, 4$ (ii) 1 (iii) $-2, \frac{2}{3}$ (iv) $\pm\sqrt{3}$

(v) 14 (vi) $(x^2-4)(x^2-9) = 0$, hence $x = \pm 3, \pm 2$

7 (i) $4(x+1)^2 + 7$ (ii) $\left(x - 3\frac{1}{2}\right)^2 - 17\frac{1}{4}$

(iii) $5\frac{1}{2} - 2\left(x + 1\frac{1}{2}\right)^2$ (iv) $14\frac{1}{3} - 3\left(x - 1\frac{1}{3}\right)^2$

8 (i) $2x^2 + 11x + 19 = 0$ (ii) $8x^2 + 6x + 5 = 0$

(iii) $4x^2 + 3x + 18 = 0$ (iv) $10x^2 + 11x + 10 = 0$

(v) $4x^2 - 63x + 250 = 0$

9 (i) $(1, 1)$ or $\left(\frac{1}{5}, 1\frac{2}{5}\right)$ (ii) $(2, 6)$

(iii) $(\pm 3, \pm 1)$ (iv) $(1, 4)$ or $(31, -16)$

Revision problems 2

1 (i) $f(1)$ and $f(-1) \neq 0$

(ii) $f(1) = 2 = a + b$, $f(-1) = -10 = -a + b$, hence $b = -4$, $a = 6$.
The remainder $= 6x - 4$.

(iv) Note (iii) shows $f(x) - 2x$ has a factor $x^2 + 1$.

Hence $$\begin{aligned} x^5 - 3x^4 + 2x^3 - 2x^2 + x + 1 &= (x^2+1)(x^3 - 3x^2 + x + 1) \\ &= (x^2+1)(x-1)(x^2 - 2x - 1) \\ \therefore \quad x &= 1, 1 \pm \sqrt{2} \end{aligned}$$

2 $(x+3)(2x-1)(x+5) > 0$, $\therefore x > \frac{1}{2}$ or $-5 < x < -3$

3 $a = 100$, $b = 10\,000$

Exercise 3(a)

1 (i) $4 + 7 + 10 + 13 + 16 + 19$ (ii) $1 + 4 + 9 + 16$

(iii) $\frac{2}{3} + \frac{3}{4} + \frac{4}{5} + \frac{5}{6} + \frac{6}{7}$ (iv) $-1 + 8 - 27 + 64$

(v) $\frac{1}{2} + \frac{1}{4} + \frac{1}{8} + \frac{1}{16} + \frac{1}{32}$

2 (i) $\sum_{r=1}^{5}(2r-1)$ (ii) $\sum_{r=1}^{4}(3r+1)$ (iii) $\sum_{r=1}^{5}\frac{r}{r+4}$

(iv) $\sum_{r=1}^{5}(-1)^{r+1}r^2$ (v) $\sum_{r=1}^{5} 45 - 5r$ (vi) $\sum_{r=1}^{4}\frac{1}{3^{r-1}}$

(vii) $\sum_{r=1}^{n} n + (r-1)k$

These expressions are not always unique. For example, (vi) could be written $\sum_{r=0}^{3}\frac{1}{3^r}$.
You should always evaluate what you have written down to check that it does give the required series.

Exercise 3(b)

1 1290

2 (i) 111 (ii) 23 (iii) 41 (iv) $3n$

3 (i) 3667 (ii) 2542 (iii) 351 (iv) 1640

4 Smallest = 207, largest = 342; 16; 4392.

5 930

6 3, 7, 11, 15

7 Solve $\frac{n}{2}(3n+1) > 2000$ to give $n > 36.3$, that is, 37 terms required.

Exercise 3(c)

1 $4(1.5^{20} - 1) = 13\,297$

2 (i) 11 (ii) 16 (iii) 10

3 1.5, 2

4 3, 4, $5\frac{1}{3}$

Exercise 3(d)

1 (i) 1 (ii) $\frac{3}{4}$ (iii) 6 (iv) $\frac{2}{15}$

2 (i) 7 (ii) 14 (iii) 9 (iv) 6

Exercise 3(e)

1 (i) $a^3 + 6a^2b + 12ab^2 + 8b^3$

(ii) $x^4 - 8x^3y + 24x^2y^2 - 32xy^3 + 16y^4$

(iii) $16a^4 + 96a^3t + 216a^2t^2 + 216at^3 + 81t^4$

(iv) $x^4 + 8x^2 + 24 + \frac{32}{x^2} + \frac{16}{x^4}$

(v) $27x^3 - 27 + \frac{9}{x^3} - \frac{1}{x^6}$

2 (i) $9120a^{17}b^3$ (ii) $-414720x^7y^3$ (iii) $448x^2$

3 (i) $8064x^5$ (ii) $-1548288x^5$ (iii) $-489888x^5$

4 ${}^{18}C_6 \times 2^{12} = 76\,038\,144$

Exercise 3(f)

1 (i) $1 - 4x + 12x^2 - 32x^3$; $-\frac{1}{2} < x < \frac{1}{2}$

(ii) $1 - x - x^2 - \frac{5}{3}x^3$; $-\frac{1}{3} < x < \frac{1}{3}$

(iii) $4 - \frac{1}{2}x - \frac{1}{32}x^2 - \frac{1}{256}x^3$; $-4 < x < 4$

(iv) $1 - x + 2x^2 - \frac{14}{3}x^3$; $-\frac{1}{3} < x < \frac{1}{3}$

(v) $1 - 3x + 5x^2 - 7x^3$; $-1 < x < 1$

(vi) $1 + \frac{3}{2}x + \frac{11}{8}x^2 + \frac{23}{16}x^3$; $-1 < x < 1$

2 0.09%

3 $1 - 1 - \frac{5}{3}x - \frac{25}{9}x^2 : x = \frac{1}{10} : \frac{36}{29}$

Miscellaneous examples 3

1 (i) $1 + 7 + 17 + 31 + 49$ (ii) $-\frac{1}{2} + \frac{2}{3} - \frac{3}{4} + \frac{4}{5} - \frac{5}{6}$

2 (i) $\sum_{r=0}^{4} 2r + 1$ (ii) $\sum_{t=1}^{8} t^2$ (iii) $\sum_{r=1}^{4} \frac{r}{r+1}$ (iv) $\sum_{k=0}^{3} n + k$

3 (i) 20 (ii) 37 (iii) 40 (iv) 201

4 (i) 12 (ii) 9 (iii) 13 (iv) 9

5 (i) 120 (ii) -115 (iii) 365 (iv) $1 - \left(\frac{1}{2}\right)^{10}$

(v) $1 - \left(\frac{1}{4}\right)^{10}$ (vi) $5\left(1 - \left(\frac{1}{5}\right)^{10}\right)$

6 (i) $81 + 432x + 864x^2 + 768x^3 + 256x^4$.

(ii) $243x^5 - 810x^4y + 1080x^3y^2 - 720x^2y^3 + 240xy^4 - 32y^5$

(iii) $16x^4 - 96x^2 + 216 - \dfrac{216}{x^2} + \dfrac{81}{x^4}$

7 $(2t)^{24} \times {}^{24}C_8 \left(\dfrac{-5}{2t^3}\right)^8 = 5^8 \times 2^{16} \times {}^{24}C_8$

8 (i) $1 + 12x + 60x^2$ all x

(ii) $\frac{1}{9}\left(1 + \frac{4}{3}x + \frac{4}{3}x^2\right)$ $-1.5 < x < 1.5$

(iii) $1 + \frac{1}{4}x + \frac{3}{32}x^2$ $-2 < x < 2$

(iv) $1 + 2x + 2x^2$ $-1 < x < 1$

(v) $2\left(1 + \frac{1}{8}x - \frac{1}{218}x^3\right)$ $-4 < x < 4$

(vi) $1 + 12x + 90x^2$ $-\frac{1}{3} < x < \frac{1}{3}$

(vii) $1 + \frac{1}{3}x - \frac{28}{9}x^2$ $-\frac{1}{4} < x < \frac{1}{4}$

(viii) $\frac{1}{3}\left(1 + \frac{2}{9}x + \frac{2}{27}x^2\right)$ $-\frac{9}{4} < x < \frac{9}{4}$

9 $1 + 2x + 6x^2$: $x = \frac{1}{100}$

$\left(\frac{96}{100}\right)^{-\frac{1}{2}} = \frac{5103}{5000}$

$\therefore \quad \dfrac{10}{4\sqrt{6}} = \frac{5000}{5103}$, that is, $\sqrt{6} = \frac{5103}{2000}$

Revision problems 3

1 (i) £29.13 (ii) 507.39

2 (i) Sum $= \frac{n}{2}(a + \ell)$ (ii) 25 (iii) 10, $12\frac{1}{2}$, 15

(iv) Using $\frac{n}{2}(2a + (n-1)d) = 500$, with $a = 6$, $d = 2$, you get the quadratic equation

$$n^2 + 5n - 500 = 0$$
$$(n + 25)(n - 20) = 0$$
$$\text{hence} \quad n = 20$$

(v) 44 cents

3 $20 \times 1.1^{n-1}$, $n - 1 > \frac{\ln 5}{\ln 1.1}$, hence $n = 18$

4 $2 + 4 \times 0.75 + 4 \times 0.75^2 + \ldots 4 \times 0.75^n$
$= 2 + 12(1 - 0.75^n) \quad n \to \infty$
distance = 14 metres.

Exercise 4(a)

1 (i) 3.16, -3 (ii) 5.83, $\frac{3}{5}$ (iii) 4.47, 2

(iv) $3.61t$, -1.5 (v) $\sqrt{p^2 + 2q^2 - 2pq}$, $\frac{q}{(p-q)}$

(v) 5.17, -3.85 (vii) 1.03, $\frac{1}{4}$ (viii) $(a + b)\sqrt{2}$, -1

2 (i) $\sqrt{6}$ (ii) $2\sqrt{3}$ (iii) $\sqrt{21}$ (iv) $\sqrt{29}t$

(v) $\frac{\sqrt{145}}{4}$ (vi) $\frac{\sqrt{806}}{60}$ (try and get this exactly) (vii) $\sqrt{14}y$

(viii) $\sqrt{17}$

3 (i) $(2, 0.5, 1.5)$ (ii) $(0, 2, -1)$ (iii) $(3.5, 1, 1)$

(iv) $(2t, 1.5t)$ (v) $(\frac{3}{8}, \frac{1}{2})$ (vi) $(\frac{11}{24}, \frac{5}{12}, \frac{27}{40})$

(vii) $(0, \frac{1}{2}y, 2\frac{1}{2}y)$ (viii) $(2.5 - t, 2 + t, 2)$

4 $(\frac{8}{5}, 1, \frac{13}{5})$

5 1 unit^2

6 (i) $\frac{5 + 2x}{7} = 1, \quad \therefore \; x = 1$

$\frac{2y}{7} = -1, \quad \therefore \; y = -3.5$

$\frac{10 + 2z}{7} = -2, \quad \therefore \; z = -12$, hence C is $(1, -3.5, -12)$

(ii) Since $x = 1$ for each point, you can consider it a two-dimensional problem, so the area $= |\frac{1}{2}(y_1(z_2 - z_3) + y_2(z_3 - z_1) + y_3(z_1 - z_2))| = 6.25$ units^2.

(iii) $CD = \sqrt{4.5^2 + 13^2} = 13.76$ $\therefore$ if h is the required distance,

$\frac{1}{2} \times 13.76h = 6.25$

$h = 0.908$

7 (i) $\dfrac{7\sqrt{13}}{13}$ (ii) 2 (iii) $\dfrac{4\sqrt{2}}{5}$

Exercise 4(b)

1 (i) $y + 2x = 7$ (ii) $y = 4x - 5$ (iii) $2y + 4x = 3$
(iv) $y = 2px - 2p^2 + q$ (v) $12y + 6x + 5 = 0$

2 (i) $2y + x = 5$ (ii) $y + 2x = 3$ (iii) $2y = 2x + 3$
(iv) $pqy + x = p + q$ (v) $2y = (t_1 + t_2)x - 2at_1t_2$

3 (i) 3 (ii) $\frac{2}{3}$ (iii) $-\frac{3}{2}$ (iv) $-\frac{5}{7}$

4 $3y + 4x = 7$

Exercise 4(c)

1 $k = 5$, $\alpha = 1.3$ **2** $a = 5$, $k = 7$ **3** $k = 10$, $q = -5$ **4** $f = 60$

Miscellaneous examples 4

1 (i) 11.2 (ii) 8.60

2 (i) -5 (ii) $-\dfrac{1}{pq}$

3 (i) $(3, -1.5)$ (ii) $(-\frac{1}{12}, \frac{7}{24})$ (iii) $(2, -0.5, 2)$

4 The sides are of length $\sqrt{2}$, $\sqrt{17}$, $\sqrt{11}$: area $= 2.12$ unit^2

5 $(3, 7)$: 1 unit^2

6 3.77

7 $C(4, 7)$, $D(5, 4)$: $C(-2, 5)$, $D(-1, 2)$

8 $a = 3$, $k = 8$

Revision problems 4

1 $n = 1.5$, $A = 5.5 \times 10^{-4}$

2 a) Plot t against $\ln N$, $k = 0.005$, $N_0 = 2.1$, b) 4.7 million

Exercise 5(a)

1 $12x^2$

2 $28x^3$

3 $3x^2 - 2$

4 $\frac{3}{2}\sqrt{x}$

5 $-\dfrac{1}{x^3}$

6 $-\frac{3}{8}x^{-\frac{3}{2}}$

7 $10(2x+1)^4$

8 $24(3x+1)$

9 $\dfrac{3x^2}{2}(1-x^3)^{-\frac{3}{2}}$

10 $1 - \dfrac{1}{x^2}$

11 $-\dfrac{3x}{2}(x^2-1)^{-4}$

12 $\frac{1}{6}x^{-\frac{1}{2}}\left(1+x^{\frac{1}{2}}\right)^{-\frac{2}{3}}$

13 $2x - \dfrac{1}{x^2}$

14 $-\dfrac{2}{x^3} + \dfrac{6}{x^4} - \dfrac{4}{x^5}$

15 $\dfrac{2(2x+1)^2(x-1)}{3x^3}$

16 18

17 1.5

18 $0, \frac{1}{2}$

19 $6q^2(1+q^2)^{-\frac{1}{2}}$

20 -22.5

Exercise 5(b)

1 $x(x+1)^2(5x+2)$

2 $(1+x)^{-2}$

3 $(x+1)(x+2)^2(5x+7)$

4 $\dfrac{1-x}{(1+x)^3}$

5 $4(x+1)^3(2x^2-1)^2(5x^2+3x-1)$

6 $\dfrac{1-x^2}{(1+x^2)^2}$

7 $\frac{1}{2}x^{-\frac{1}{2}} + 1 + \frac{3}{2}x^{\frac{1}{2}}$

8 $\dfrac{2+\sqrt{x}-x^{-\frac{1}{2}}}{2(1+\sqrt{x})^2}$

9 $2\left(x+1-\dfrac{1}{x^2}-\dfrac{1}{x^3}\right)$

10 $\dfrac{2}{\sqrt{x}(1-\sqrt{x})^3}$

11 $\dfrac{6x^2+3x-1}{2x^2}\sqrt{\dfrac{x}{x+1}}$

12 $\dfrac{x^2(x^2+3)}{(1-x^2)^3}$

13 $-\dfrac{2(x+2)}{(x-3)^3}$

14 $\dfrac{2+x^{\frac{1}{2}}-2x^{-\frac{1}{2}}}{(1+\sqrt{x})^2}$

15 $-\dfrac{1}{(1+x)^{\frac{3}{2}}\sqrt{1-x}}$

Exercise 5(c)

1 (i) $\dfrac{dy}{dx} = 3x^2 = 3 \times 3^2 = 27,\ (y-28) = 27(x-3),\ \therefore\ \ y = 27x - 53$

(ii) $\dfrac{dy}{dx} = \frac{1}{6},\ 6y = x + 9$

(iii) $\dfrac{dy}{dx} = 12,\ y = 12x - 40$

(iv) $\dfrac{dy}{dx} = -\frac{3}{4},\ 4y + 3x = 22$

(v) $\dfrac{dy}{dx} = -\frac{1}{4},\ 4y + x = 3$

(vi) $\dfrac{dy}{dx} = -27,\ y + 27x + 81 = 0$

2 (i) $\frac{dy}{dx} = 6x = 12$, hence the gradient of the normal $= -\frac{1}{12}$, $(y - 10) = -\frac{1}{12}(x - 2)$,

$12y + x = 122$

(ii) $y = 3x - 2$

(iii) $y = x + 1$

(iv) $2y + 3x + 5 = 0$

3 (i) 3.0001 (ii) 3.9997

4 $4y + 9x - 16 = 0$

Exercise 5(d)

1 (i) $(\frac{5}{12}, \frac{23}{24})$; minimum

(ii) $(0, 1)$; inflexion

(iii) $(0, 0)$ maximum, $(-\sqrt{2}, -4)$ and $(\sqrt{2}, -4)$ minima

(iv) $(1, 2)$ minimum, $(-1, -2)$; maximum

(v) $x = 2 + \sqrt{5}$ minimum, $x = 2 - \sqrt{2}$; maximum

(vi) $(0.08, 1.96)$; minimum, $(-1.08, -5.59)$; maximum

2 1.5 rev/sec

3 $S = 2\pi r^2 + 2\pi rh, \pi r^2 h = 200$

$\therefore \quad S = 2\pi r^2 + \frac{400}{r}, r = 3.17$

4 $-\frac{2}{27}$

5 1.97 cm × 2.36 cm × 2.15 cm

Exercise 5(e)

1 1.2π cm^2/s **2** 20cm^2/s **3** (a) 70 (b) 10

4 $\frac{7}{3}$ cm × 7 cm × $\frac{7}{2}$ cm

5 (i) 12 m (ii) $4\frac{2}{3}$ sec (iii) $-\frac{49}{3}$ m/s at $t = 2\frac{1}{3}$

Miscellaneous examples 5

1 (i) $\frac{-3}{4x^4}$ (ii) $\frac{-6}{(x+1)^3}$

(iii) $\frac{-2x}{\sqrt{1 - 2x^2}}$ (iv) $\frac{2 + x}{2(1 + x)^{\frac{3}{2}}}$

(v) $\frac{4x^3 - 3x^2 + 2}{(1 - x^3)^2}$ (vi) $-\frac{1}{2}x^{-\frac{3}{2}} + 2x^{-3} + \frac{5}{2}x^{-\frac{7}{2}}$

2 $y + 9x - 8 = 0$, tangent, $9y - x + 10 = 0$ normal

3 $2x + 3y = 18\frac{1}{9}$

4 $(\frac{1}{2}, -8)$; minimum, $(-\frac{1}{2}, 8)$; maximum

5 $46.4\ \text{cm}^3/\text{s}$

6 $\sqrt{\dfrac{100}{3\pi}}$ cm

7 At B, $v = \dfrac{ds}{dt} = 0, \quad \therefore \quad t = 12$, hence $AB = 288$

Revision problems 5

1 (b) $100x - \dfrac{4x^3}{3}$ (c) $333\frac{1}{3}\ \text{cm}^3$

2 (i) $8x + \dfrac{1}{x^2}, 8 - \dfrac{2}{x^3}$ (ii) $-\frac{1}{2}$, 24 (iii) $-\frac{1}{2}, 1$

(iv) (a) behaves like $\dfrac{1}{x}$; (b) behaves like $4x^2 - 3$

Exercise 6(a)

1

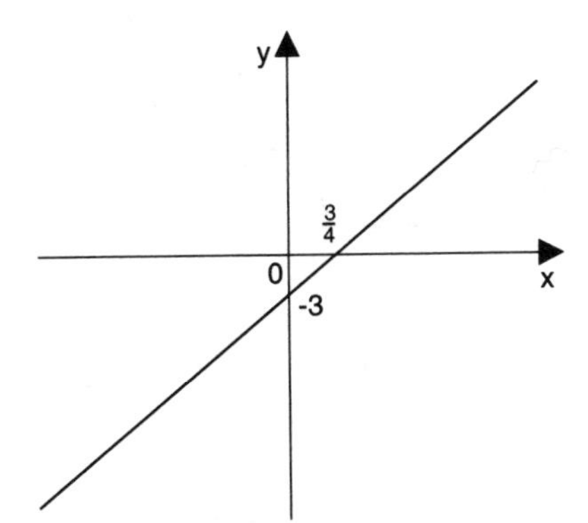

2

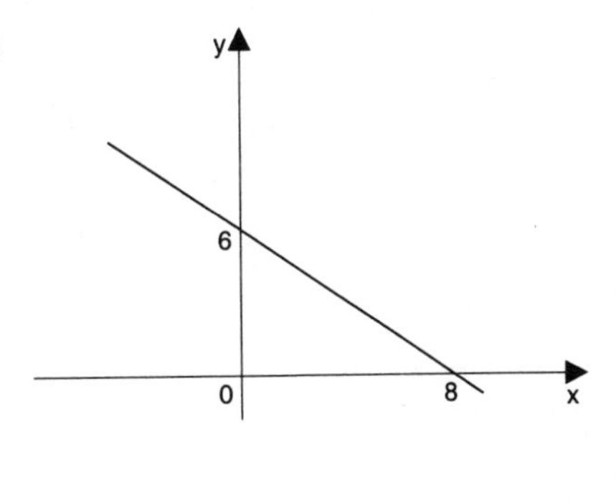

3

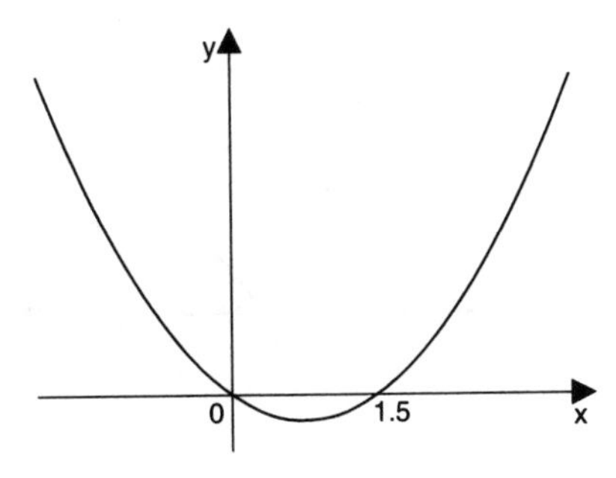

4

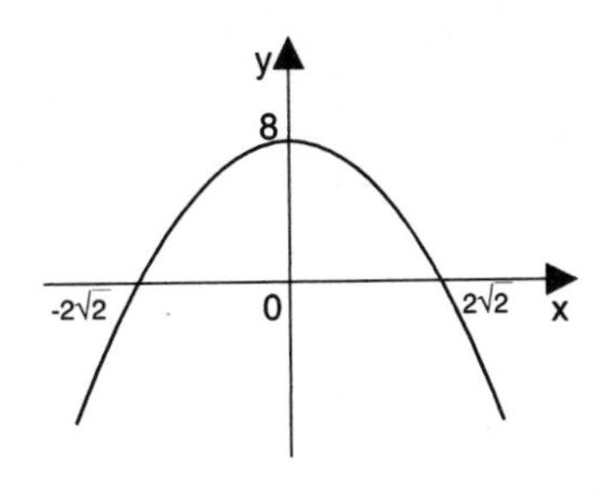

5

y

8

0 2 4 x

6

y

3

0 x

7

y

0 x

8

y

0 1.5 x

9

y

1

0 x

10

y

7

0 x

11

12

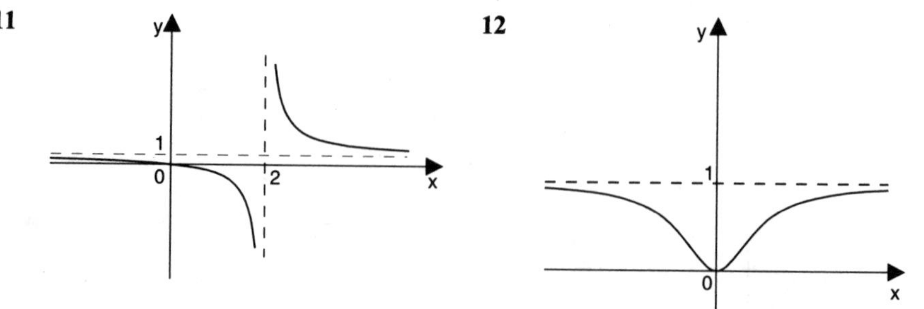

Exercise 6(b)

1 (i)

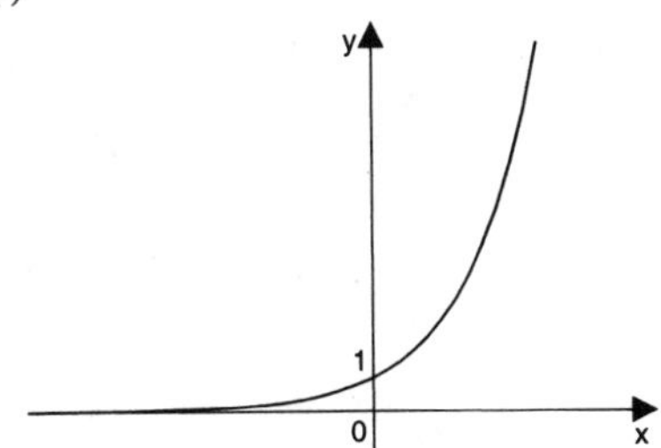

(ii)

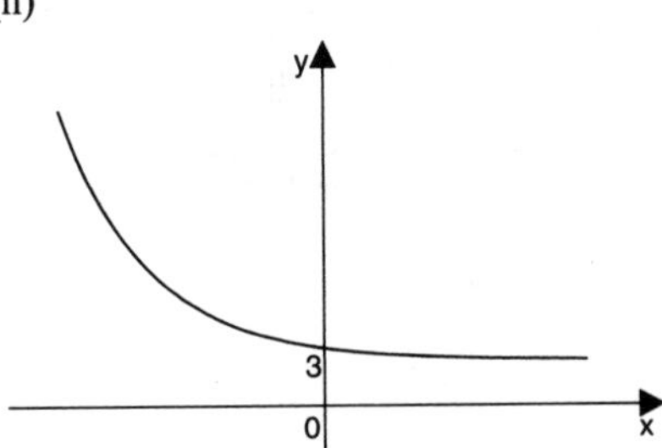

(iii)

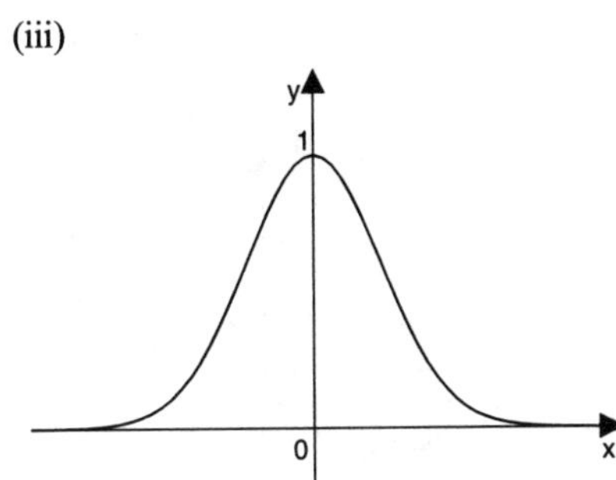

(iv)

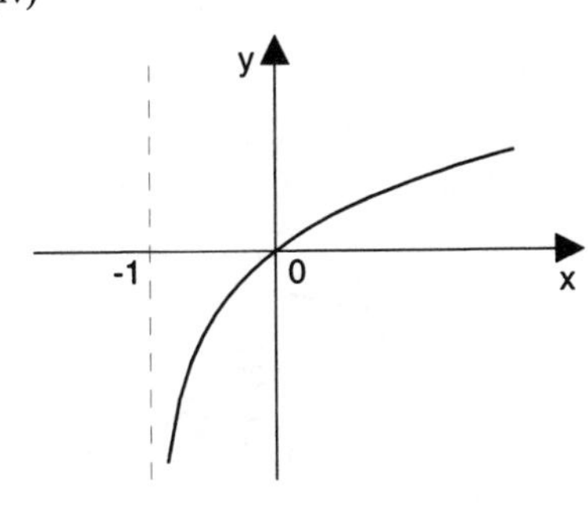

(v)

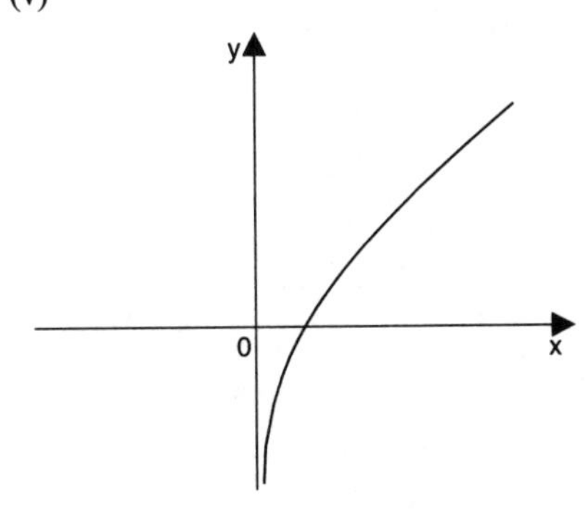

(vi)

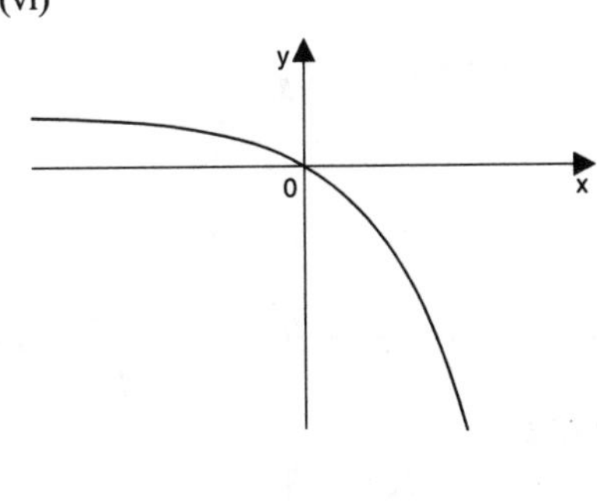

2 (i)

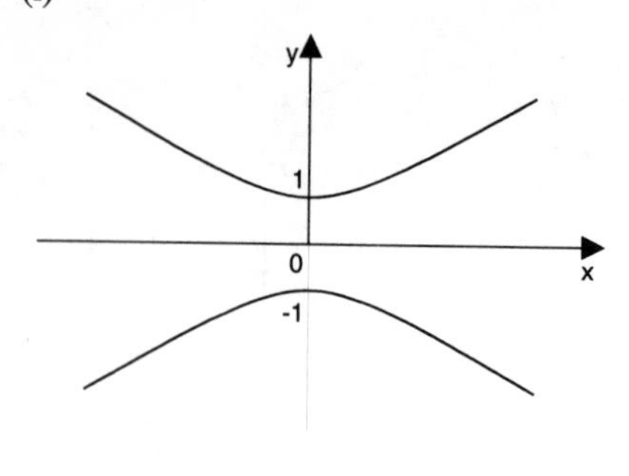

(ii)

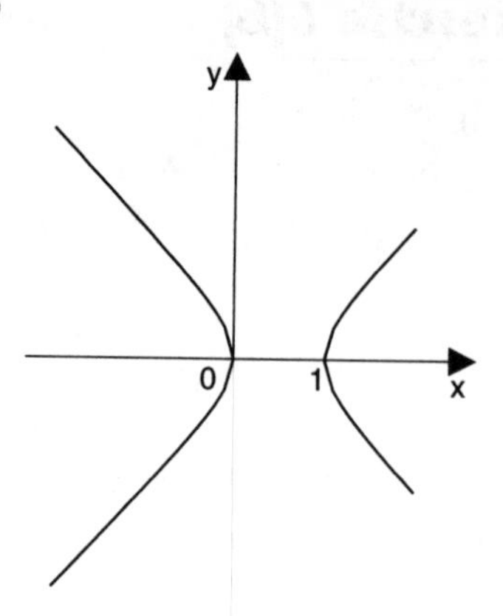

(iii)

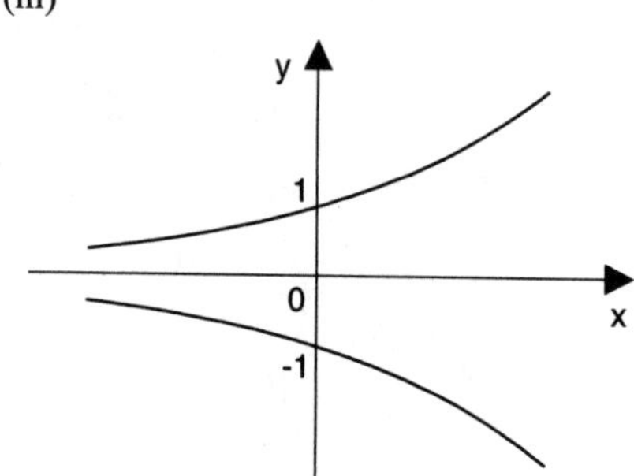

(iv)

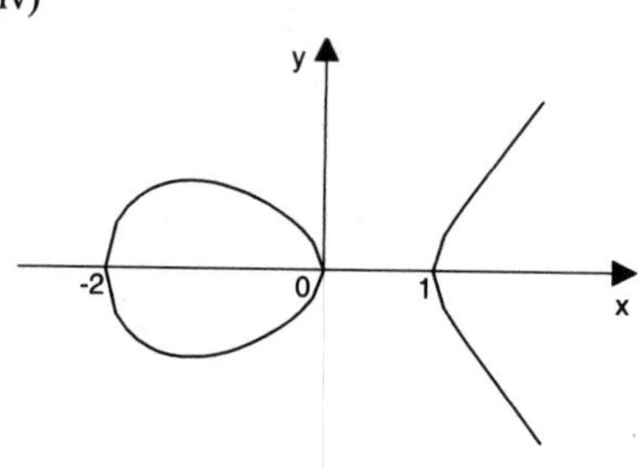

(v)

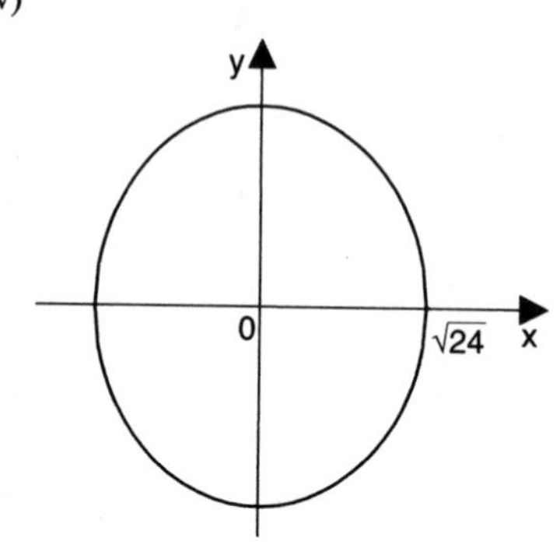

(vi)

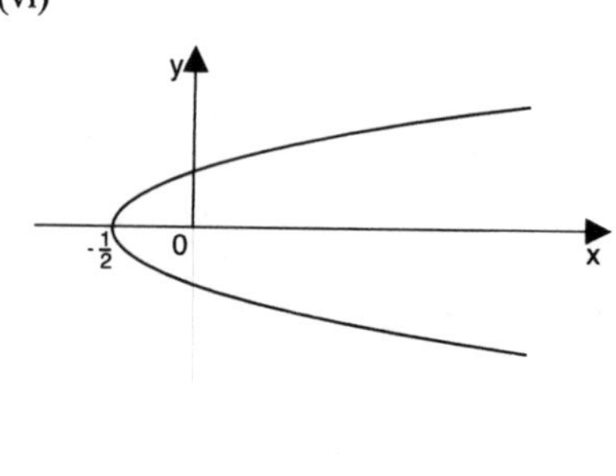

3 (i) $x^2 + y^2 = 16$ (ii) $x^2 + y^2 - 2y = 8$

(iii) $x^2 - 2x + y^2 + 4y = 31$ (iv) $16x^2 + 64x + 16y^2 + 32y + 79 = 0$

(v) $x^2 + 6x + y^2 - 12y = 19$

4 (i) $(0, 0)$; 6 (ii) $(0, 0)$; $\sqrt{8}$ (iii) $(0, 0)$; $\sqrt{12.5}$

(iv) $(-1, 2)$; 5 (v) $(2, 3)$; 4

5 (i) 10π unit^2

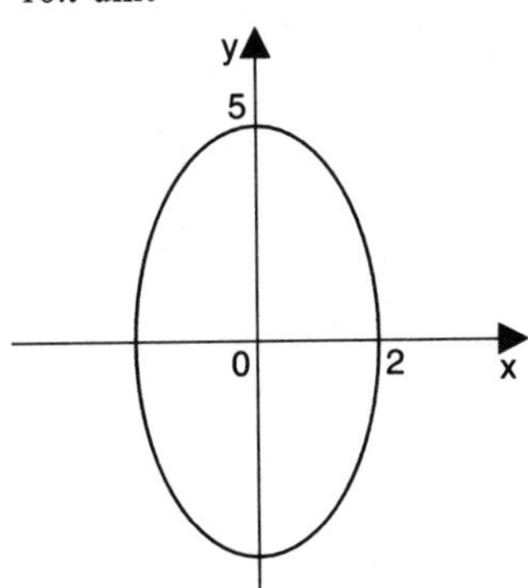

(ii) 6π unit^2

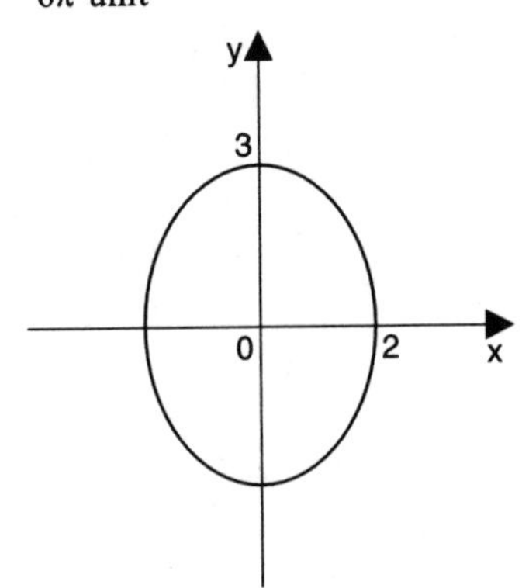

(iii) $2\sqrt{6}\pi$ unit^2

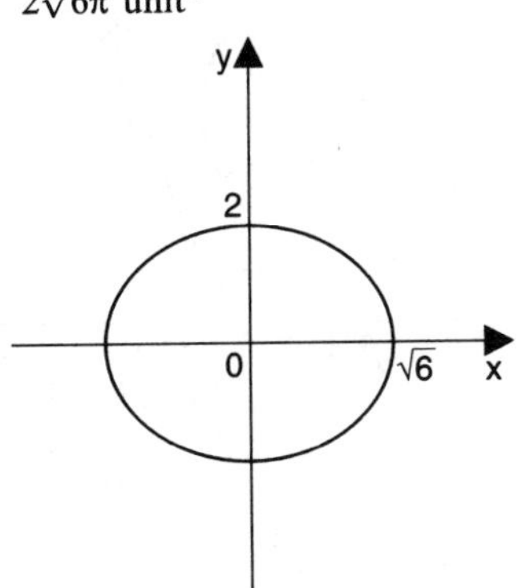

(iv) π unit^2

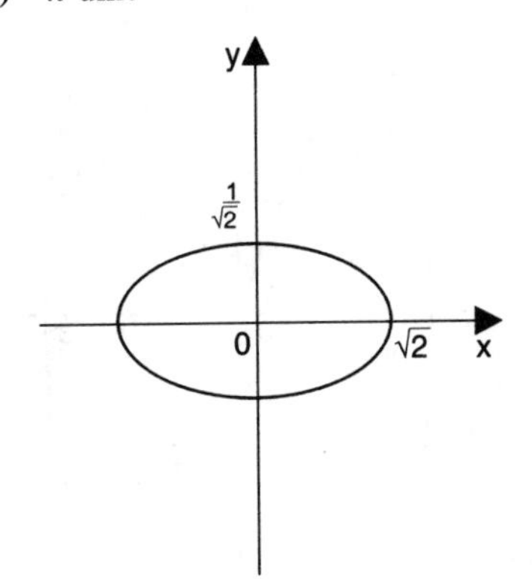

Exercise 6(c)

1

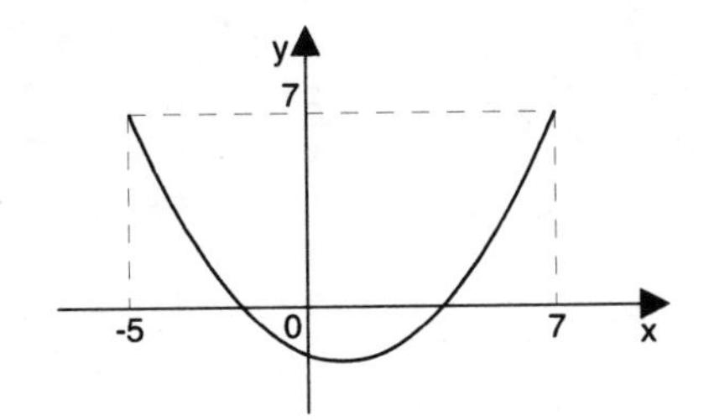

2

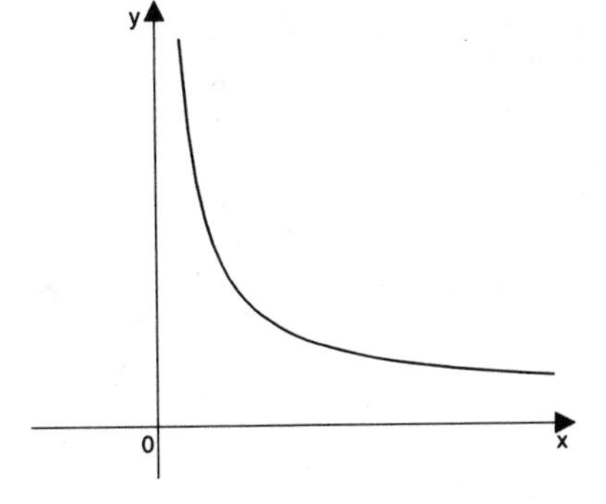

3

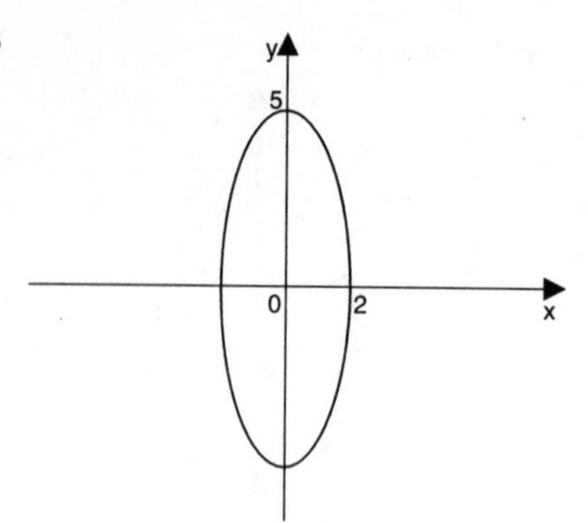

4

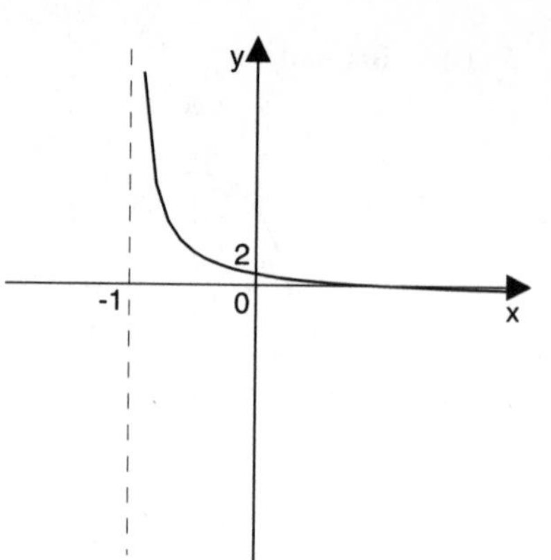

5

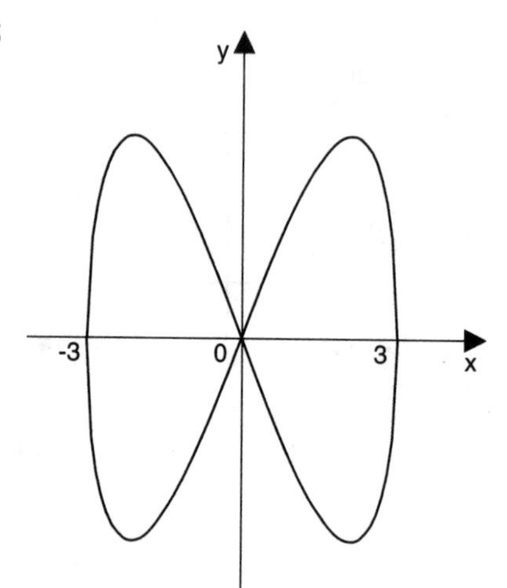

6

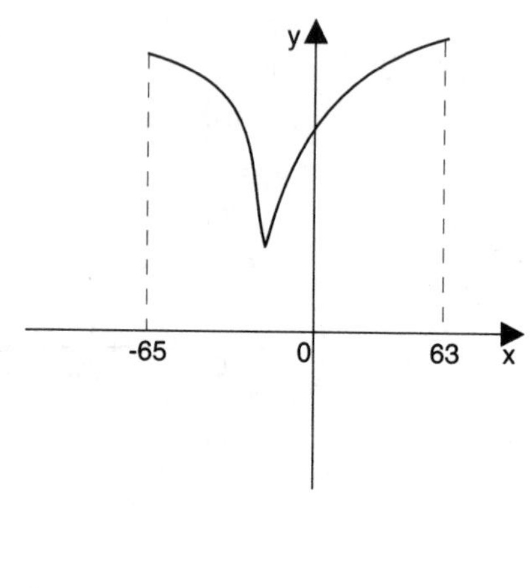

Miscellaneous examples 6

1 (i)

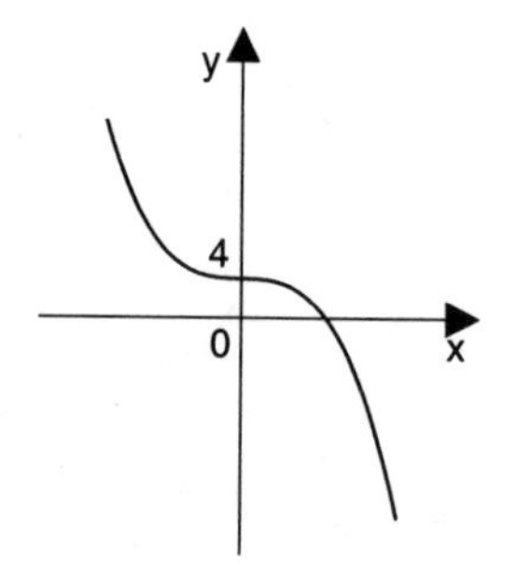

(ii)

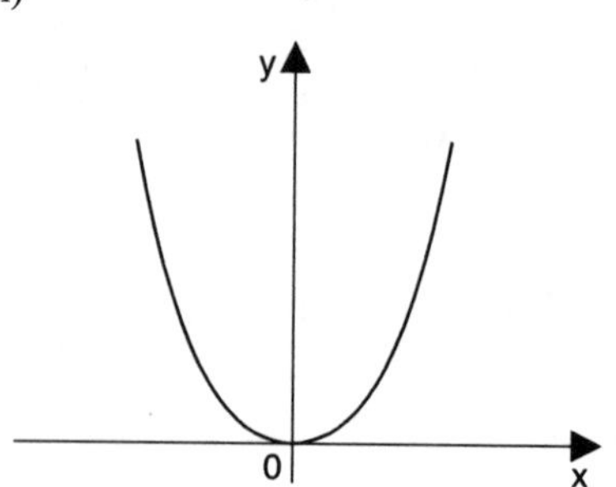

(iii)

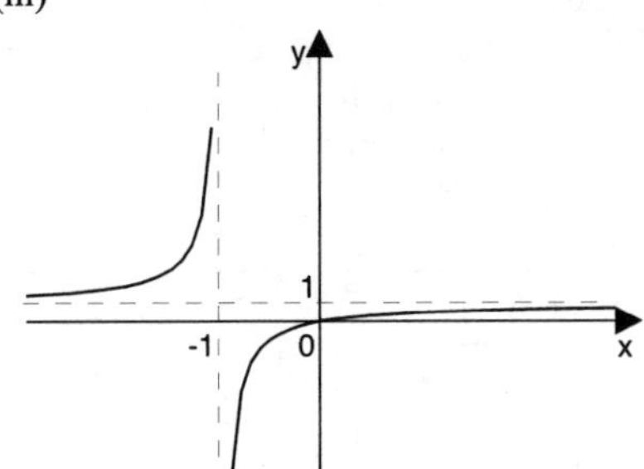

(iv)

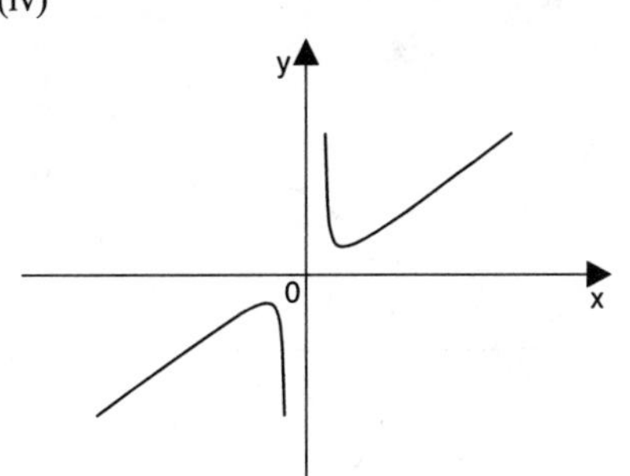

(v)

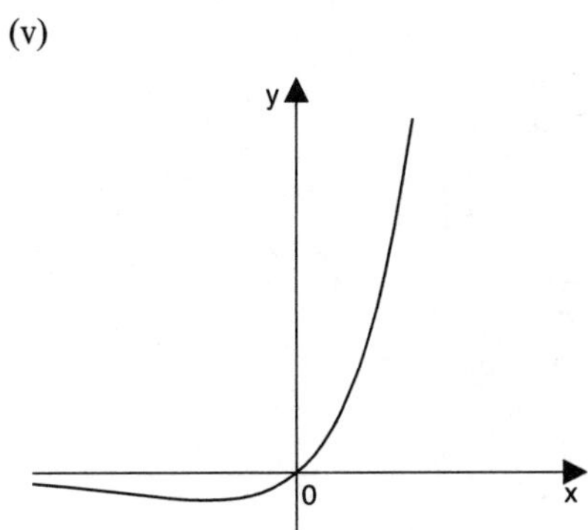

(vi)

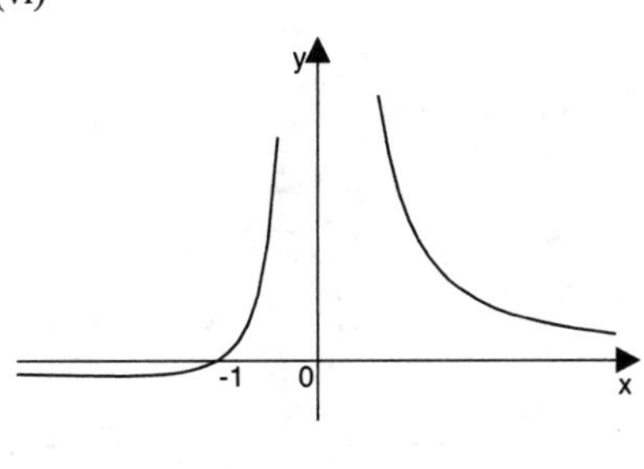

2 (i)

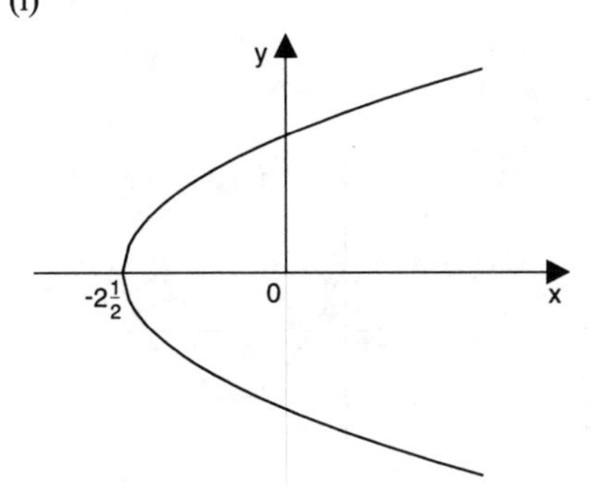

(ii)

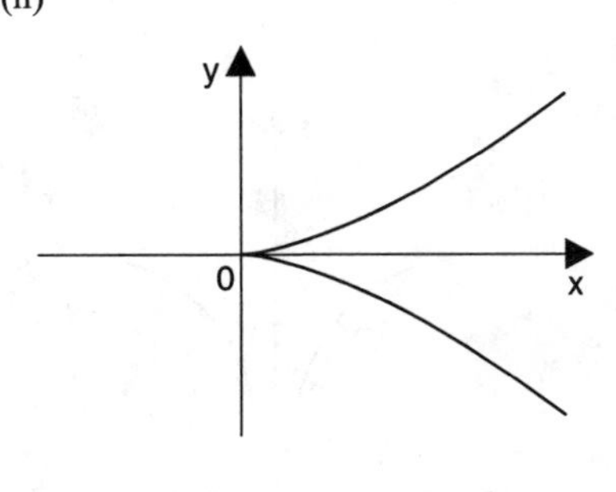

(iii)

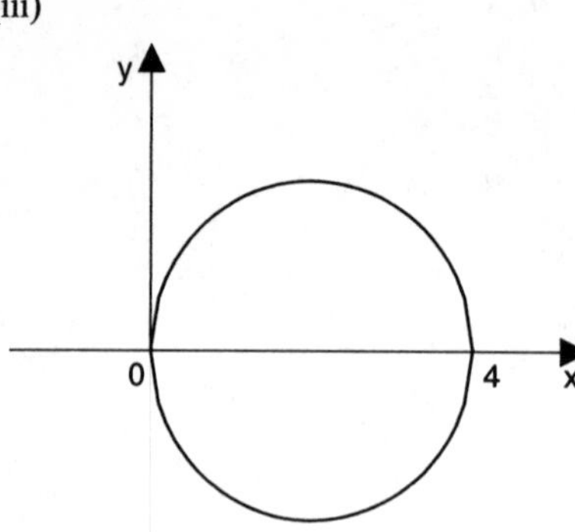

(iv)

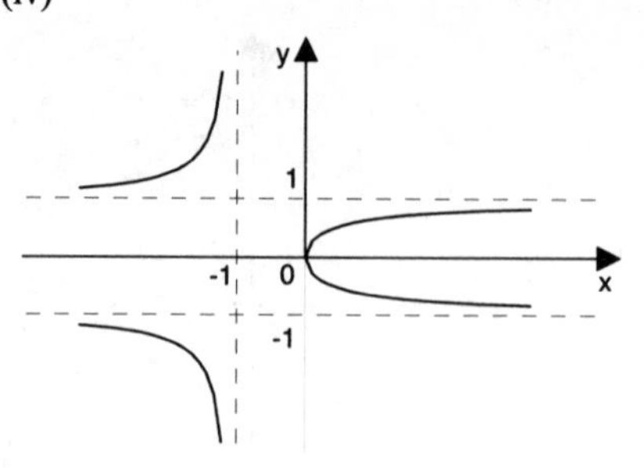

(v)

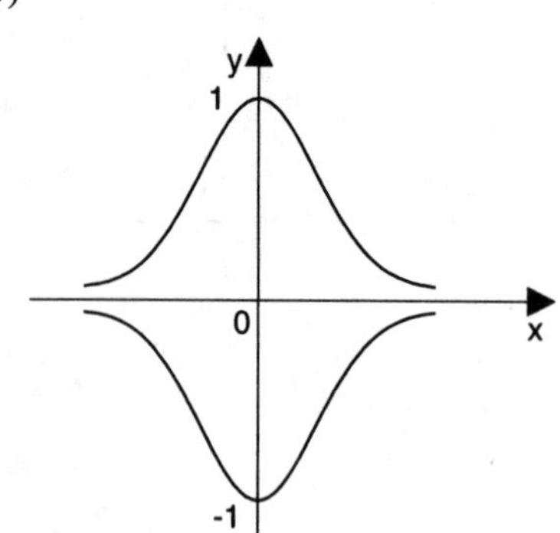

3 (i)

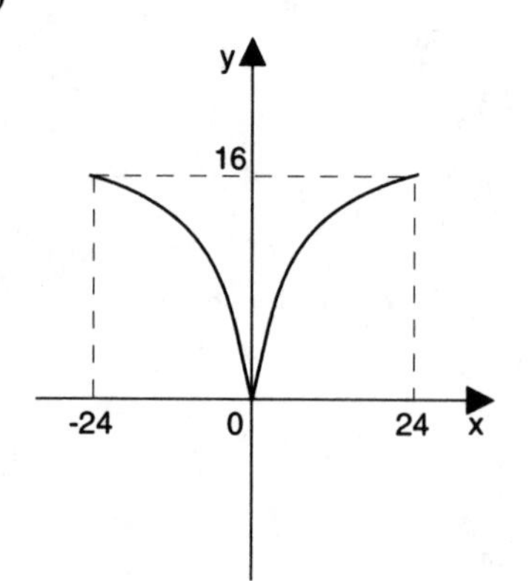

(ii)

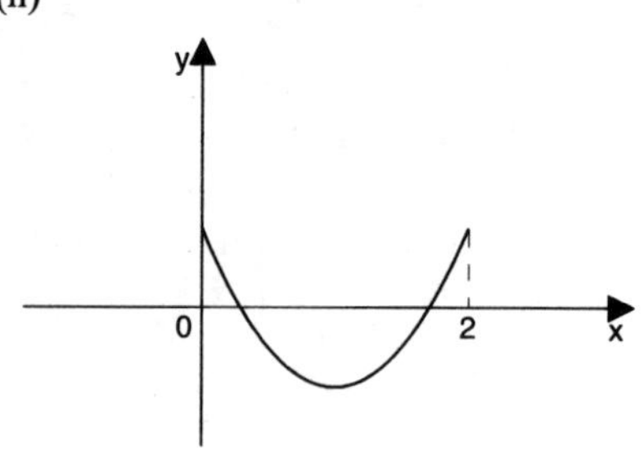

(iii)

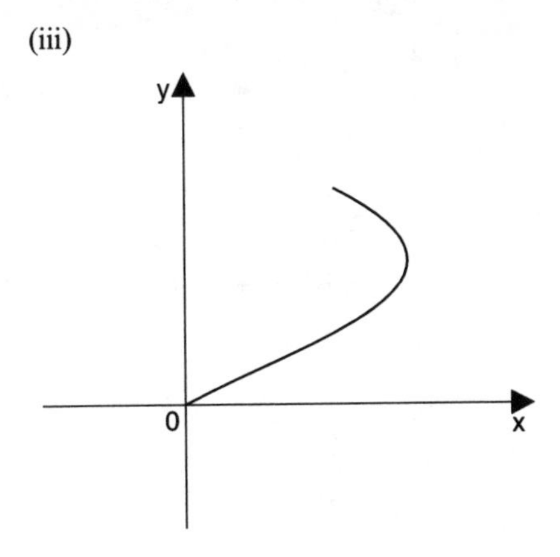

Revision Problems 6

1 y= 0, $x = -2$; $\left(2, \frac{3}{8}\right)$ is a maximum

(i)

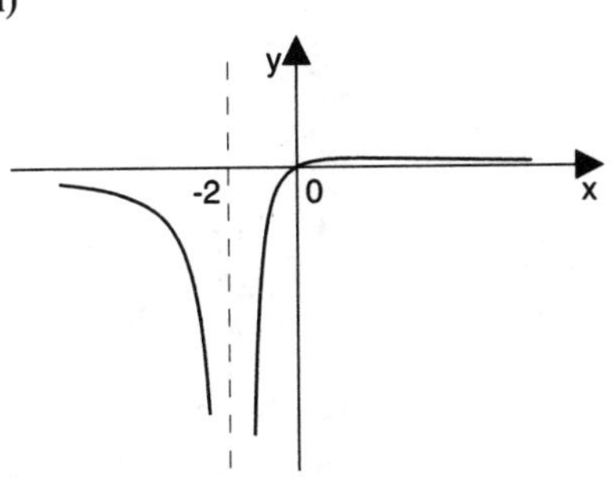

(ii)

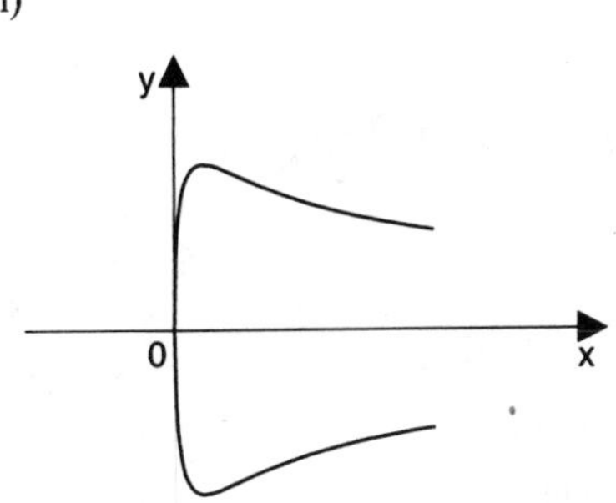

2 (i)

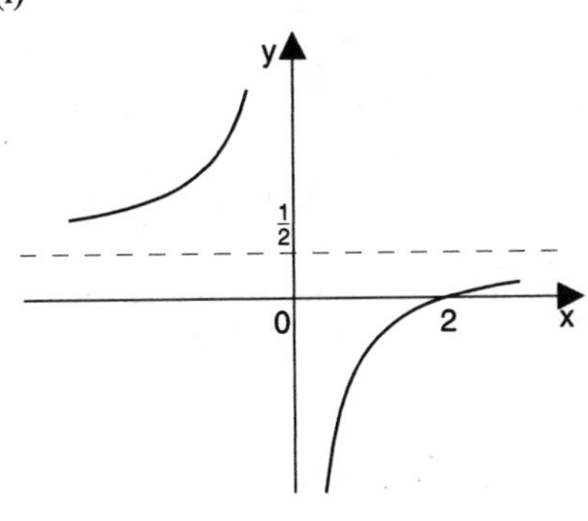

(ii)

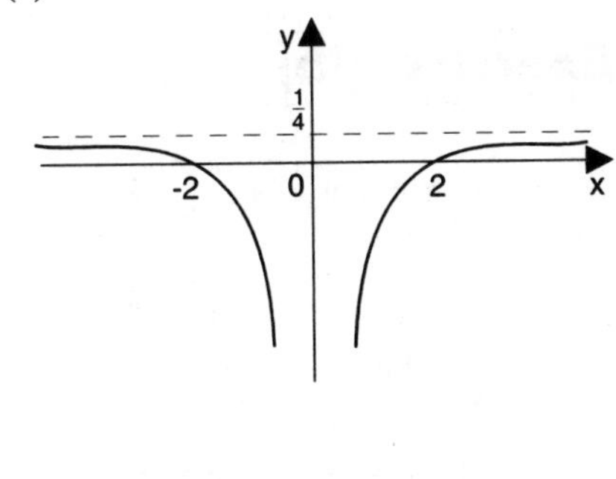

(iii)

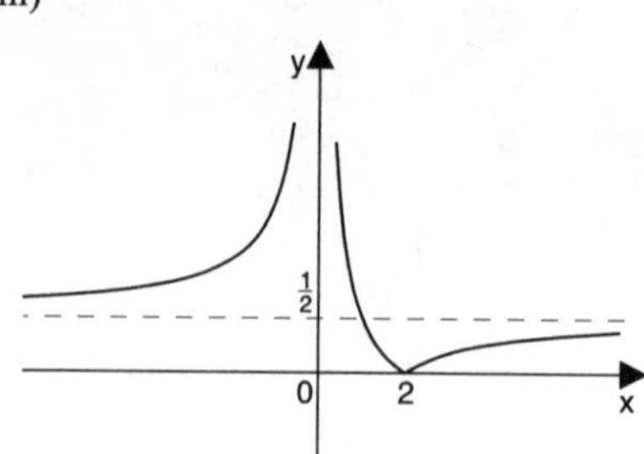

(iv)

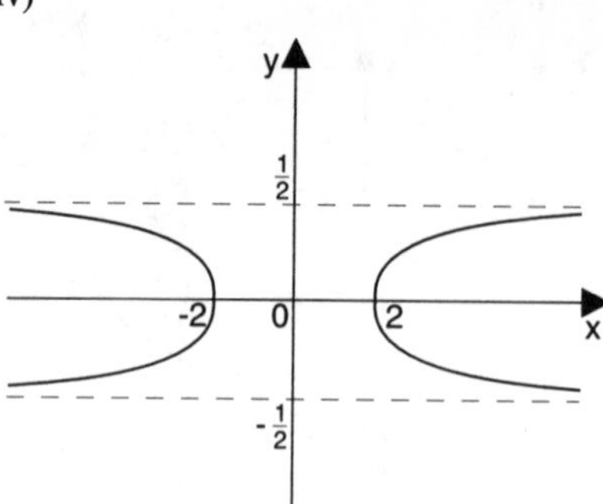

3 $(x^2 + y^2 - 4x - 6) - (x^2 + y^2 + 2x + 4y - 8) = 0$
that is, $-6x - 4y + 2 = 0$
or $3x + 2y = 1$

4 $x^2 = 36\sin^2\theta\cos^2\theta = \dfrac{36y^2}{16}\left(1 - \dfrac{y^2}{16}\right)$
that is, $256x^2 = 36y^2(16 - y^2)$
or $64x^2 = 9y^2(16 - y^2)$

Exercise 7(a)

1 $-1 \leq y \leq 7$
2 $4 \leq y \leq 20$
3 $-4 \leq y \leq 12$
4 $\frac{1}{4} \leq y \leq 1$
5 $-44 \leq y \leq 4$
6 $-4 \leq y \leq 1$
7 $0 \leq y \leq 15$
8 $-3 \leq y \leq -1\frac{2}{3}$
9 Neither
10 neither
11 Even
12 Neither
13 Odd
14 Odd
15 Even

Exercise 7(b)

1 No inverse, not 1 : 1
2 $g^{-1}(x) = \dfrac{x-4}{3}$
3 $h^{-1}(x) = \sqrt[3]{x-1}$
4 $k^{-1}(x) = \dfrac{1}{x} - 1$
5 $\ell^{-1}(x) = \sqrt{\dfrac{4-x}{2}}$
6 No inverse, not 1 : 1
7 $n^{-1}(x) = \dfrac{\ln(1-x)}{\ln 2}$
8 $p^{-1}(x)$ exists, but not possible to write down an explicit function.
9 $q^{-1}(x) = \ln(x)$
10 $r^{-1}(x) = e^x - 1$

Exercise 7(c)

1 $(2x+1)^2$

2 fg(9) does not exist.

3 g gives values greater than -1

4 $2(2-x^2)+1$

5 $(x+1)$

6 $\dfrac{1}{(1-x)}$

Exercise 7(d)

1 f$(2x)$ halves all x-values

f$(x-2)$ translates curve $+2$ parallel to $0x$

2f(x) stretches curve by factor 2 parallel to $0y$

For example

(vii)

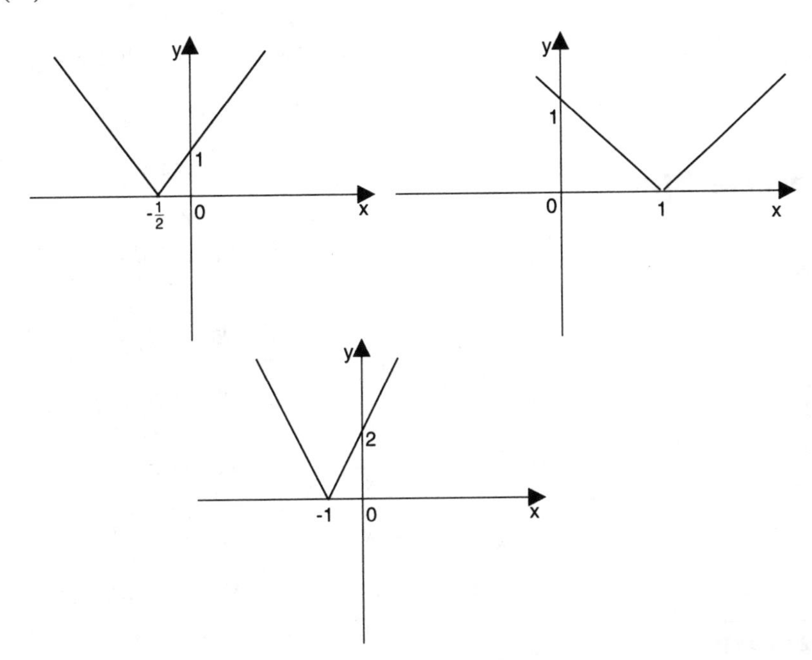

2 (i) $(-1,1)$, $(0,-1)$, $(1,0)$

(ii) $(2,1)$, $(3,-1)$, $(4,0)$

(iii) $(1,3)$,$(2,-3)$, $(3,0)$

(iv) $(1,-1)$, $(2,-3)$, $(3,-2)$

Exercise 7(e)

1 $x < -1.5$

2 $2 < x < 3$

3 $-1.5 \leq x \leq 2$

4 $-4 < x < 0$

5 $x > 2$ or $x < 0$

6 $-1 < x < 3$ or $x < -5$

7 $x \geq 1$ or $-2 \leq x \leq -1$

8 $x > -1$

Miscellaneous examples 7

1 $y \geq -4$, $x = \pm\sqrt{14}$

2 (i) $y \geq \ln 2$ (ii) $y \geq 0$
(iii) $-24 \leq y \leq 24$ (iv) $0 < y \leq e^{-1}$

3 (i) $f^{-1} : x \mapsto \frac{1}{2}x + 3$ (ii) $g^{-1} : x \mapsto -\ln x$
(iii) No inverse (iv) $k^{-1} : x \mapsto \left(\frac{x}{3}\right)^{\frac{1}{3}}$

4 $fg : x \mapsto 4x^2 + 4x$, $gf : x \mapsto 2x^2 - 1$, $-1 \pm \frac{1}{2}\sqrt{2}$

5 $x > 4$ or $x < \frac{2}{3}$

6 $-1 < x < \frac{1}{3}$ or $x > 3$

Revision Problems 7

1 (a) $\pm 2, \pm 4$ (b) 3, 4

3 (i) $\ln\frac{4}{x}, \frac{1}{2}e^x$ (ii) Reflection in the line $y = x$
(iii) $q^{-1}(x) = \frac{1}{2}e^{\sqrt{x/3}}$

Exercise 8(a)

1 $\frac{x^5}{5} + c$ **2** $\frac{2}{3}x^6 + c$ **3** $-\frac{3}{x} + c$

4 $\frac{4}{3}x^{\frac{3}{2}} + c$ **5** $-\frac{x^2}{2} - \frac{1}{x} + c$ **6** $-\frac{1}{x} - \frac{1}{2x^2} + c$

7 $\frac{1}{5}(x-1)^5 + c$ **8** $\frac{x^3}{3} - \frac{1}{x} + 2x + c$ **9** $6x^{\frac{2}{3}} + c$

10 $\frac{2\sqrt{2}}{3}x^{\frac{3}{2}} + c$ **11** $-\frac{1}{x} - \frac{1}{x^2} - \frac{1}{3x^3} + c$ **12** $-6x^{-\frac{1}{2}} + c$

13 $\frac{8}{3}x^{\frac{3}{4}} + c$ **14** $\frac{7}{4}x^4 + 2x^3 - x + c$ **15** $\frac{x^4}{4} - \frac{2}{x} + c$

16 $\frac{14}{9}x^{\frac{9}{2}} + c$ **17** $\frac{2}{3}x^{\frac{3}{2}} + 2x^{\frac{1}{2}} + c$ **18** $\frac{2}{3}x^{\frac{3}{2}} + 2x^{\frac{1}{2}} + c$

19 $\frac{1}{8}(2x-1)^4 + c$ **20** $\frac{1}{3a}(ax+b)^3$

Exercise 8(b)

1 $\frac{1}{6}(1+x^3)^6 + c$ **2** $\frac{1}{40}(4x+1)^{10} + c$ **3** $-(x^3-1)^{-1} + c$

4 $2\sqrt{x^2+x-5}$ **5** $(3x-1)^{\frac{1}{3}} + c$ **6** $\frac{1}{6}(1+2x^2)^{\frac{3}{2}} + c$

7 $-\frac{1}{3}(1+x^3)^{-3} + c$ **8** $-\frac{3}{4}(x^2-4x+1)^{-2} + c$ **9** $-\frac{1}{6}(1-3x^2)^5 + c$

10 $\frac{1}{6}(1+x^{\frac{3}{2}})^4 + c$

Exercise 8(c)

1 $\frac{7}{3}$	**2** $\frac{14}{3}$	**3** $-\frac{1}{2}$	**4** 2	**5** $\frac{2}{3}$
6 10	**7** $\frac{55}{24}$	**8** $1\frac{1}{2}$	**9** $1\frac{13}{24}$	**10** 1
11 0.609	**12** $\frac{31}{15}$	**13** $\frac{5}{48}$	**14** −0.848	**15** 0.166

Exercise 8(d)

1 (i) $2\frac{1}{3}$ unit^2 (ii) $10\frac{2}{3}$ unit^2 (iii) $56\frac{3}{4}$ unit^2 (iv) $\frac{1}{4}$ unit^2

2 (i) 1.22 unit^2 (ii) 1 unit^2 (iii) 1.52 unit^2 (iv) 6.93 unit^2

3 9 unit^2

4 36 unit^2

5 2 unit^2 (there are two parts)

6 (i) 19.5 unit^3 (ii) 107 unit^3 (iii) 10600 unit^3 (iv) 0.115 unit^3

7 (i) 4.71 unit^3 (ii) 2.09 unit^3 (ii) 7.24 unit^3 (iv) 52.2 unit^3

Exercise 8(e)

1 (i) 6.6 seconds (ii) 61.2 metres

2 (i) 43.2 metres (ii) 32 m/s^2

Exercise 8(f)

1 (i) 0.435 (ii) 1.89 (iii) 2.37 (iv) 0.443

2 (i) 0.441 (ii) 1.84 (iii) 2.36 (iv) 0.441

Miscellaneous examples 8

1 (i) 0.542 (ii) $\frac{2}{3}$ (iii) 0.989 (iv) $\frac{1}{3}$ (v) 0.0156

2 1.6 unit^2 **3** $\frac{1}{3}\pi r^2 h$ **4** 11.2 **5** 0.587

Revision Problems 8

1 $n \to \infty$, $A = \frac{1}{3}$ **2** (i) 0.792 (iii) 0.748

Exercise 9(a)

1	$\sin 40°$	**2**	$-\cos 20°$	**3**	$-\tan 45°$	**4**	$\tan 60°$
5	$-\sin 80°$	**6**	$-\cos 40°$	**7**	$-\tan 60°$	**8**	$-\tan 1°$
9	$-\sin 60°$	**10**	$\cos 10°$	**11**	$\cos 80°$	**12**	$\tan 70°$
13	$\tan 20°$	**14**	$\cos 41°$	**15**	$-\cos 16°$	**16**	$-\tan 7°$
17	$\sin 84°$	**18**	$-\cos 60°$	**19**	$\sin 0°$	**20**	$-\cos 87°$

Exercise 9(b)

1	$-\sec 80°$	**2**	$\operatorname{cosec} 40°$	**3**	$\cot 60°$	**4**	$\sec 60°$
5	$-\cot 70°$	**6**	$-\operatorname{cosec} 80°$	**7**	$-\sec 20°$	**8**	$\cot 10°$
9	$\sec 20°$	**10**	$\cot 0°$				

Exercise 9(c)

1 (i) $22\frac{1}{2}°$ (ii) $540°$ (iii) $120°$ (iv) $57.3°$
(v) $75°$ (vi) $149°$ (vii) $315°$ (viii) $270°$
(ix) $1080°$ (x) $482°$

2 (i) $\frac{\pi}{12}$ (ii) 0.873 (iii) 0.34 (iv) $\frac{7\pi}{3}$
(v) $-\frac{3\pi}{4}$ (vi) $\frac{10\pi}{3}$ (vii) $\frac{3\pi}{8}$ (viii) 0.14
(ix) 5π (x) 1.86

Exercise 9(d)

1 1.03 cm^3 **2** 7.89 cm **3** 0.16 rad **4** $\frac{\pi}{2}$ rad

Exercise 9(f)

1 $\sec x$ **2** $\sec^2 x$ **3** 1 **4** $\cot^4 x$

5 $\frac{\sin x}{1-\cos x}$ or $\frac{1+\cos x}{\sin x}$ **6** $\sec^2 x$

Exercise 9(g)

1 (i) 36.9°, 143.1° (ii) 134.4°, 225.6°
(iii) 17.6°, 107.6°, 197.6°, 287.6° (iv) 75°, 135°, 255°, 315°
(v) 26.5°, 153.5°, 206.5°, 333.5° (vi) 14°, 194°

2 (i) $-112.5°, -22.5°, 67.5°, 157.5°$

(ii) $-142.3°, -37.7°, 37.7°, 142.3°$

(iii) $-165°, -135°, -45°, -15°, 75°, 105°$

(iv) $-23°, 83°$

3 (i) $\frac{\pi}{9}, \frac{4\pi}{9}, \frac{7\pi}{9}$ (ii) 0.32, 1.25, 3.46, 4.39

(iii) 0.23, 1.35, 3.37, 4.49 (iv) $\frac{5\pi}{4}$

(v) $\frac{\pi}{6}, \frac{5\pi}{18}, \frac{5\pi}{6}, \frac{17\pi}{18}, \frac{3\pi}{2}, \frac{29\pi}{18}$

4 (i) $30° + 120°n$ (ii) $33.75° + 45°n$

(iii) $26.6° + 180°n$ (iv) $41° + 180°n$ or $94° + 180°n$

5 (i) $54.7°, 125.3°, 234.7°, 305.3°$

(ii) $16.9°, 43.1°, 76.9°, 103.1°, 136.9°, 163.1°, 196.9°, 223.1°, 256.9°, 283.1°, 316.9°, 343.1°$

(iii) $30°, 150°$

(iv) $38.2°, 141.8°$

Exercise 9(h)

2 (i) $60°, 120°, 240°, 300°$ (ii) $0°, 60°, 120°, 180°, 240°, 300°, 360°$

(iii) $179.5°, 359.5°$

Exercise 9(i)

1 (i) $\sqrt{13}\cos(x - 33.7°)$ (ii) $13\cos(x - 157.3°)$

(iii) $\sqrt{2}\cos(x - 135°)$ (iv) $\sqrt{13}\cos(x - 326.3°)$

2 (i) $107.6°, 319.8°$ (ii) $114.3°, 335.7°$

(iii) $79.1°, 263.5°$ (iv) $25.4°, 228.4°$

Miscellaneous examples 9

1 (i) $-\frac{56}{33}$ (ii) $-\frac{16}{65}$

(iii) $\cos A \cos B - \sin A \sin B = -\frac{4}{5} \times -\frac{12}{13} - \frac{3}{5} \times \frac{5}{13} = \frac{33}{65}$

2 0.869, 3.45

3 $\tan 3x = \tan(x + 2x)$; $\frac{3\tan x - \tan^3 x}{1 - 3\tan^2 x}$

4 $\sqrt{52}$

5 $52.8° + 180°n$ or $157.3° + 180°n$

7 $90°n$ or $\pm 120° + 360°n$

8 $\frac{\tan\theta - 4}{1 - 4\tan\theta}$

Revision Problems 9

1 0.0753

2 (i) 0.916 seconds (ii) −1.9 cm (cork is above fixed line)

3 Solve $8\sin^2 x = 3\sin x + 1$, −168°, −12°, 36°, 144°

4 Remember to work in radians, there are 12 occasions

5 (i) 75° (ii) $\dfrac{1}{(2-\sqrt{3})}$

Exercise 10(a)

1 (i) 4θ (ii) $2 - \dfrac{9\theta^2}{2}$ (iii) $1 + \theta^2$ (iv) θ

(v) 6θ (vi) 4θ

Exercise 10(b)

1 $4\cos 4x$

2 $-\sin\left(x + \dfrac{\pi}{4}\right)$

3 $2\sec^2 x \tan x$

4 $-8\sin 2x$

5 $12\sec^2 4x$

6 $-30\sin 3x \cos 3x = -15\sin 6x$

7 $2\sec^2\left(2x - \dfrac{\pi}{4}\right)$

8 $3\tan^2 x \sec^2 x$

9 $4\sin 2x \cos 2x = 2\sin 4x$

10 $x\cos x + \sin x$

11 $\cos^2 x(\cos x - 3x\sin x)$

12 $\dfrac{1 + \sin x - x\cos x}{(1 + \sin x)^2}$

13 $2x\cos\left(2x + \dfrac{\pi}{4}\right) + \sin\left(2x + \dfrac{\pi}{4}\right)$

14 $x(3x\cos 3x + 2\sin 3x)$

15 $\dfrac{(1 + x)(2\sin x - (1 + x)\cos x)}{\sin^2 x}$

16 $8\sin 2x\cos 2x(1 + \sin^2 2x) = 4\sin 4x(1 + \sin^2 2x)$

17 $\dfrac{2(\sin x - x\cos x)}{(x + \sin x)^2}$

18 $x^2(2x\sec^2 2x + 3\tan 2x)$

19 $2x\tan x(x\sec^2 x + \tan x)$

20 $36x^2\sin^2(x^3 + 1)\cos(x^3 + 1)$

Exercise 10(c)

1 $2\sec 2x\tan 2x$

2 $\cot x - x\mathrm{cosec}^2 x$

3 $-6\,\mathrm{cosec}^3 2x\cot 2x$

4 $-\mathrm{cosec}\left(x + \dfrac{\pi}{4}\right)\cot\left(x + \dfrac{\pi}{4}\right)$

5 $2\sec\left(2x-\frac{\pi}{3}\right)\tan\left(2x-\frac{\pi}{3}\right)$

6 $3x^2\sec 3x(x\tan 3x+1)$

7 $\dfrac{2\sec x\tan x}{(1-\sec x)^2}$

8 $\sec^2 x(\cos x+2\sin x\tan x)$

9 $-6\,\text{cosec}^2 2x(1+\cot 2x)^2$

10 $-\dfrac{2}{(1-x)^2}\,\text{cosec}\left(\dfrac{1+x}{1-x}\right)\cot\left(\dfrac{1+x}{1-x}\right)$

Exercise 10(d)

1 (i) $\frac{1}{4}\sin 4x+c$ (ii) $-\frac{3}{2}\cos 2x$ (iii) $\frac{1}{2}x+\dfrac{\sin 8x}{16}+c$

(iv) $\frac{1}{3}\tan 3x-x+c$ (v) $\frac{1}{2}\sec 2x+c$ (vi) $\frac{1}{2}\tan 2x+c$

(vii) $\frac{1}{2}\,\text{cosec}\,2x+c$

2 (i) 0.433 (ii) 3 (iii) 0.5 (iv) 1 (v) 0.845

3 $\frac{1}{2}-\dfrac{\pi\sqrt{3}}{12}=0.0466$

4 1

5 1

Miscellaneous examples 10

1 $\frac{1}{2}+\frac{3}{2}\theta-\frac{1}{8}\theta^2$

2 (i) $\sin x(\sin x+2x\cos x)$ (ii) $\sec x\tan x(2\sec^2 x+\tan^2 x)$

(iii) $\dfrac{-2\sin x}{(1-\cos x)^2}$ (iv) $\sec x(x\tan x+1)$

(v) $-2\,\text{cosec}^2 x\cot x$

(vi) $\dfrac{4(\cot 4x\sec^2 4x-\text{cosec}^2 4x+\tan 4x\,\text{cosec}^2 4x)}{(1-\tan 4x)^2}$

3 $\dfrac{dy}{dx}=x\cos x+\sin x=1.26$ at the point; $y=1.26x-0.43$; $1.26y+x=1.49$

4 (i) 0.5 (ii) 0.1 (iii) $\int_0^{\frac{\pi}{6}} 2(1-\cos 6x)\,dx=\dfrac{\pi}{3}$

5 $\displaystyle\int_{\frac{\pi}{6}}^{\frac{5\pi}{6}} \sin x-0.5\,dx=0.685$

Revision Problems 10

1 0.5

2 $\dfrac{\pi^2}{12}$

3 $\left(\dfrac{\pi}{2},\pi\right)$ point of inflexion

4 (i) $\frac{\pi t}{6}$ (ii) 2.21

(iii) Minimum depth occurs when $\cos(30t)^\circ = -1$, $t = 6$, depth $= 0.3$ m

(iv) $h = 2.3 + 2\cos\frac{\pi t}{6}$ in radians

$\therefore \quad \frac{dh}{dt} = \frac{-\pi}{3}\sin\frac{\pi t}{6}$ (*A*) 0.91 mh^{-1} falling (*B*) 0.91 mh^{-1} rising

(v) $t = 9$

Exercise 11(a)

1 (i) $-\mathbf{i} - 5\mathbf{j}$ (ii) $2\mathbf{i} - 3\mathbf{j} - \mathbf{k}$ (iii) $-2\mathbf{i} - 2\mathbf{k}$

(iv) $-\mathbf{i} + 4\mathbf{j} - 4\mathbf{k}$ (v) $(2t-1)\mathbf{i} + (1-t)\mathbf{j} - (t+3)\mathbf{k}$

2 (i) $\sqrt{26}$ (ii) $\sqrt{14}$ (iii) $\sqrt{8}$

(iv) $\sqrt{33}$ (v) $\sqrt{(2t-1)^2 + (1-t)^2 + (t+3)^2} = \sqrt{6t^2 + 11}$

3 1.22 unit^2

4 4 unit^3

Exercise 11(b)

1 (i) $\mathbf{r} = \begin{pmatrix} 1 \\ 2 \\ 1 \end{pmatrix} + \lambda \begin{pmatrix} 3 \\ -2 \\ 1 \end{pmatrix}$ (ii) $\mathbf{r} = \begin{pmatrix} 2 \\ -1 \\ -3 \end{pmatrix} + \lambda \begin{pmatrix} 4 \\ -1 \\ -2 \end{pmatrix}$

(iii) $\mathbf{r} = \begin{pmatrix} 1 \\ 0 \\ 1 \end{pmatrix} + \lambda \begin{pmatrix} 3 \\ 0 \\ -2 \end{pmatrix}$

2 (i) $\mathbf{r} = \begin{pmatrix} 1 \\ 3 \\ 1 \end{pmatrix} + \lambda \begin{pmatrix} 1 \\ -2 \\ 1 \end{pmatrix}$ (ii) $\mathbf{r} = \begin{pmatrix} 4 \\ -1 \\ 0 \end{pmatrix} + \lambda \begin{pmatrix} -1 \\ 1 \\ 2 \end{pmatrix}$

(iii) $\mathbf{r} = \begin{pmatrix} t \\ 2t \\ 3t \end{pmatrix} + \lambda \begin{pmatrix} 3t \\ 5-2t \\ 3t \end{pmatrix}$

3 (i) $(3, -1, 3)$ (ii) $(\frac{3}{2}, -\frac{1}{2}, 2)$ (iii) $(2, 5, -3)$

Exercise 11(c)

1 (i) 80.7° (ii) 79.5°

2 (i) 19.5° (ii) 79.5°

Exercise 11(d)

1 $2x + 4y - 5z = 9$

2 $\mathbf{r} = \begin{pmatrix} 6 \\ -2 \\ 0 \end{pmatrix} + \lambda \begin{pmatrix} 19 \\ -11 \\ 1 \end{pmatrix}$

3 37.6°

4 $\dfrac{1}{\sqrt{6}}$

Miscellaneous examples 11

1 23.4°

2 135.6°

3 $3x + 3y + 2z = 6$

4 $\dfrac{2}{\sqrt{14}}$

5 $(-2, -3, -2.5)$

6 3.84 unit^2

Revision Problems 11

1 (a) $(2, 0, -1)$

(b) $\mathbf{r} = \begin{pmatrix} 2 \\ -1 \\ -3 \end{pmatrix} + s \begin{pmatrix} 3 \\ 2 \\ 6 \end{pmatrix}$

2 (a) $-4\mathbf{i} + 2\mathbf{j} + 6\mathbf{k}$

(c) 6.76

(d) $\mathbf{r} = \begin{pmatrix} 8 \\ 4 \\ 0 \end{pmatrix} + \lambda \begin{pmatrix} 4 \\ 2 \\ -6 \end{pmatrix}$

(e) $\begin{pmatrix} 8 + 4\lambda \\ 4 + 2\lambda \\ -6\lambda \end{pmatrix} \cdot \begin{pmatrix} 4 \\ 2 \\ -6 \end{pmatrix} = 0$ if point X is such that OX is perpendicular to BD, $\lambda = -\frac{5}{7}$, the point is $\left(5\frac{1}{7}, 2\frac{4}{7}, 4\frac{2}{7}\right)$

Exercise 12(a)

1 $42.95 \le H \le 43.65$

2 $1.08 \le T \le 1.64$

3 $18.4 \le T \le 18.5$

4 $0.348 \le f \le 0.447$

5 10.5% maximum error

6 $0.867 \le \dfrac{v}{s} \le 0.899$

Exercise 12(b)

1 1.4

2 3.45

3 2.24

4 1.8

5 2.3

Exercise 12(c)

1 3.03 **2** 1.95

Exercise 12(d)

1 1.21 **2** 0.25 **3** 1.25 **4** 1.59 **5** 1.21

Miscellaneous examples 12

1 2% error **2** 4.71 **3** −0.3 **4** 0.77

Revision problems 12

1 (c) 1.76, 1.61, 1.97, 0.596, not converging; 1.72

2 (b) 2.79

(c) You did not have information about the second decimal place before; and so cannot guarantee this is correct

(d) One root

Exercise 13(a)

1 $\frac{6!}{2!2!} = 180$

2 ${}^{250}P_6$

3 Consider the number of As and Es in the word.

A	0	0	1	1	1	2	2	2	Total
E	1	2	0	1	2	0	1	2	
Number	24	36	24	72	36	36	36	6	260

4 $\frac{9!}{4!3!2!} = 1260$, 140

5 $5! - 2 \times 4! = 72$

6 (i) $6 \times 5 \times 4 \times 3 = 360$ (ii) 180 (iii) 60

7 One

8 $7! - 2 \times 6! = 3600$

Exercise 13(b)

1 210

2 $^{30}C_4 \times {}^{25}C_5$

3 $^{40}C_{12}$

4 $^5C_2 \times {}^4C_2 = 60,\ 60 - {}^4C_1 \times {}^3C_1 = 48$

5 (i) 135751 (ii) 131875

7 $4 \times 6 \times 5 \times 3 \times {}^{14}C_1 = 5040$

Exercise 13(c)

1 (i) $\frac{1}{20}$ (ii) $\frac{3}{20}$ (iii) $\frac{7}{20}$

2 (i) $\frac{1}{4}$ (ii) $\frac{1}{12}$

3 (i) 0.24 (ii) 0.76

4 $1 - (0.8)^3 = 0.488$

5 (i) $\frac{5}{6}$ (ii) $\frac{2}{3}$

6 (i) $\frac{2}{3}$ (ii) $\frac{16}{37}$

7 (i) $\frac{4}{5}$ (ii) $\frac{1}{16}$ (iii) $\frac{3}{4}$

Miscellaneous examples 13

1 2520, $\frac{1}{2}$ **2** $\frac{2}{7}$ **3** $^7C_2 \times {}^8C_2 = 588$

4 $n - 1$ **5** 120, $\frac{3}{5}$

6 (i) $\frac{18}{25}$ (ii) $\frac{9}{50}$ (iii) $\frac{3}{5}$

Revision problems 13

1 (i) (a) $\frac{1}{4}$ (ii) (b) $\frac{4}{5}$

2 (i) (a) A and C are mutually exclusive

(b) A and B are independent since $P(A \cap B) = P(A)P(B)$

(ii) (a) 0.45 (b) $0.4p + 0.2(1 - p) = 0.2p + 0.3(1 - p)$ hence $p = \frac{1}{3}$

Exercise 14(a)

1 (i) $\dfrac{-y}{4y + x}$

(ii) $\dfrac{-(x + y)}{(x + 6y)}$

(iii) $-\dfrac{(2x+\sin 2y)}{2x\cos 2y}$

(iv) $-\dfrac{\tan y}{x\sec^2 y} = -\dfrac{\sin y\cos y}{x} = -\dfrac{\sin 2y}{2x}$

(v) $4(x-y)(1-y')+1=0,\ y' = \dfrac{4x-4y+1}{4(x-y)}$

(vi) $\dfrac{(y-3x^2)}{(3y^2-x)}$

(vii) $-\dfrac{\tan y}{2x}$

(viii) $3(2x+3y)^2.(2+3y')=0$

$(2x+3y)^2=1$, so $2+3y'=0, y'=-\frac{2}{3}$

(ix) $\dfrac{\tan^2 y}{1-2x\tan y\sec^2 y}$

(x) $\dfrac{6x\cos y}{(2y+3x^2y)}$

2 (i) 1 (ii) $-\dfrac{1}{\pi}$ (iii) -2

(iv) 1 (v) ∞ (vi) $\dfrac{1}{2\sqrt{\pi}}$

3 (i) $x+y=2,\ y=x$ (ii) $y=1,\ x=2$

(iii) $y+2x=3,\ 2y=x-4$

Exercise 14(b)

1 (i) t (ii) $-\dfrac{1}{t}$ (iii) $-\frac{1}{2}\cot 3\theta$

(iv) $-\dfrac{1}{t^2}$ (v) $\frac{3}{4}\operatorname{cosec}\alpha$ (vi) 1

(vii) $\dfrac{2-6t}{2+6t}$ (viii) $-\left(\dfrac{1+\cos\theta}{1+\sin\theta}\right)$

2 $t=2,\ 3y=8x-17,\ 8y+3x=125$

3 $y=tx-at^2,\ \frac{1}{2}a^2t^3$

4 $4y=3x+15$, solve $4(t^3+1)=3(4t-1)+15$, $(7,9)$

Exercise 14(c)

1 (i) $3e^{3x}$ (ii) $e^{-2x}(1-2x)$

(iii) $2(x\cos x+\sin x)$ (iv) $e^x\cos x(\cos x-2\sin x)$

(v) $\dfrac{e^x(1-x)^2}{(1+x^2)^2}$ (vi) $e^x(2\sec^2 2x+\tan 2x)$

(vii) $\dfrac{x^2(2x^3+x^2+2x+3)e^{2x}}{(1+x^2)^2}$ (viii) $-\cos x e^{-\sin x}$

(ix) $6e^{2x}(1+e^{2x})^2$

(x) Write $y=\frac{1}{2}e^{-x}\sin 4x$ first, then $y'=\frac{1}{2}e^{-x}(4\cos 4x-\sin 4x)$

2 (i) $-e^{-4t}$

(ii) $\dfrac{\cos t-\sin t}{\cos t+\sin t}=\dfrac{1-\tan t}{1+\tan t}=\tan\left(\dfrac{\pi}{4}-t\right)$

(iii) $-\frac{2}{3}e^{-5\theta}$

(iv) $\dfrac{\cos t+\sin t}{\cos t-\sin t}=\tan\left(\dfrac{\pi}{4}+t\right)$

3 $y=2ex-e$, $2ey+x=1+2e^2$

4 $(\ln e, \dfrac{1}{e}+\ln e)$ minimum

5 0.655

Exercise 14(d)

1 $\dfrac{1}{x}$ **2** $1+\ln x$ **3** $-\tan x$ **4** $\tan x$

5 $\dfrac{2x}{1+x^2}$ **6** $\dfrac{-2}{(1-x^2)}$ **7** $\dfrac{e^x}{1+e^x}$ **8** $e^x\left(\dfrac{1}{x}+\ln x\right)$

9 $2\cot x$ **10** $\dfrac{2\ln x}{x}$ **11** $2t^2+2$

12 $y=0.0677x+0.125$ **13** $(0.606, -0.184)$, minimum

Exercise 14(e)

1 $\dfrac{2}{1+4x^2}$ **2** $\dfrac{-3}{\sqrt{1-9x^2}}$ **3** $\dfrac{4}{\sqrt{1-16x^2}}$

4 $\dfrac{1}{2x^2+4x+4}$ **5** $\dfrac{1}{\sqrt{2x-x^2}}$ **6** $\dfrac{3}{\sqrt{-9x^2-6x}}$

7 $\dfrac{x}{\sqrt{1-x^2}}+\sin^{-1}x$ **8** $\dfrac{-e^x}{\sqrt{1-x^2}}+e^x\cos^{-1}x$ **9** $\dfrac{2x}{\sqrt{2x^2-x^4}}=\dfrac{2}{\sqrt{2-x^2}}$

10 $\dfrac{3x^2}{x^6+2x^3+2}$

Miscellaneous examples 14

1 $54y=17x-31$

2 Tangent is $y=2ex-e+1$, area $=\dfrac{1}{4e}(1-e)^2$

3 $(\pm 1.32, \mp 1.73)$

4 $(0.354, 2.23)$, minimum

6 π

Revision Problems 14

1 (i) $\dfrac{dy}{dx} = 0$ if $y = -2x$, substitute into equation to give:

$$x = \pm\frac{2}{\sqrt{3}},\ y \mp \frac{4}{\sqrt{3}};\ \frac{dy}{dx} = \infty \text{ if } x = -2y, \text{ hence } y = \pm\frac{2}{\sqrt{3}},\ x = \mp\frac{4}{\sqrt{3}}$$

(ii) $(0, \pm 2)$, $2y + x = \pm 4$; $(\pm 2, 0)$, $y + 2x = \pm 4$

2 (a) 4.58

3 (ii) $-\frac{1}{2}t^2$ (iii) $2y = x - 6$ (iv) $(8, 1)$

4 (a) 1 (c) $\left(e^{\frac{1}{2}}, \frac{1}{2}e^{-1}\right)$, $-2e^{-2}$ (d) $\frac{1}{2}(1 - \ln 2)$

(e) 0.133

5 (ii) $y = a\sin\theta - a\theta\cos\theta$ (iv) $\tan\theta$ (v) $a\left(1 + \frac{1}{2}\theta^2\right)$

Exercise 15(a)

1 (i) $\frac{1}{2}e^{2x} + c$ (ii) $-\frac{3}{2}e^{-2x} + c$ (iii) $\dfrac{2^x}{\ln 2} + c$

(iv) $\frac{1}{2}\ln x + c$ (v) $\frac{1}{3}\ln(3x - 1) + c$ (vi) $4\ln(x + 6) + c$

(vii) $\dfrac{-1}{2(x+2)^2} + c$ (viii) $-\frac{1}{4}e^{-4x} + c$ (ix) $\frac{1}{3}e^{3x} + c$

(x) $\frac{1}{2}e^{x^2} + c$

2 (i) 3.19 (ii) 8.56 (iii) 0.231 (iv) -0.575

(v) 1.15 (vi) 0.28 (vii) 0.0585 (viii) 0.316

(ix) $2(e - 1)$ (x) 0.0151 (xi) $\frac{2}{3}\ln 4$ (xii) 2.16

3 12.6 unit^2 **4** 2520 unit^3 **5** 46.1 unit^3

6 (i) 3230 m/s (ii) 10 300 metres

7 0.313 unit^2

Exercise 15(b)

1 (i) $\frac{1}{20}\left(x^2 + 1\right)^{10} + c$ (ii) $\dfrac{-3}{4((1 + x^2)^2} + c$

(iii) $\dfrac{(x + 10)^{11}(11x - 16)}{66} + c$ (iv) $\frac{1}{3}\cos^3 x - \cos x + c$

(v) $\frac{2}{3}e^{3x^3} + c$

(vi) $\ln \sin x + c$

(vii) $\frac{2}{3}(\ln x)^2 + c$

(viii) $\frac{-1}{(1+e^x)} + c$

(ix) $\frac{1}{3(1-x^2)^3} + c$

(x) $\frac{1}{4}(1 + \tan x)^4 + c$

2 (i) $\frac{1}{3}\ln(1 + x^3) + c$ (ii) 2.34 (iii) -6307

(iv) $\frac{2}{3}$ (v) $\frac{7}{72}$ (vi) The limit 1 means 1 radian, 2.06

Exercise 15(c)

1 $\frac{1}{3}\tan^{-1}\frac{x}{3} + c$ **2** $\sin^{-1}\frac{x}{\sqrt{2}} + c$ **3** $\frac{1}{3}\sin^{-1}\frac{3x}{2} + c$

4 $\frac{1}{2}\tan^{-1}\frac{(x+3)}{2} + c$ **5** 0.253 **6** 0.0583

Exercise 15(d)

1 (i) $\frac{1}{2}xe^{2x} - \frac{1}{4}e^{2x} + c$

(ii) $x^2 \sin x + 2x \cos x - 2 \sin x + c$

(iii) $\frac{1}{3}x^3 \ln x - \frac{1}{9}x^3 + c$

(iv) $-x^2e^{-x} - 2xe^{-x} - 2e^{-x} + c$

(v) $-\frac{2}{5}e^{-x}\cos 2x - \frac{1}{5}e^{-x}\sin 2x + c$

(vi) $\frac{1}{2}x^2 \tan^{-1} x - \frac{1}{2}x + \frac{1}{2}\tan^{-1} x + c$

(vii) $x \tan^{-1} x - \frac{1}{2}\ln(1 + x^2)$

(viii) $x \ln(x+2) - x + 2\ln(x+2) + c$

2 (i) 1.3 (ii) 2.32 (iii) 1.84

Exercise 15(e)

1 (i) $\frac{-1}{3(1+x)} + \frac{2}{3(1-2x)}$

(ii) $\frac{-1}{x} + \frac{4}{1+2x}$

(iii) $\frac{1}{4(1+x)} - \frac{1}{2(1+x)^2} + \frac{1}{4(1-x)}$

(iv) $\frac{2}{5(x-3)} - \frac{(2x+1)}{5(x^2+1)}$

(v) Divide first, $x - 1 + \frac{1}{3(x-1)} + \frac{8}{3(x+2)}$

(vi) $\frac{1}{9(x-1)} + \frac{1}{3(x-1)^2} - \frac{(x+4)}{9(x^2+2)}$

2 (i) $\ln\frac{8}{3}$ (ii) $\frac{1}{2} + \ln\frac{2}{3}$ (iii) 0.000437

Exercise 15(f)

1 $y = \frac{x^3}{3} - \frac{5}{3}$

2 $y = \frac{2}{3}x$

3 $\ln y = \frac{1}{\sqrt{2}} - \cos x$

4 $\tan^{-1} y = -\frac{1}{x} + \frac{1}{4}$

5 $\ln y = \frac{x^2}{2} - x + \ln 7 - \frac{3}{2}$

6 $y = \frac{\pi}{6} - \ln \cos x$

Miscellaneous examples 15

1 $\frac{-3^{-x}}{\ln 3} + c$

2 $\frac{1}{3}x^2 \cos 3x + \frac{2}{9}x \sin 3x + \frac{2}{27} \cos 3x + c$

3 $\ln(1 + x^2) - \tan^{-1} x + c$

4 $\frac{1}{2}e^{x^2} + c$

5 $\frac{1}{2}\sin^{-1} \frac{2x}{3} + c$

6 $\frac{x^4}{4} \ln 2x - \frac{x^4}{16} + c$

7 $\frac{1}{3}$

8 $\frac{1}{4}\ln\frac{3}{2} + \frac{1}{4}$

9 6.7

10 0.138

11 0.283

12 −1.05

13 0.00234

14 $\frac{1}{3}$

15 0.376

16 $-\frac{1}{2}x + \frac{1}{4}x^2$

17 $-1 + 3x - 5x^2 + 9x^3$, $-\frac{1}{2} < x < \frac{1}{2}$

Revision Problems 15

1 $300 - 1.5h = 300e^{-1.5t}$, 11 months

2 (i) (1, 0) (ii) $-\frac{1}{x} + \frac{4}{x+3}$ (iv) 3 (vi) 5.1

Index